MW01633492

Industrial Power Distribution and Illuminating Systems

ELECTRICAL ENGINEERING AND ELECTRONICS

A Series of Reference Books and Textbooks

1. Rational Fault Analysis, *edited by Richard Saeks and S. R. Liberty*
2. Nonparametric Methods in Communications, *edited by P. Papantoni-Kazakos and Dimitri Kazakos*
3. Interactive Pattern Recognition, *Yi-tzuu Chien*
4. Solid-State Electronics, *Lawrence E. Murr*
5. Electronic, Magnetic, and Thermal Properties of Solid Materials, *Klaus Schröder*
6. Magnetic-Bubble Memory Technology, *Hsu Chang*
7. Transformer and Inductor Design Handbook, *Colonel Wm. T. McLyman*
8. Electromagnetics: Classical and Modern Theory and Applications, *Samuel Seely and Alexander D. Poularikas*
9. One-Dimensional Digital Signal Processing, *Chi-Tsong Chen*
10. Interconnected Dynamical Systems, *Raymond A. DeCarlo and Richard Saeks*
11. Modern Digital Control Systems, *Raymond G. Jacquot*
12. Hybrid Circuit Design and Manufacture, *Roydn D. Jones*
13. Magnetic Core Selection for Transformers and Inductors: A User's Guide to Practice and Specification, *Colonel Wm. T. McLyman*
14. Static and Rotating Electromagnetic Devices, *Richard H. Engelmann*
15. Energy-Efficient Electric Motors: Selection and Application, *John C. Andreas*
16. Electromagnetic Compossibility, *Heinz M. Schlicke*
17. Electronics: Models, Analysis, and Systems, *James G. Gottling*
18. Digital Filter Design Handbook, *Fred J. Taylor*
19. Multivariable Control: An Introduction, *P. K. Sinha*
20. Flexible Circuits: Design and Applications, *Steve Gurley, with contributions by Carl A. Edstrom, Jr., Ray D. Greenway, and William P. Kelly*
21. Circuit Interruption: Theory and Techniques, *Thomas E. Browne, Jr.*
22. Switch Mode Power Conversion: Basic Theory and Design, *K. Kit Sum*
23. Pattern Recognition: Applications to Large Data-Set Problems, *Sing-Tze Bow*
24. Custom-Specific Integrated Circuits: Design and Fabrication, *Stanley L. Hurst*
25. Digital Circuits: Logic and Design, *Ronald C. Emery*
26. Large-Scale Control Systems: Theories and Techniques, *Magdi S. Mahmoud, Mohamed F. Hassan, and Mohamed G. Darwish*
27. Microprocessor Software Project Management, *Eli T. Fathi and Cedric V. W. Armstrong (Sponsored by Ontario Centre for Microelectronics)*
28. Low Frequency Electromagnetic Design, *Michael P. Perry*
29. Multidimensional Systems: Techniques and Applications, *edited by Spyros G. Tzafestas*
30. AC Motors for High-Performance Applications: Analysis and Control, *Sakae Yamamura*

31. Ceramic Materials for Electronics: Processing, Properties, and Applications, *edited by Relva C. Buchanan*
32. Microcomputer Bus Structures and Bus Interface Design, *Arthur L. Dexter*
33. End User's Guide to Innovative Flexible Circuit Packaging, *Jay J. Miniet*
34. Reliability Engineering for Electronic Design, *Norman B. Fuqua*
35. Design Fundamentals for Low-Voltage Distribution and Control, *Frank W. Kussy and Jack L. Warren*
36. Encapsulation of Electronic Devices and Components, *Edward R. Salmon*
37. Protective Relaying: Principles and Applications, *J. Lewis Blackburn*
38. Testing Active and Passive Electronic Components, *Richard F. Powell*
39. Adaptive Control Systems: Techniques and Applications, *V. V. Chalam*
40. Computer-Aided Analysis of Power Electronic Systems, *Venkatachari Rajagopalan*
41. Integrated Circuit Quality and Reliability, *Eugene R. Hnatek*
42. Systolic Signal Processing Systems, *edited by Earl E. Swartzlander, Jr.*
43. Adaptive Digital Filters and Signal Analysis, *Maurice G. Bellanger*
44. Electronic Ceramics: Properties, Configuration, and Applications, *edited by Lionel M. Levinson*
45. Computer Systems Engineering Management, *Robert S. Alford*
46. Systems Modeling and Computer Simulation, *edited by Naim A. Kheir*
47. Rigid-Flex Printed Wiring Design for Production Readiness, *Walter S. Rigling*
48. Analog Methods for Computer-Aided Circuit Analysis and Diagnosis, *edited by Takao Ozawa*
49. Transformer and Inductor Design Handbook, Second Edition, Revised and Expanded, *Colonel Wm. T. McLyman*
50. Power System Grounding and Transients: An Introduction, *A. P. Sakis Meliopoulos*
51. Signal Processing Handbook, *edited by C. H. Chen*
52. Electronic Product Design for Automated Manufacturing, *H. Richard Stillwell*
53. Dynamic Models and Discrete Event Simulation, *William Delaney and Erminia Vaccari*
54. FET Technology and Application: An Introduction, *Edwin S. Oxner*
55. Digital Speech Processing, Synthesis, and Recognition, *Sadaoki Furui*
56. VLSI RISC Architecture and Organization, *Stephen B. Furber*
57. Surface Mount and Related Technologies, *Gerald Ginsberg*
58. Uninterruptible Power Supplies: Power Conditioners for Critical Equipment, *David C. Griffith*
59. Polyphase Induction Motors: Analysis, Design, and Application, *Paul L. Cochran*

60. Battery Technology Handbook, *edited by H. A. Kiehne*
61. Network Modeling, Simulation, and Analysis, *edited by Ricardo F. Garzia and Mario R. Garzia*
62. Linear Circuits, Systems and Signal Processing: Advanced Theory and Applications, *edited by Nobuo Nagai*
63. High-Voltage Engineering: Theory and Practice, *edited by M. Khalifa*
64. Large-Scale Systems Control and Decision Making, *edited by Hiroyuki Tamura and Tsuneo Yoshikawa*
65. Industrial Power Distribution and Illuminating Systems, *Kao Chen*
66. Distributed Computer Control for Industrial Automation, *edited by D. Popovic and Vijay P. Bhatkar*

Additional Volumes in Preparation

Computer-Aided Analysis of Active Circuits, *Adrian Ioinovici*

Electrical Engineering-Electronics Software

1. Transformer and Inductor Design Software for the IBM PC, *Colonel Wm. T. McLyman*
2. Transformer and Inductor Design Software for the Macintosh, *Colonel Wm. T. McLyman*
3. Digital Filter Design Software for the IBM PC, *Fred J. Taylor and Thanos Stouraitis*

Industrial Power Distribution and Illuminating Systems

Kao Chen
Carl-sons Consulting Engineers
San Diego, California

MARCEL DEKKER, INC. New York and Basel

Library of Congress Cataloging-in-Publication Data

Chen, Kao

Industrial power distribution and illuminating systems / Kao Chen.
p. cm. -- (Electrical engineering and electronics ; 65)
Includes bibliographical references.
ISBN 0-8247-8237-2 (alk. paper)
1. Electric engineering. 2. Industrial buildings--Lighting.
I. Title. II. Series.
TK146.C347 1990
621.319'24--dc20

89-71511
CIP

This book is printed on acid-free paper.

MARCEL DEKKER, INC.
270 Madison Avenue, New York, New York 10016

Current printing (last digit):
10 9 8 7 6 5 4 3 2 1

PRINTED IN THE UNITED STATES OF AMERICA

Preface

The past three decades have witnessed a technological explosion in the fields of solid-state electronics, microprocessors, lasers, fiber optics, exotic signal processing, and space science, fields that have come to dominate the technical interests of college faculty and students. Today, power engineering courses are almost excluded from the curriculum of most engineering colleges and universities. As a result, the shortage of electrical engineers with adequate knowledge of industrial power distribution or illuminating theory and practice has now reached a critical point.

This decline in the supply of power engineers is causing great concern to industry. Many industrial firms are forced to hire graduate electronics or mechanical engineers for positions responsible for power distribution and illumination design work. Normally, a college graduate with basic knowledge in physics and engineering principles should be able to master such subjects as industrial power distribution, application of protective relaying, principles of grounding and shielding, and fundamentals of illumination design within two to three semesters, given that academic institutions offer these subjects in a continuing or evening education program. Unfortunately, this availability is not the case.

Industry needs a continuing supply of qualified power engineers because there are many problem areas that require sophisticated engineering analysis. A near-term and perhaps effective way to overcome this shortage might begin with continuing education programs to be offered at local colleges and universities. This book provides a useful reference for practicing engineers and designers. It can also serve as an excellent text for a continuing education program in the field of industrial power distribution and illuminating engineering.

Chapters 1–10 cover all important elements of industrial power distribution—system planning; selection of distribution voltages and systems; principles of grounding; methods of fault current calculations; system protective relaying; power, switching, and carrying equipment and devices; and, finally, special power distribution requirements for computers and sensitive electronic equipment. Chapters 11–18 cover the illuminating engineering and design principles based on the latest concepts and approaches; a review of light sources, luminaires, and ballasts; applications and installations of industrial illuminating systems; floodlighting design procedures and examples; energy conservation practices and current codes and standards; and a thorough review of the up-to-date lighting controls and guidance for selection and implementation of optimum controls for specifically chosen system. There is an abundance of tables and diagrams giving practical values and limits to aid in the design and evaluation of systems.

The long-range correction for the power engineer shortage will require the concerted efforts of colleges and universities throughout the nation to reorient their curricula and revive the interests of faculty and students. Manufacturing industries are just as important to our national economic health and defense posture as the service industries. Continuing availability of industrial power engineers is essential to our nation's ability to survive and excel in the coming century. I hope that I have made some contribution toward this majestic goal.

Special acknowledgment and thanks are extended to Rene Castenschiold, Rudy Elam, Ed Palko, Dan DelBianco, Gary Forcey, and Jim Edwards for their photographic and additional technical assistance.

My gratitude also goes to Dr. Eileen Gardiner, Ms. Beth Wooster, Ms. Barbara Zeiders, Ms. Ruth Dawe, Ms. Rhonda Lillianthal and Messrs. John L. Bottomley, Jon Tell, and Graham Garratt of Marcel Dekker, Inc., who most patiently encouraged and supported this effort.

Kao Chen

Contents

Industrial Power Distribution and Illuminating Systems

Part I

1

Introduction

1.1 PURPOSE

This book, which has been prepared by the author after more than 30 years' experience in designing and developing industrial power distribution and energy-efficient illuminating systems, will hopefully be of great value to those of today's power-oriented engineers who have limited industrial plant experience. It can also be an aid to all engineers responsible for the electrical design of industrial plants. However, it is not intended as a replacement for many excellent engineering texts and handbooks commonly in use today, nor is it detailed enough to be a design manual. It should be used as a guide and up-to-date general reference on power and illumination facilities design for industrial plants.

Today's industrial plants, because of their increasing size and complexity, and their extensive use of computer systems and programmable controllers, have become more and more dependent on reliable and clean power systems. One can better provide such required systems by understanding the principles of system protection and grounding, basic power distribution systems, the available power equipment and carrying devices, and the specific requirements for computer systems. It is the author's hope that this book will fill the needs of those engineers responsible for designing and building demanding power distribution systems aimed at higher productivity and attendant higher profits from the products manufactured.

1.2 IEEE STANDARDS

The Industry Applications Society of the Institute of Electrical and Electronics Engineers has published the following "color book" standards:

1. ANSI/IEEE Standard 141-1986 (Red Book), Recommended Practice for Electric Power Distribution for Industrial Plants
2. ANSI/IEEE Standard 241-1983 (Gray Book), Recommended Practice for Electric Power Systems in Commercial Buildings
3. ANSI/IEEE Standard 242-1986 (Buff Book), Recommended Practice for Protection and Coordination of Industrial and Commercial Power Systems
4. ANSI/IEEE Standard 739-1984 (Bronze Book), Recommended Practice for Energy Conservation and Cost-Effective Planning in Industrial Facilities
5. ANSI/IEEE Standard 142-1982 (Green Book), Recommended Practice for Grounding of Industrial and Commercial Power Systems
6. ANSI/IEEE Standard 399-1980 (Brown Book), Recommended Practice for Industrial and Commercial Power Systems Analysis
7. ANSI/IEEE Standard 446-1987 (Orange Book), Recommended Practice for Emergency and Standby Power for Industrial and Commercial Applications
8. ANSI/IEEE Standard 602-1986 (White Book), Recommended Practice for Electric Systems in Health Care Facilities
9. ANSI/IEEE Standard 493-1980 (Gold Book), Recommended Practice for the Design of Reliable Industrial and Commercial Power Systems

The present author has been either the chapter author or contributing author to various chapters in six of the above-listed standards during a span of 15 years. His deep involvement in preparing materials for chapters on power distribution, power systems protection and grounding, power system analysis, and energy-efficient illumination for industrial environment has given him a solid foundation for constructing this timely book.

1.3 CODES AND STANDARDS

1.3.1 National Fire Protection Association Standards

The electrical wiring and design recommendations in the National Electrical Code (NEC), ANSI/NFPA 70-1987, are vitally important guidelines for industrial plant engineers. The NEC is published by and available from the National Fire Protection Association (NFPA). It does not represent a design specification, but simply minimum requirements. The NEC is also available from the American National

Standards Institute (ANSI) and from each state's Board of Fire Underwriters.

Other NFPA publications cited throughout this book are:

1. NFPA 75, Electronic Computer/Data Processing Equipment
2. NFPA 77, Recommended Practice on Static Electricity
3. NFPA 78, Lightning Protection Code
4. NFPA 110, Emergency and Standby Power Systems

1.3.2 National Electrical Manufacturers Association Standards

The National Electrical Manufacturers Association (NEMA) prepares standards that establish dimensions, ratings, and performance requirements for electric equipment for manufacturers. Their standards are widely used in the preparation of purchase specifications.

1.3.3 Underwriters Laboratories, Inc. Standards

Underwriters Laboratories (UL) prepares safety standards for electric equipment, including appliances and test equipment for compliance with these standards. Manufacturers that have products approved by UL as meeting the standards are authorized to use the UL label on the equipment. UL periodically publishes lists of approved equipment.

1.3.4 American National Standards Institute

The American National Standards Institute (ANSI) does not write standards. It promotes and coordinates the development of American National Standards and approves as American National Standards those documents that have been prepared in accordance with ANSI regulations.

Standards that have been approved by other organizations and then approved as American National Standards carry the identification numbers of both organizations and may be purchased from either. The sponsoring organization retains the responsibility for keeping the standards current.

Standards carrying only an ANSI number were prepared by American National Standards Committees organized and administered by other organizations in accordance with ANSI regulations. These ANSI committees are used to coordinate participation by a large number of organizations.

ANSI standards of interest to industrial plant engineers include the following:

1. ANSI Y 1.1-1972 (R1984), American National Standard Abbreviations for Use on Drawings and in Text

2. ANSI Y 32.9-1972, American National Standard Graphic Symbols for Electrical Wiring and Layout Diagrams Used in Architecture and Building Construction

1.4 SAFETY AND ENVIRONMENTAL CONSIDERATIONS

1.4.1 Occupational Safety and Health Administration

Legislation by the U.S. federal government has had the effect of giving standards, such as those of ANSI, the impact of law. The Occupational Safety and Health Act, administered by the U.S. Department of Labor, permits federal enforcement of codes and standards. The Occupational Safety and Health Administration (OSHA) established Design Safety Standards for Electrical Systems, published in the *Federal Register* as 29 CFR Part 1910, Subpart S. The regulation became effective April 16, 1981, and some articles and sections apply to all electrical installations and utilization equipment, and thus are retroactive.

1.4.2 Environmental Considerations

The limited availability of energy sources and the steadily increasing cost of electrical energy require that engineers be concerned with energy conservation. Electrical engineer should participate in studies such as total energy required compared to utility power available, electric heating versus fossil fuel, and the like. In these studies the effects of noise, vibration, exhaust gases, cooling methods, and energy requirements must be considered in relation to the immediate and sometimes general environment.

1.5 EDISON ELECTRIC INSTITUTE

The Edison Electric Institute (EEI), the trade association of privately owned electric utilities, publishes the following handbooks, which are of interest to electrical engineers:

1. A Planning Guide for Architects and Engineers
2. Industrial and Commercial Power Distribution
3. Industrial and Commercial Lighting
4. Underground Systems Reference Book

1.6 HANDBOOKS AND PERIODICALS

The following handbooks and periodicals have, over the years, established reputations in the electrical field. The list is not intended to be all-inclusive; other excellent references are also available. The list is limited primarily to subjects of special relevance to this book.

Handbooks

1. Electrical Transmission and Distribution Reference Book, Westinghouse Electric Corp., Trafford, Pa., 1964.
2. Electrical Utility Engineering Reference Book, Vol. 3: Distribution Systems, Westinghouse Electric Corp., Trafford, Pa., 1965.
3. Beeman, D. L., editor, Industrial Power Systems Handbook, McGraw-Hill, New York, 1955.
4. Lighting Handbook, Illuminating Engineering Society, New York, Application Volume, 1987, Reference Volume, 1984.
5. Underground Systems Reference Book, Edison Electric Institute, Washington, D.C., 1957.
6. Fink, D. G. and Beaty, H. W., editors, Standard Handbook for Electrical Engineers, McGraw-Hill, New York, 12th Edition.

Periodicals

1. Consulting-Specifying Engineer, 270 St. Paul Street, Denver, CO 80206
2. Electrical Construction and Maintenance (EC&M), 1221 Avenue of the Americas, New York, NY 10020
3. Electrical Systems Design, 5123 West Chester Pike, Edgemont, PA 19028
4. Transmission and Distribution, 5123 West Chester Pike, Edgemont, PA 19028
5. Plant Engineering, 1301 South Grove Avenue, Barrington, IL 60010
6. Power Engineering, 1301 South Grove Avenue, Barrington, IL 60010
7. Lighting Design and Application (IES), 345 East 47 Street, New York, NY 10017
8. Architectural Lighting, P.O. Box 10955, Eugene, OR 97440
9. Lighting Dimensions, 135 Fifth Avenue, New York, NY 10010

1.7 MANUFACTURERS' DATA

Manufacturers' catalogs are a valuable source of equipment information. Some of the larger manufacturers' complete catalogs are very extensive, covering dozens of volumes; however, these companies may also issue abbreviated or condensed versions that are adequate for most applications. Data sheets referring to specific items are always available from the sales offices. Manufacturers' representatives, both sales and technical, can do much to provide complete information on their products.

2
System Planning

2.1 INTRODUCTION

A typical power distribution system is not adaptable to all industrial plants because different plants have different specific requirements. Analyses must be made of specific requirements before the engineer is able to design the system that will meet the requirements of a particular plant with due consideration given to both present and future operating conditions.

2.2 BASIC DESIGN CONSIDERATIONS

Several basic considerations should be carefully given before a power distribution system is chosen.

2.2.1 System Reliability

The degree of importance in service reliability is dependent on the type of manufacturing or process operation of the plant. Some plants can tolerate momentary outages, whereas others require a high degree of service continuity. The system should be designed with the capability to isolate faults with minimum disturbance and should have features to produce the maximum dependability consistent with specific requirements. Many factors influence the reliability of a plant's power supply. The most important factors are the following:

1. Reliability of the bulk power supply from utility and/or local generation. Faults in the supply system can cause momentary

voltage depressions in the supply voltage which may adversely affect industrial equipment.
2. Plant distribution system arrangement.
3. Simplicity of system.
4. Simplicity in operation and maintenance.
5. Reliability of equipment and installation.

Various recent studies of equipment reliability are reported in the Bibliography at the end of the chapter. System design engineers should pay special attention to equipment with high failure rates.

2.2.2 Economics

In general, the power distribution cost represents only 2 to 10% of plant investment. Economics is a very important consideration in the design of a power distribution system; engineers must compare systems on the basis of cost as well as other factors. In making economic comparisons, it is important to include all parts of the system, from the power source down to and including the utilization equipment. Economics studies should include installation as well as equipment and operating costs. Selection of power incoming voltage and utilization voltages has a serious impact on the economics of system operation. The costs of switchgear, transformers, and cables are varied according to the voltages selected for the system. The fundamental consideration in selecting equipment is to choose the optimum equipment consistent with the requirements of the plant. Some widely accepted guidelines are:

1. Choose metal-clad or metal-enclosed equipment.
2. Choose nonflammable or dry-type transformers for indoor applications.
3. Use factory-assembled equipment for easier field installation.
4. Be sure that all equipment ratings are adequate, such as voltage, current, momentary and interrupting ratings, and basic insulation level.

2.2.3 Flexibility

Plants change manufacturing processes from time to time. Both process and product may change as demands and style change. The power distribution system should be flexible enough so that complete new process layouts can be made without requiring major changes in the distribution system. Flexibility for future expansion should be considered. The engineer should strive for a system design that will permit reasonable expansion with minimum investment and minimum downtime to existing production. Two types of equipment contribute significantly to flexibility: (1) the load center system and (2) plug-in busway that permits the installation of equipment with a great degree of flexibility.

2.2.4 System Quality

System stability and decent voltage regulation are important factors in a well-designed power distribution system. Surges, transients, and dips occurring on the distribution line can result in poor or even damaged products. Design must be aimed at minimizing any kind of disturbances.

2.2.5 Operation and Maintenance

In planning the distribution system, the accessibility and availability for inspection and repair should be given careful consideration. Proper maintenance is as important to successful performance of the system as is selection of the system and its components. Design engineers can aid in maintenance by designing systems that can be maintained economically with a minimum of downtime, or by providing alternate power circuits. It should be possible to remove one circuit for maintenance without dropping essential load. Drawout equipment should be very easy to maintain. All this is possible by careful equipment location and provision of convenient auxiliary services.

2.2.6 Future Expansion

Plant loads increase through the years. Selection of plant voltages, equipment ratings, space for additional equipment, and capacity to be allowed for expansion must be planned carefully. Figure 2.1 shows a typical industrial unit substation incorporating the foregoing considerations.

2.3 LOAD SURVEY

Determination of the load is the design engineer's first problem and is difficult to solve. The size and number of primary and secondary substations; the size, number, and arrangement of primary feeders; and the type of secondary distribution are largely dependent on the amount and nature of the load.

2.3.1 Preliminary Load Survey

How to make preliminary estimates of loads is a problem that deserves careful study. These estimates may have to be used as the basis for major decisions. At this stage design engineers often have only a few scanty building layout drawings. The location of major pieces of equipment may not be known. Starting with the scanty information on hand, engineers must depend on their knowledge and experience to enable them to arrive at estimates that will stand up at a later date. It is better to consider the lighting and power loads separately and combine them later to determine the demand in any one area, since the usual practice is to supply these loads from a common load center substation. Several factors frequently used in determining distribution load are listed below:

1. *Demand factor:* the ratio of the maximum demand on a system to the total connected load of the system. (The maximum demand is usually the integrated maximum kilowatt demand over a 15- or 30-min interval rather than the instantaneous or peak demand.)

Figure 2.1 Typical industrial unit substation.

2. *Diversity factor:* the ratio of the sum of the individual maximum demands of the various parts of a system to the maximum demand of the entire system.
3. *Load factor:* the ratio of the average load over a designated period to the peak load occurring in that period.

Information on the demand and diversity factors for various loads and groups of loads is needed for system design. For example, the sum of the connected loads on a branch load circuit multiplied by the demand factor of these loads gives the maximum demand that the branch circuit must carry. The sum of the maximum demands of the branch circuits associated with a subload center or panelboard divided by the diversity factor of these branch circuits gives the maximum demand at the subload center and on the circuit supplying it. The sum of the maximum demands of the circuits radiating from a load center, divided by the diversity factor of these circuits, gives the maximum demand on the transformer at the load center. The sum of the maximum demands of the load center transformers divided by the diversity factor of the transformer loads gives the maximum demand on their primary feeder. By using the proper demand and diversity factors, maximum demands on the various parts of the system from the branch load circuits to the power source can be determined.

2.3.2 Lighting Loads

To estimate lighting loads, the general type of construction must be known as well as mounting height and space and location of roof trusses and columns so that the optimum arrangement can be made against physical requirements. The intensity of illumination and the type of lighting system [fluorescent or high-intensity discharge (HID)] desired, together with general construction features, will make possible computation of load using up-to-date recommended practice or data from fixture manufacturers.

For a quick estimate of lighting load today, the engineer must not ignore various mandatory lighting energy conservation guidelines adopted by different states which indicate unit power density figures to be observed. Also, many new and efficient light sources are available for selection and application. Once followed, these practices tend to reduce the lighting load and yet to produce satisfactory results. Table 2.1 shows the IES recommended power limits for a number of typical industries. Information on outdoor lighting is readily available. Fence lighting can be estimated at 200 W per 100 ft. Total outdoor lighting will seldom exceed 25% of all lighting load and may be as low as 5%. Table 2.2 shows estimated lighting requirements for various industries which can be used as a guide. The diversification of the lighting load will be low and the demand factor of the

Table 2.1 IES Recommended Lighting Power Limits

Industry	Power limit[a] (W/ft^2)
Aircraft manufacturing	2.5–6.9
Automobile manufacturing	1.8–7.1
Brewing	1.4
Cement	1.0–11.0
Dairy products	0.8–3.8
Electrical equipment manufacturing	3.3
Electric generating station	0.7–2.5
Explosive manufacturing	0.9
Foundries	0.8–4.5
Glass works	1.4–6.4
Iron and steel manufacturing	0.8–7.4
Machine shop	1.7–3.7
Paper manufacturing	1.1–4.4
Petroleum and chemical plants	0.5–4.0
Printing	2.4–7.8
Rubber tire manufacturing	1.3–3.3
Sheet metal works	1.5–6.8
Textile mills	1.4–3.8
Welding	1.8
Woodworking	1.8

[a]The range given is used to indicate that certain functions will require a higher lighting level than the others for a typical industry. For detailed information, reference should be made to IES publications.

Table 2.2 Estimated Lighting Requirements for Various Industries

Industry	Lighting (% of total connected load)
Steel foundries	1–3
Rolling mills, oil refining	3–5
Heavy electric equipment	5–8
Auto equipment, baking	8–10
Machine parts	10–15
Auto assembly parts	15–25

lighting load connected to a lighting panelboard may vary as shown in Table 2.3.

Table 2.3 Typical Lighting Demand Factors

Distribution system component	Demand factor
Lighting panelboard bus and main overcurrent device	1.0
Lighting panelborad feeder and feeder overcurrent device	1.0
Distribution panel board bus and main overcurrent device	
First 50,000 W or less	0.5
All over 50,000 W	0.4
Remaining components	0.4

2.3.3 Power Loads

Estimating the power loads is considerably more difficult than estimating the lighting loads. Load densities for various industries are difficult to obtain, and if obtained, could be misleading. The technology advancement and recent energy conservation efforts have had an impact on density figures published previously. However, some demand factors established through long years of experience may still serve as a useful guide in estimating load requirements. Table 2.4 shows some typical demand factors. When the loads of individual

Table 2.4 Typical Power Load Demand Factors

Load	Estimated demand factor
Arc furnaces	1.00
Arc welders	0.30
Induction furnaces	0.80
Lighting	1.00
Motors	
General-purpose application	0.30
Semicontinuous process	0.60
Continuous process	0.90
Resistance ovens, heaters, and furnaces	0.80

machines or areas are known, it is necessary to combine them to obtain the maximum demand that determines system capacity. Applying demand factors such as those listed in Table 2.4 to the total connected load will yield a total plant demand assuming a unity diversity factor. If the diversity factor is also known, the demand thus obtained, divided by the diversity factor, gives the actual demand.

The National Electrical Code contains several tables of demand factors on lighting loads for dwelling units, hospitals, hotels, warehouses, and so on; and on nondwelling receptacle loads, electric clothes dryers, and electric ranges. Some of these demand figures may be used to estimate industrial loads.

2.3.4 Electronic Data Processing

Large computers require that special consideration be given to the electric distribution systems that supply them. Special requirements are usually stipulated, but the requirements vary with the computer design. A more detailed treatment is presented in Chapter 10. Power requirements of typical large computer systems range from 200 to 500 kVA, with an air-conditioning load of 25 to 75 kVA.

2.3.5 Conclusion

After a survey or estimate of loads is made, the design engineer will be able to predict the required system capacity. At this point before a complete power distribution system is determined, analyses must be made concerning the characteristics of various loads that will be supplied by the system. Impact loads and loads that when energized can cause surges and/or dips in the system can be handled advantageously with separate load centers, not mixed with the same load center supplying dedicated electronic equipment and/or computer systems. This can be accomplished easily at the planning stage. Purifying power supplies for different loads will be much more difficult and costly once the plant is in operation.

2.4 SYSTEM AND VOLTAGE SELECTION

At the planning stage the design engineer must have a well-established concept in mind about the selection of a system and voltage ratings for various components that are to make up the entire distribution system to meet various requirements of the loads from the survey.

2.4.1 Selection of System

Many types of circuit arrangements are commonly used: radial, secondary selective, primary selective, looped primary, and secondary network. Detailed discussions on each system and its advantages or disadvantages are given in Chapter 4. Although the various systems may be compared on several bases, the choice usually resolves itself into the selection of a system that will provide the required degree of service reliability at minimum cost.

In controlling initial investment, far more can be accomplished by proper selection of the system than by economizing on equipment details. Cost reductions should be obtained by using a less expensive distribution system, with some sacrifice in reserve capacity and reliability, instead of using inferior apparatus. The degree to which extra expenditure should be added to a plant distribution system to increase its service reliability depends on (1) the characteristics of the manufacturing process and (2) the reliability of the power source. For many manufacturing plants, a short power outage can be tolerated. In other plants, a short outage can result in spoilage of considerable material in process. In the latter case, it might be possible to justify an extra expenditure to increase the reliability of the distribution system.

If the source circuits are underground cable, the probability of faults will be much less than in overhead, open wire circuits. Although the frequency of occurrence is low, it usually takes as long as 24 hours to locate and reapir the fault. On the other hand, faults on overhead wire circuits normally require much less time to locate and repair. Historical records of similar installation under similar conditions will provide a reasonable guide as to what to expect.

2.4.2 Selection of System Voltage

The selection of utilization and distribution voltage levels is one of the most important considerations in the power distribution system design. System voltage usually affects the economics of equipment selection and plant expansion more than does any other single factor. The various voltage levels may be classified broadly as follows:

Low voltage: 1,000 V or less
Medium voltage: 1,000–100,000 V
High voltage: above 100,000 V

Factors affecting voltage selection are as follows:

1. *Service voltage available from utility.* When the best distribution voltage for application is not available from the utility, the design engineer must determine the relative costs of using the service voltage for distribution or providing substation transformer capacity to obtain the desired distribution voltage.

2. *Magnitude of the load.* This is also an important factor in selecting distribution voltage. It is usually the determining factor for selecting distribution voltage when the plant is compact and where the load is concentrated in one area.

3. *Distance over which power is to be carried.* Distance is another important consideration in the selection of distribution voltages. It becomes particularly important when sizable loads are located at a distance from the main plant.

4. *Ratings of utilization devices.* Lamps, small motors, business machines, appliances, and so on, are generally available with 115- or 120-V ratings. A 120-V single-phase supply must be made available for service. Three phase motors are available for voltages from 208 V to 13.2 kV. Large arc furnace transformers are generally limited to a maximum of 23 to 34.5 kV because of switching equipment limitations.

5. *Safety.* Safety is a major consideration in selecting system voltage in the area of 120 V or below. A voltage of 32 V has been selected for some ungrounded frame portable tools because voltages above 50 V to ground can be lethal. For safety reasons work on current-carrying parts should be done only with the circuit deenergized.

6. *Codes and standards.* The National Electrical Code places limitations on the voltage ratings of equipment and distribution circuits within buildings and should be consulted to assure conformance.

2.5 POWER SYSTEMS STUDIES

Besides being familiar with the characteristics of loads, effects of loads on the power system, and effects of system irregularities on individual loads, system design engineers must incorporate the findings from the power systems studies in the design of new industrial power distribution system. The principal objective of power system studies is to provide design engineers with a convenient and relatively inexpensive source of information needed to make a system fulfill its goals. These goals are outlined in Section 2.2. Principal types of studies include short circuit, protective device coordination, load flow, transient overvoltage, power-factor improvement, transient stability, and motor starting and grounding.

The transient stability study is most useful to industries that have their generation tied in with utility supplies. Many industrial systems are equipped with under-frequency-load shedding relays that can detect drops in frequency due to severe overload and trip the breakers feeding noncritical loads. This can help to prevent complete system collapse. The proper delay settings for these relays are determined by a transient stability study which calculates the rate of frequency decrement upon loss of utility tie.

A load flow study is primarily an aid to system planning. It can accurately predict changes in voltage level, power factor, and line and transformer loading in an existing system due to a projected increase in system load, permitting the required modifications in the system to be planned for. Another important study is the motor starting study. Starting large motors across the line on a relatively weak system can cause severe voltage dips. Whether the voltage at critical load locations will fall to dangerously low levels or whether voltage will be adequate to accelerate the load can be answered by such a study.

2.5.1 Some Practical Solutions to the New Load Requirements

To maintain the quality level of power to computers and other solid-state equipment, it is essential to minimize transients, voltage spikes, and dips caused either externally or internally. A typical externally caused transient is lightning or the result of capacitor and regulator switching by the utility. Internal transients can be caused by the operation of compressor motors. Voltage spikes from radio-frequency sources within a plant can cause not only power problems but can also ruin equipment. A solution to this type of problem is to ground equipment separate from the neutral grounding of the electrical system. The neutral should be isolated from the building ground all the way back to the transformer. Other means to eliminate transients are by use of rotating equipment (motor-generator set with flywheel), diesel-driven ac generators, or static equipment (rectifiers and inverters).

In view of the inadequacy of utility systems in meeting the stringent requirements of new equipment and devices, whether or not an uninterruptible power supply is included as part of the system design should be given due consideration at an early stage. Some fundamental questions should be answered before a decision can be reached about incorporating a proper interface in the system design:

1. What tangible consequences result from computer failures ascribed to voltage transients?
2. What degree of system reliability is warranted by the consequences of system failures?
3. Which alternative power source can supply the required level of reliability most economically?
4. What voltage and frequency variations can the important sections of the information system tolerate? What are their transient starting characteristics?
5. What are tolerable levels of noise, vibration, and radio-frequency interference (RFI) from the power interface?

2.6 CODES AND STANDARDS

Safety to life and protection of property from electric arcing, fire, and explosions that may result from power distribution system installation and use of electrical equipment should be considered during the design stages. State and municipal regulations will vary to some extent, and fire and casualty insurance carriers usually have specific requirements for safeguarding from certain hazards associated with the use of electrical equipment. Details of the proposed system should be reviewed with representatives of these groups. In some cases, the only authorities having any jurisdiction are the insurance carriers.

In general, all distribution system construction should conform to the latest requirements of the National Electrical Code and the National Electrical Safety Code. Where the requirements are not covered by code, the design of the system and selection of the distribution equipment should conform to all applicable ANSI/IEEE Standards, particularly C37-1986 and C57-1986. The National Electrical Code is published by the National Fire Protection Association (NFPA). It has generally been accepted by most states, municipalities, and insurance authorities as comprising the minimum requirements necessary for safety.

2.7 MAINTENANCE

Dirt is probably the greatest enemy of all electric equipment. It results in greater heating of parts, with resulting lowered efficiency and therefore progressively increased heating. At the same time, this shortens the life of the insulation, because the life expectancy of most insulating materials decreases rapidly as the temperature rises. Insulation usually depreciates with age. A sudden drop in the insulation resistance of any piece of equipment is an indication of developing trouble. This can be prevented by immediate corrective measures. Protective devices will function properly only if regularly examined and checked. The essence of good maintenance can be summarized as follows:

1. Cleanliness of equipment
2. Regular inspections
3. Lubrication of rotating equipment
4. Repairs and adjustments
5. Record keeping

BIBLIOGRAPHY

Chen, Kao, Power Systems Analysis: An Essential Step in the Planning of a Future Industrial System, IEEE-IGA Conference Record 71C1-IGA, pp. 459–463.

Guideline on Electric Power for ADP Installations, Federal Information Processing Standards Publication 94, U.S. Department of Commerce, Sept. 21, 1983.

IEEE Committee Report, Reliability of Electric Utility Supplies of Industrial Plants, IEEE-IC&PS Technical Conference Record, 75-CHO947-1-1A, Toronto, Canada, May 5–8, 1975, pp. 131–133.

IEEE Committee Report, Report of Switchgear Bus Reliability Survey of Industrial Plants, *IEEE Transactions on Industry Applications*, Mar./Apr. 1979, pp. 141–147.

IEEE Committee Report, Report on Reliability Survey of Industrial Plants, Parts 1–6, *IEEE Transactions on Industry Applications*, Mar./Apr., pp. 213–252; July/Aug., pp. 456–476; Sept./Oct., p. 681; 1974.

IEEE Standard 141-1976, Recommended Practice for Electric Power Distribution for Industrial Plants.

IEEE Standard 493-1980, Recommended Practice for Design of Reliable Industrial and Commercial Power Systems.

Recommended Procedure for Lighting Power Limit Determination, Illuminating Engineering Society Publication, LEM-1-1982.

Regotti, A. R., and Trasky, J. G., What to Look for in a Low-Voltage Unit Substation, *IEEE Transactions on Industry and General Applications*, vol. IGA-5, Nov./Dec. 1969, pp. 710–719.

3

Distribution System Voltage Standards and Selections

3.1 INTRODUCTION

An understanding of system voltage nomenclature and the preferred voltage ratings of distribution apparatus and utilization equipment provides a solid foundation for design engineers to make proper selections of industrial power distribution systems and equipment. Satisfactory power supply voltage levels must be maintained under all conditions of plant operation to ensure good-quality production.

3.2 VOLTAGE STANDARDS FOR THE UNITED STATES

The voltage standard for electric power systems in the United States is ANSI C84.1-1982, Voltage Ratings for Electric Power Systems and Equipment (60 Hz). This standard lists all the standard nominal system voltages at the point of delivery by a utility and the point of utilization. Two sets of tolerance limits are provided: range A, which specifies the limits under most operating conditions, and range B, which allows minor excursions outside the range A limits. Table 3.1 shows the Standard Nominal System Voltages and Voltage Ranges A and B at three different voltage classes: low voltage (600 V and below), medium voltage (2.4 to 69 kV), and high voltage (115 kV and above).

Table 3.1 Standard Nominal System Voltages and Voltage Ranges

Standard Nominal System Voltages and Voltages Ranges

VOLTAGE CLASS	NOMINAL SYSTEM VOLTAGE (Note a)			VOLTAGE RANGE A			VOLTAGE RANGE B		
				Minimum		Maximum	Minimum		Maximum
	Two-wire	Three-wire	Four-wire	Utilization Voltage	Service Voltage	Utilization and Service Voltage (Note d)	Utilization Voltage	Service Voltage	Utilization and Service Voltage
Low Voltage (Note c)				**Single-Phase Systems**					
	120			110	114	126	106	110	127
		120/240		**110/220**	**114/228**	**126/252**	**106/212**	**110/220**	**127/254**
				Three-Phase Systems					
			208Y/120 (Note e)	**191Y/110**	**197Y/114**	**218Y/126**	**184Y/106** (Note f)	**191Y/110** (Note f)	**220Y/127**
			240/120	**220/110**	**228/114**	**252/126**	**212/106**	**220/110**	**254/127**
		240		220	228	252	212	220	254
			480Y/277	**440Y/254**	**456Y/263**	**504Y/291**	**424Y/245**	**440Y/254**	**508Y/293**
		480		**440**	**456**	**504**	**424**	**440**	**508**
		600 (Note g)		550	570	630 (Note g)	530	550	635 (Note g)
Medium Voltage		2 400		2 160	2 340	2 520	2 080	2 280	2 540
			4 160Y/2 400	3 740Y/2 160	4 050Y/2 340	4 370/2 520	3 600Y/2 080	3 950Y/2 280	4 400Y/2 540
		4 160		**3 740**	**4 050**	**4 370**	**3 600**	**3 950**	**4 400**
		4 800		4 320	4 680	5 040	4 160	4 560	5 080
		6 900		6 210	6 730	7 240	5 940	6 560	7 260
			8 320Y/4 800	(Note h)	8 110Y/4 680	8 730Y/5 040	(Note h)	7 900Y/4 560	8 800Y/5 080
			12 000Y/6 930		11 700Y/6 760	12 600Y/7 270		11 400Y/6 580	12 700Y/7 330
			12 470Y/7 200		**12 160Y/7 020**	**13 090Y/7 560**		**11 850Y/6 840**	**13 200Y/7 620**
			13 200Y/7 620		**12 870Y/7 430**	**13 860Y/8 000**		**12 540Y/7 240**	**13 970Y/8 070**
			13 800Y/7 970		13 460Y/7 770	14 490Y/8 370		13 110Y/7 570	14 520Y/8 380
		13 800		**12 420**	**13 460**	**14 490**	**11 880**	**13 110**	**14 520**
			20 780Y/12 000	(Note h)	20 260Y/11 700	21 820Y/12 600	(Note h)	19 740Y/11 400	22 000Y/12 700
			22 860Y/13 200		22 290Y/12 870	24 000Y/13 860		21 720Y/12 540	24 200Y/13 970
		23 000			22 430	24 150		21 850	24 340
			24 940Y/14 400		**24 320Y/14 040**	**26 190Y/15 120**		**23 690Y/13 680**	**26 400Y/15 240**
			34 500Y/19 920		**33 640Y/19 420**	**36 230Y/20 920**		**32 780Y/18 930**	**36 510Y/21 080**
		34 500			33 640	36 230		32 780	36 510

VOLTAGE CLASS	Two-wire	Three-wire	Four-wire	Maximum Voltage
Medium Voltage		46 000		(Note i) 48 300
		69 000		**72 500**
High Voltage		**115 000**		**121 000**
		138 000		**145 000**
		161 000		169 000
		230 000		**242 000**
Extra-High Voltage (Note j)		345 000		**362 000**
		500 000		**550 000**
		765 000		**800 000**
Ultra-High Voltage (Note j)		1 100 000		**1 200 000**

NOTES:

(a) Three-phase three-wire systems are systems in which only the three-phase conductors are carried out from the source for connection of loads. The source may be derived from any type of three-phase transformer connection, grounded or ungrounded. Three-phase four-wire systems are systems in which a grounded neutral conductor is also carried out from the source for connection of loads. Four-wire systems in Table 1 are designated by the phase-to-phase voltage, followed by the letter Y (except for the 240/120 volt delta system), a slant line, and the phase-to-neutral voltage. Single-phase services and loads may be supplied from either single-phase or three-phase systems.

(c) Minimum utilization voltages for 120-600 volt circuits not supplying lighting loads are as follows:

Nominal System Voltage	Range A	Range B
120	108	104
120/240	**108/216**	**104/208**
(Note f) **208Y/120**	**187Y/108**	**180Y/104**
240/120	**216/108**	**208/104**
240	216	208
480Y/277	**432Y/249**	**416Y/240**
480	**432**	**416**
600	540	520

(d) For 120-600 volt nominal systems, voltages in this column are maximum service voltages. Maximum utilization voltages would not be expected to exceed 125 volts for the nominal system voltage of 120, nor appropriate multiples thereof for other nominal system voltages through 600 volts.

(e) A modification of this three-phase, four-wire system is available as a 120/208Y volt service for single-phase, three-wire, open-wye applications.

(f) Many 220 volt motors were applied on existing 208 volt systems on the assumption that the utilization voltage would not be less than 187 volts. Caution should be exercised in applying the Range B minimum voltages of Table 1 and Note (c) to existing 208 volt systems supplying such motors.

(g) Certain kinds of control and protective equipment presently available have a maximum voltage limit of 600 volts; the manufacturer or power supplier or both should be consulted to assure proper application.

(h) Utilization equipment does not generally operate directly at these voltages. For equipment supplied through transformers, refer to limits for nominal system voltage of transformer output.

(i) For these systems Range A and Range B limits are not shown because, where they are used as service voltages, the operating voltage level on the user's system is normally adjusted by means of voltage regulation to suit their requirements.

(j) Information from American National Standard C92 2-1978 Nominal voltages above 230,000 volts are not standardized. The nominal voltages listed are typically used with the associated preferred standard maximum voltages.

3.2.1 Power Transmission and Distribution in the United States

A general understanding of the principles of power transmission and distribution in utility systems is essential for a design engineer, since most industrial plants obtain their power from the local utility. Most generating stations are located near a source of fuel and water. The electric power generated is transformed in a substation to a transmission voltage, generally 69 kV or higher. This power is classed as unregulated. Note in Table 3.1 that ANSI specifies only the nominal and maximum values for systems over 34.5 kV. Transmission lines supply a distribution substation with transformers to step down to a primary distribution voltage, generally in the range 4.16 to 34.5 kV, with 12.47 kV and 13.2 kV as the most widely used voltages. In recent years there has been an increasing trend toward the use of 34.5 kV as the primary distribution voltage. This is where voltage control is normally applied when necessary for the purpose of maintaining satisfactory voltage to the terminals of utilization equipment. Substation transformers are generally equipped with tap-changing-underload devices in order to maintain a more constant voltage. The regular controls are provided with compensators that raise the voltage as the load increases and lower the voltage as the load decreases, to compensate for any voltage drop in the primary distribution system. Thus a fixed average voltage can be maintained along the primary distribution system. Figure 3.1 illustrates the compensation effect of regulator.

In general, plants close to the substation will receive voltages higher than those received by plants remote from the substation. The primary distribution system supplies power to distribution transformers, which then step it down to utilization voltages in the range 120 to 600 V. Supply voltages to an industrial plant are often varied depending on the size of the plant load. Large plants with loads greater than a few thousand kVA are connected to the transmission system and will have to provide the primary distribution system, the distribution transformers, the secondary distribution system, and may even provide the substation. Small plants with loads only up to several hundred kVA may be supplied from low-voltage secondary distribution systems. If power is supplied by in-plant generation, the generators will replace the utility supply system wholly or in part. Generators may replace the distribution transformers if they deliver power at 600 V or below.

3.3 VOLTAGE TOLERANCE LIMITS

The voltage tolerance limits for ANSI C84.1-1982 are based on NEMA MG1-1972, Motors and Generators, which established the voltage tolerance limits of the standard induction motor at ±10% of nameplate ratings of 230 V and 460 V. Since motors represent the major portion of utilization equipment, they were given primary consideration in the establishment of the voltage standard.

One convenient way to show the voltage in a distribution system is in terms of a 120 V base. The tolerance limit of the 460 V motor in terms of 120 V base becomes 115 V plus 10% or 126.5 V, and 115 V minus 10%, or 103.5 V. This tolerance range of 23 V should be divided among the primary distribution system, the distribution transformer, and the secondary distribution system. Table 3.1, range B, shows a maximum service voltage of 127 V, allowing a 13 V drop in the primary distribution system and a 4 V drop in the distribution transformer, resulting in 110 V as a minimum service voltage. A 4 V drop is also allotted in the plant wiring; thus a minimum utilization voltage of 106 V is established. The range A limits for the standard were established by reducing the maximum tolerance limits from 127 V to 126 V and increasing the minimum tolerance limits from 106 V to 110 V. The spread of 16 V was then allotted as follows: 9 V for the voltage drop in the primary distribution system to provide a minimum primary service voltage of 117 V; 3 V for the voltage drop in the distribution transformer and the secondary connections to provide a minimum utility secondary service voltage of 114 V; and 4 V for the voltage drop in the plant wiring to provide a minimum utilization voltage of 110 V. Table 3.2 shows a standard voltage pro-

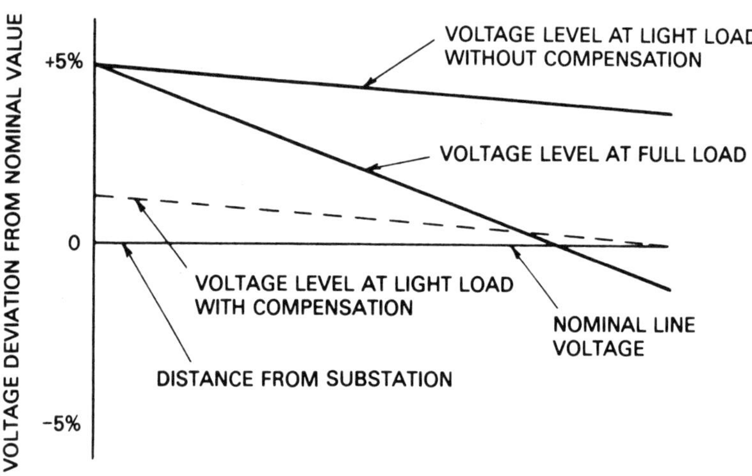

Figure 3.1 Effect of regulator compensation on primary distribution system voltage.

file for a low-voltage regulated power distribution system (120 V base) which summarizes the discussions above.

Electric supply systems are to be designed and operated so that most service voltages fall within the range A limits. User systems are also to be designed and operated within range A. Range B is to allow limited excursion of voltage outside the range A limits which result from practical design and operating conditions. The supplying utility is expected, within a reasonable period, to take action to restore service voltage to range A limits. The user is also expected to take action to restore utilization voltage to range A limits. In general, utilization equipment may be expected to give acceptable performance at a voltage outside range A but within range B.

3.4 DEFINITION OF NOMINAL SYSTEM VOLTAGE

The term "nominal system voltage" designates not a single voltage, but a range of voltage over which the actual voltages at any point in the system may vary and still provide satisfactory operation for equipment connected to the system. The voltage range for all standard nominal system voltages in the utilization and distribution range 120 V to 34.5 kV is specified in Table 3.1 for two critical points in

Table 3.2 Standard Voltage Profile for Low-Voltage Regulated Power Distribution System with 120-V Base[a]

	Range A (V)	Range B (V)
Maximum allowable voltage	126(125*)	127
Voltage drop allowance for primary distribution line	9	13
Minimum primary service voltage	117	114
Voltage drop allowance for distribution transformer	3	4
Minimum secondary service voltage	114	110
Voltage drop allowance for plant wiring	6(4+)	6(4+)
Minimum utilization voltage	108(110+)	104(106+)

[a]*, For utilization voltage of 120 to 600 V; +, for building wiring circuits supplying lighting equipment.

the distribution system: the point of delivery by the supply utility and the point of connection to utilization equipment. For transmission voltages over 34.5 kV, only the maximum voltage is specified, because the voltages are normally unregulated and only a maximum voltage is required to establish the design insulation level for the line and associated apparatus.

3.5 SYSTEM VOLTAGE NOMENCLATURE

The nominal system voltages in Table 3.1 are designated in the same way as on the nameplate of the transformer for the winding or wingings supplying the system.

1. Single-phase systems
   a. 120: a single-phase two-wire system in which the nominal voltage between the two wires is 120 V.
   b. 120/240: a single-phase three-wire system in which the nominal voltage between two phase conductors is 240 V, and between each phase conductor and the neutral is 120 V.
2. Three-phase systems
   a. 240/120: a three-phase four-wire system supplied from a delta-connected transformer. The midtap of one winding is connected to a neutral. The three-phase conductors provide a nominal 240 V three-phase system, and the neutral and the two adjacent phase conductors provide a nominal 120/240 V single-phase system.
   b. Single number: a three-phase three-wire system in which the number designates the nominal voltage between phases.
   c. Two numbers separated by a Y/: a three-phase four-wire system from a wye-connected transformer in which the first number denotes the nominal phase-to-phase voltage and the second number denotes the nominal phase-to-neutral voltage.

Figure 3.2 shows principal transformer connections that supply the various nominal system voltages as described above.

3.5.1 Standard Nominal System Voltages in the United States

The nominal system voltages listed in the left-hand column of Table 3.3 are officially designated as "standard nominal system voltages" in the United States by ANSI C84.1-1982. In addition, those shown in boldface type in Table 3.1 are designated as preferred standards to provide a long-range plan for reducing the multiplicity of voltages.

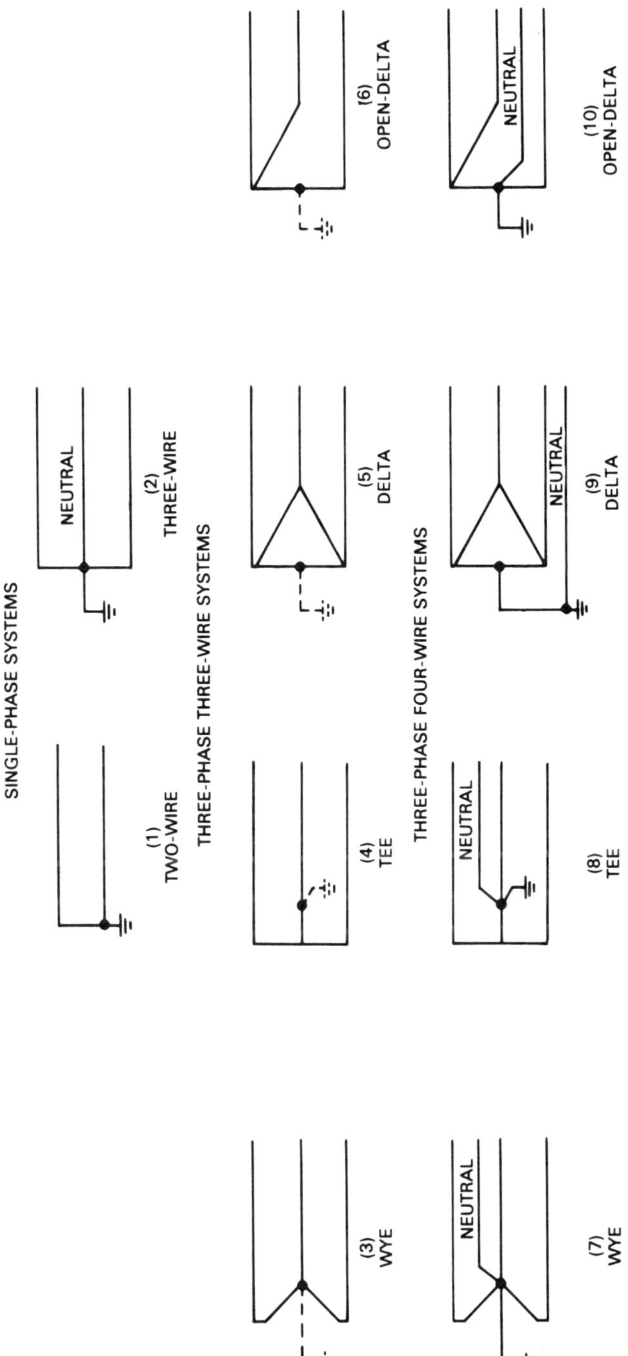

Figure 3.2 Principal transformer connections to supply various nominal voltages.

Table 3.3 Nominal System Voltages

Standard Nominal System Voltages	Associated Nonstandard Nominal System Voltages
Low voltages	
120	110, 115, 125
120/240	110/220, 115/230, 125/250
208Y/120	216Y/125
240/120	
240	230, 250
480Y/277	460Y/265
480	440
600	550, 575
Medium Voltages	
2400	2200, 2300
4160Y/2400	
4160	4000
4800	4600
6900	6600, 7200
8320Y/4800	11 000, 11 500
12 000Y/6930	
12 470Y/7200	
13 200Y/7620	
13 200	
13 800Y/7970	14 400
13 800	
20 780Y/12 000	
22 860Y/13 200	
23 000	
24 940Y/14 400	
34 500Y/19 920	
34 500	33 000
46 000	44 000
69 000	66 000
High Voltages	
115 000	110 000, 120 000
138 000	132 000
161 000	154 000
230 000	220 000
Ultra-High Voltages	
345 000	
500 000	
765 000	
1 100 000	

For the utilization voltages of 600 V and below, the associated nominal system voltage in the right-hand column are obsolete and should not be used. In the case of medium distribution voltages, the numbers in the right-hand column may designate an older system in which the voltage tolerance limits are maintained at a different level than the standard nominal system voltage, and special consideration must be given to the distribution transformer ratios, taps, and tap settings.

There are two situations that necessitate the use of taps:

1. Taps are needed if the primary voltage has a nominal value that is slightly different from the transformer primary nameplate rating. For example, if a 13,200/480 V transformer is connected to a nominal 13,800 V system, the nominal secondary voltage would be 502 V.
2. Taps are used to adjust the utilization voltage spread band to provide a closer fit to the tolerance limits of the utilization equipment.

Table 3.4 shows the voltage tolerance limits of standard 460 V and 440 V three-phase induction motors. Table 3.5 shows the tolerance limits for 277 V and 265 V fluorescent lamp ballasts. A study of these two tables shows that a tap setting of "normal" will provide the best fit with tolerance limits of the 460 V motor and 277 V ballast. Note that the examples above assume that the tolerance limits of the supply and utilization voltages are within the tolerance limits specified in ANSI C84.1-1982. This may not be true, so the actual voltages should be measured with a recording chart. The actual voltages can then be used to assist the selection of transformer taps.

When a plant has not yet been built, the supply utility should be requested to provide the expected spread band. Recommendations should be obtained from the supply utility on transformer ratios, taps, and tap settings. If the supply voltage offered by the utility is one of the associated nominal system voltages listed in Table 3.3, the taps on a standard distribution transformer will generally be sufficient to

Table 3.4 Tolerance Limits for Low-Voltage Three-Phase Motors

Motor rating (V)	Tolerance limit (V)	
	-10%	+10%
460	414	506
440	396	484

Table 3.5 Tolerance Limits for Low-Voltage Standard Fluorescent Lamp Ballasts

Ballast rating (V)	Tolerance limit (V)	
	-10%	+10%
277	249	305
265	238	292

adjust the distribution transformer ratio to provide a satisfactory utilization voltage range. Taps serve only to move the secondary voltage spread band up and down in the steps of the taps. They cannot correct for excessive spread in the supply voltage or excessive drop in the plant wiring system.

3.6 VOLTAGE RATINGS FOR LOW-VOLTAGE UTILIZATION EQUIPMENT

Utilization equipment is defined as electric equipment that uses electric power by converting it into another form of energy: light, heat, or mechanical motion. Most utilization equipment carries a nameplate rating which is the same as that of the voltage system. The major exception is that of motors and equipment containing motors. Single-phase motors for use on a 120-V system have been rated 115 V for many years. Single-phase motors for use on 208 V single-phase systems are rated 200 V, and for 240 V single-phase systems, they are rated 230 V.

Prior to the late 1960s, low-voltage three-phase motors were rated 220 V for use on both 208 V and 240 V systems, 440 V for use on 480 V systems, and 550 V for use on 600 V systems. The reason was that in large industrial plants, relatively long circuits resulted in considerable voltage drop at the end of the circuits. Also, the utility supply system had a limited capacity and low voltages were fairly common. By the mid-1960s surveys indicated that the average voltage supply for 440 V motors on a 480 V system was 460 V. At about the same time MEMA decided that the improvements in motor design and insulation systems would allow a reduction of two frame sizes for standard induction motors rated 600 V and below. Consequently, the nameplate voltage rating of the new T-frame motor was raised from 220/440 V (for U-frame motors) to 230/460 V. A motor rated 200 V is to be used on a 208-V system. Table 3.6 shows the nameplate voltage ratings of standard induction motors.

Table 3.6 Motor and Motor Control Equipment Nameplate Voltage Ratings

	Nameplate voltage rating (V)			
	Integral horsepower		Fractional horsepower	
Voltage (V)	Three-Phase	One-Phase	Three-Phase	One-Phase
120	—	115	—	115
208	200	—	200	—
240	230	230	230	230
480	460	—	460	—
600[a]	575	—	575	—
2400	2300	—	—	—
4160	4000	—	—	—
4800	4600	—	—	—
6900	6600	—	—	—
13800	13200	—	—	—

[a]Certain kinds of control and protective equipment presently available have a maximum voltage limit of 600 V; the manufacturer or power supplier, or both, should be consulted to ensure proper application.

3.6.1 Effect of Voltage Variations on Utilization Equipment

When the voltage at the terminals of utilization equipment deviates from the value of the nameplate of the equipment, the performance and operating life of the equipment will be affected.

Induction Motor

The variation in characteristics as a function of the applied voltage is shown in Table 3.7. It might appear that motors operating on a voltage over 110% of the motor nameplate rating would run cooler, but it does not work that way. Above 110% voltage can cause severe overheating. Motor current should be analyzed into two components:

1. *Load current:* varies almost directly with load and inversely with voltage

Table 3.7 General Effect of Voltage Variations on Induction Motor Characteristics

	Voltage Variation		
Characteristic	*Function of Voltage*	90 *Percent Voltage*	110 *Percent Voltage*
Starting and maximum running torque	$(\text{Voltage})^2$	Decreases 19%	Increase 21%
Synchronous Speed	Constant	No Change	No Change
Percent Slip	$1/(\text{Voltage})^2$	Increase 23%	Decrease 17%
Full-Load Speed	(Synchronous Speed-Slip)	Decrease 1½%	Increase 1%
*Efficiency**			
Full Load	—	Decrease 2%	Increase ½-1%
¾ Load	—	Practically No Change	Practically No Change
½ Load	—	Increase 1-2%	Decrease 1-2%
Power Factor			
Full Load	—	Increase 1%	Decrease 3%
¾ Load	—	Increase 2-3%	Decrease 4%
½ Load	—	Increase 4-5%	Decrease 5-6%
Full-Load Current	—	Increase 11%	Decrease 7%
Starting Current	Voltage	Decrease 10-12%	Increase 10-12%
Temperature Rise,* Full Load	—	Increase 6-7C	Decrease 1-2C
Maximum Overload Capacity	$(\text{Voltage})^2$	Decrease 19%	Increase 21%
Magnetic Noise—No Load in particular	—	Decrease Slightly	Increase Slightly

* This data applies to motors of over 25 horsepower.

2. *Magnetizing current:* is essentially constant with load, but increases with voltage

Because these two are in quadrature relationship, they must be added vectorially to arrive at total motor current. Magnetizing current changes linearly with load up to about 110% of voltage, and rises sharply beyond 110% volts if the motor goes into saturation. Power factor and efficiency go down. Therefore, excess voltage can make the motor hot.

Lamps

The light output and life of incandescent filament lamps are critically affected by the supply voltage. The variation of life and light output with voltage is shown in Table 3.8. Lamp ratings of 125 V and 130 V are also included because these ratings are useful in signs and other applications where long life is the predominant consideration. Maintenance of the proper voltage is therefore an important factor in obtaining good performance from lamps.

Fluorescent Lamps. The voltage at the luminaire should be kept well within the normal operating range for the ballast. Both low and high voltage reduce efficiency and shorten lamp life. This is in contrast with filament lamps, where low voltage reduces efficiency but prolongs life. Low voltage may also cause starting difficulty. Slow or delayed starting results in shortened lamp life. High voltage causes overheating of the ballast and premature end blackening and early lamp failure. The normal operating range for "low-voltage" ballasts is 110 to 125 V; for "high-voltage" equipment, 220 to 250 V.

High-Intensity-Discharge Lamps. Mercury lamps using conventional unregulated ballast will experience a 30% decrease in light output for a 10% decrease in terminal voltage. If a constant-wattage ballast is used, the decrease in light output for a 10% decrease in terminal voltage will be about 2%. Low-voltage conditions that require repeated starting will shorten lamp life. Excessive high voltage raises

Table 3.8 Effect of Voltage Variations on Incandescent Lamps

	Lamp rating					
	120 V		125 V		130 V	
Applied voltage (V)	Percent Life	Percent Light	Percent Life	Percent Light	Percent Life	Percent Light
105	575	64	880	55	—	—
110	310	74	525	65	880	57
115	175	87	295	76	500	66
120	100	100	170	88	280	76
125	58	118	100	100	165	88
130	34	132	59	113	100	100

the arc temperature, which could damage the glass enclosure. High-pressure-sodium and metal halide lamps have characteristics similar to those of mercury lamps, although the starting and operating voltages may be somewhat different. Ballasts for these lamps are not interchangeable except in special cases.

Solid-State Equipment

Thyristors, transistors, and other solid-state devices do not have thermionic heaters. They are not nearly as sensitive to voltage variations as are electronic tubes. Internal voltage regulators are normally provided for sensitive equipment. An individual study of the maximum voltage of the equipment, including surge characteristics, is essential to determine the effect of maximum system voltage or abnormal low voltage in terms of possible malfunction.

3.7 PHASE VOLTAGE UNBALANCE AND HARMONICS

3.7.1 Phase Voltage Unbalance

Most utility lines are four-wire grounded-wye primary distribution systems that provide power to both single-phase and three-phase loads. Variations in single-phase loading cause the currents in three-phase conductors to differ, causing the phase voltages to become unbalanced. Sometimes blown fuses on a three-phase capacitor bank can cause phase voltage unbalance. Industrial plants make extensive use of either 208Y/120 V or 480Y/277 V utilization voltage to supply lighting loads connected across phase to neutral. It is always desirable to keep the load unbalance and the corresponding phase voltage unbalance within reasonable limits. The amount of voltage unbalance can better be expressed in symmetrical components as the negative-sequence component of the voltage:

$$\text{voltage unbalance factor} = \frac{\text{negative-sequence voltage}}{\text{positive-sequence voltage}}$$

3.7.2 Effect of Phase Voltage Unbalance

When unbalanced phase voltage is applied to three-phase motors, the phase voltage unbalance causes additional negative-sequence currents to circulate in the motor, increasing the heat losses in the rotor. Table 3.9 shows the effect of phase voltage unbalance on motor temperature rise for both U-frame and T-frame motors. Although phase voltage unbalance will cause an increase in the motor lead current, the increase is generally insufficient to indicate actual temperature rise. Some electronic equipment, such as computers, may also be affected by phase voltage unbalance of more than 2 to 2.5%. This is discussed in more detail in Chapter 10. In general, single-phase loads should not be connected to three-phase circuits supplying equipment sensitive to phase voltage unbalance. A separate circuit should be provided.

3.7.3 Harmonics

Harmonics are integral multiples of the fundamental frequency. For example, for 60-Hz power systems, the second harmonic would be 120 Hz and the third harmonic would be 180 Hz. Harmonics are caused by devices that change the shape of the normal sine wave of voltage or current in synchronism with the 60-Hz supply. Any distorted wave must be made up of a fundamental and harmonics of various frequencies and magnitudes. The harmonics content and magnitude existing in any power system is largely unpredictable and effects will vary widely in different parts of the same system. Harmonics may be transferred from one circuit or system to another by direct connection or by inductive or capacitive coupling. Since 60-Hz harmonics are in the low-frequency audio range, the transfer of these frequencies into communication, signaling, and control circuits employing frequencies in the same range may cause objectionable interference. Studies have identified the following areas where harmonics can cause operating problems:

1. Interference with ripple control and power-line carrier systems, causing misoperation of systems that accomplish remote switching, load control, and metering
2. Excessive losses, resulting in heating of induction and synchronous machines

Table 3.9 Effect of Phase Voltage Unbalance on Motor Temperature Rise

Motor type	Load	Percent voltage unbalance	Percent added heating	Insulation system class	Temperature rise (°C)
U frame	Rated	0	0	A	60
	Rated	2	8	A	65
	Rated	3½	25	A	75
T frame	Rated	0	0	B	80
	Rated	2	8	B	86.4
	Rated	3½	25	B	100

3. Overvoltages and excessive currents on the system from resonance to harmonic voltages or currents on the network
4. Dielectric breakdown of insulated cables or capacitor bank, resulting from harmonic overvoltages on the system
5. Inductive interference with telecommunications system
6. Signal interference and relay malfunction, particularly in solid-state and microprocessor-controlled systems
7. Interference with large motor controllers and power plant excitation systems
8. Unstable operation of firing circuits based on zero-voltage-crossing detection or latching

How to Reduce Harmonics Effects

Where harmonics interference exists, the regular measures of increasing the separation between the power and communication conductors and the use of shielded communication conductors should be considered. Where resonant conditions exist, the capacitor bank should be changed in size to shift the resonant point. Where harmonics pass from a power system to a communications, signal, or control circuit through a direct connection such as a power supply, filters may be required to suppress the harmonic frequencies. During preliminary meetings with the supplying utility, the anticipated harmonics problems should be identified. This information, together with other information provided by manufacturers of any equipment to be installed that may generate a voltage distortion, should be used to govern the specifications for other equipment that may be exposed to the harmonic conditions.

BIBLIOGRAPHY

ANSI C84.1-1982, Voltage Ratings for Electric Power Systems and Equipment (60 Hz).

Frank, Jerome M., and Luebke, C. R., Transients and Hermonics in Industrial and Commercial Electrical Systems, IAS Conference Record, Part II, 86CH2272-3, pp. 974–981.

IEEE Standard 141-1976, Recommended Practice for Electric Power Distribution for Industrial Plants.

Lighting Handbook, Westinghouse Electric Corporation, Bloomfield, N.J., Jan. 1976.

NEMA MG1-1972, Motors and Generators.

4

Power Distribution Systems for Industrial Plants

4.1 INTRODUCTION

A variety of basic circuit arrangements are available for industrial plant power distribution. Selection of the best system or combination of systems will depend on the needs of the manufacturing process. In general, system costs increase with the system reliability. Assuming that component quality is equal, maximum reliability per unit investment can be obtained by using properly applied and well-designed components. For processes that are little affected by power interruptions, a simple radial system is satisfactory. For other processes that may sustain long-term damage by even a brief interruption, a more complex system with an alternate power source for critical load may be justified.

For the majority of applications today, the radial and secondary selective arrangements are preferred, with the radial system accounting for 55 to 65% of industrial plant installation. With modern equipment properly installed and maintained, these two arrangements offer optimum reliability, flexibility, and expendability consistent with minimum cost. The secondary network system is used only in a small percentage of the total. In the following sections we discuss in detail several of the most commonly used distribution systems for modern industrial plants.

4.2 SIMPLE RADIAL SYSTEM

In this system a given secondary feeder is fed from only one transformer and one primary cable. Each substation operates independently and there is no duplication of equipment. No extra tie cables, circuit breakers, or transformers with large reserve capacity are needed. System investment is usually the lowest of all circuit arrangements. Despite this, for a well-planned and well-maintained system, the service reliability of the radial system is usually high. Figure 4.1 shows a typical simple radial system. It is true that the loss of a primary cable or transformer will cut off service to the affected loads until repairs can be made. This is often the reason for design engineers to select one of the other circuit arrangements, which can minimize production shutdown.

4.2.1 Expanded Radial System

The advantage of the radial system may be applied to larger loads by using a radial primary distribution system to supply a number of unit substations located near the centers of load, supplying the load through radial secondary systems. This will result in better voltage regulation and in lower power loss than in a system using heavy low-voltage feedres of extensive length. Figure 4.2 shows a single medium-voltage service entrance circuit breaker supplying a number of unit substations. The National Electrical Code permits the installation of transformers without individual primary protection under specified conditions. If the transformers are installed without individual

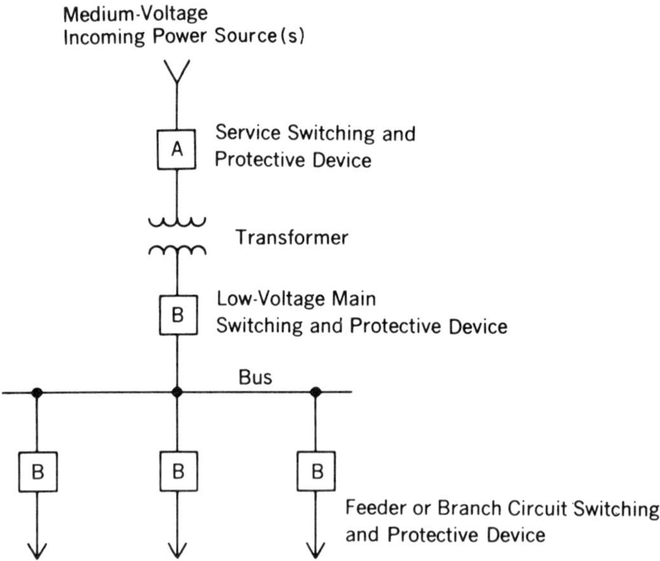

Figure 4.1 A typical simple radial system.

primary protection, a fault in any one transformer will result in loss of the entire system. Since the transformers are very dependable devices, this arrangement has been used in many installations. In Figure 4.3 the transformers are shown with primary fuses. In this arrangement, a fault on the primary system up to the transformer fuses will result in a complete loss of power. A fault in a transformer or on the secondary bus will cause loss of power to the load served by the transformer only. In Figure 4.4 a fault in the supply circuit or on the medium-voltage bus will result in a complete loss of

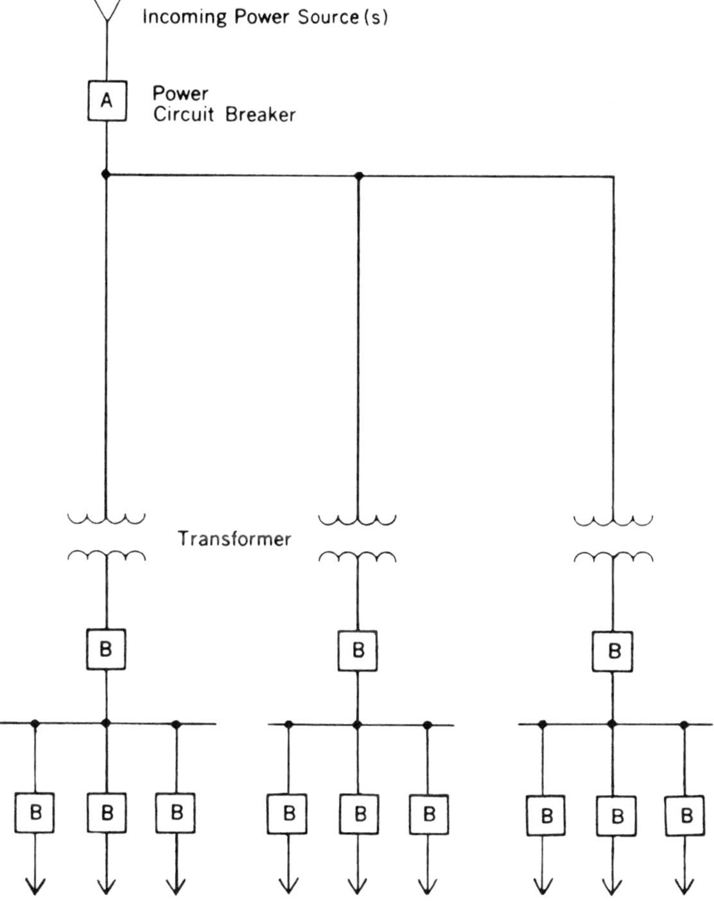

Figure 4.2 Expanded radial system—transformers without individual primary protection.

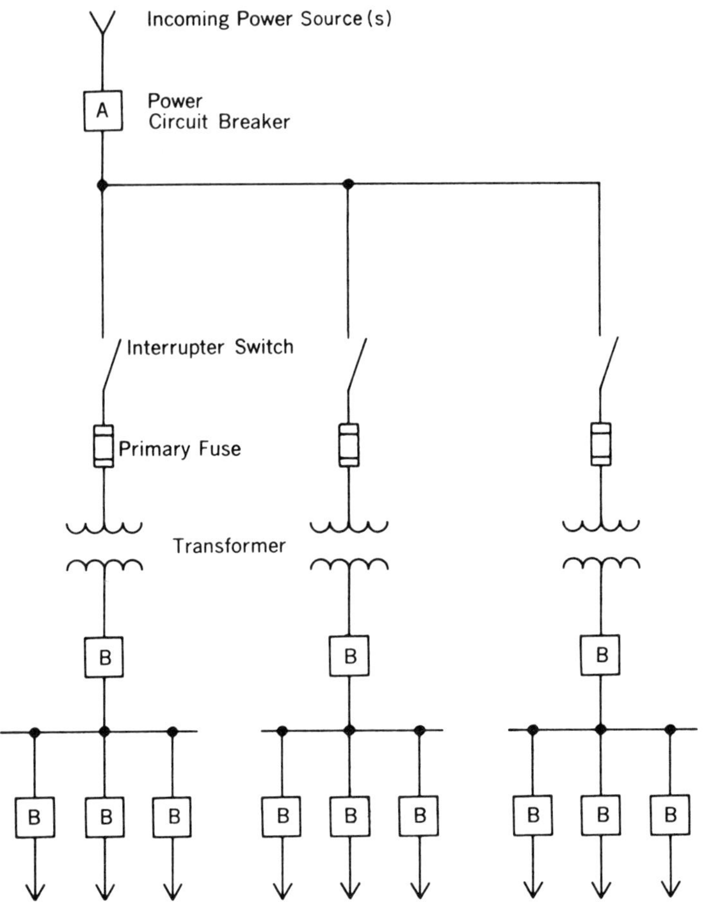

Figure 4.3 Expanded radial system—transformers with individual primary protection.

power. A fault in one of the medium-voltage feeders will result in loss of the load connected to that feeder only. The first cost for the arrangement in Figure 4.4 would be higher than that of the system shown in Figure 4.3, due to the additional length of the medium-voltage cable.

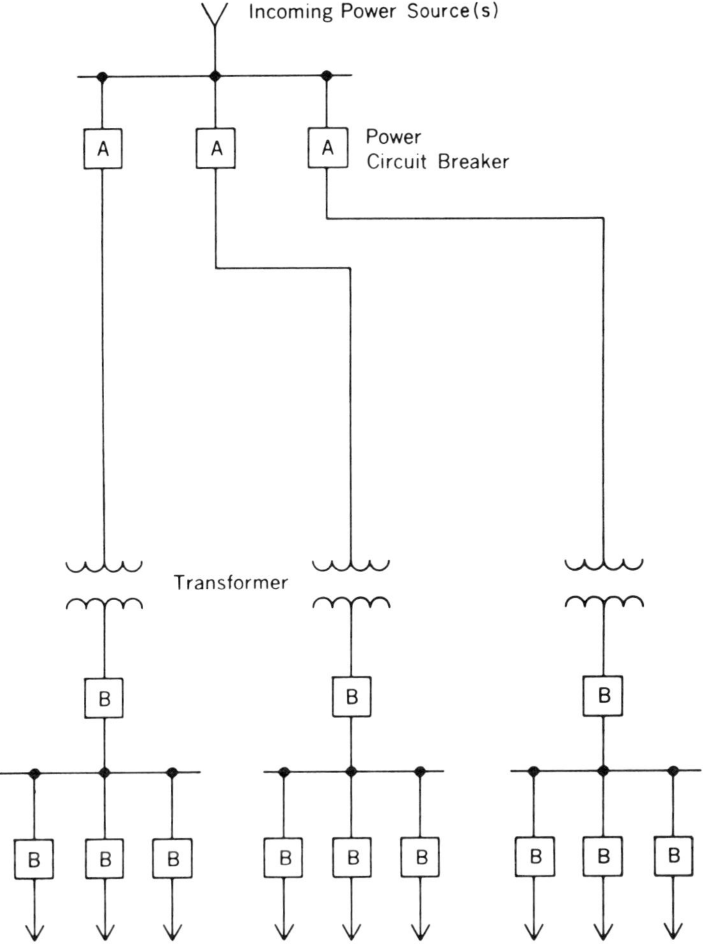

Figure 4.4 Expanded radial system—independently protected medium-voltage supply for each transformer.

4.3 SECONDARY SELECTIVE SYSTEM

The next widely used system in industrial power installation is the secondary selective system. Two varieties of the secondary selective system are shown in Figure 4.5. The system is somewhat similar to the radial system. If pairs of unit substations are connected through a normally open secondary tie circuit breaker, the result is a secondary system. Use of this system increases reliability by reducing

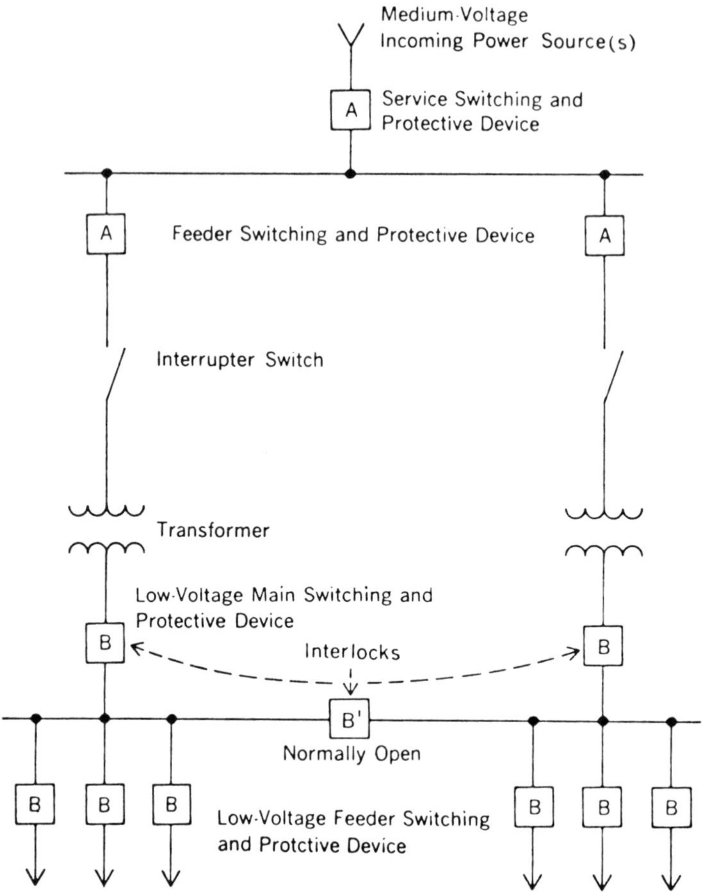

Figure 4.5 (a) Secondary selective system.

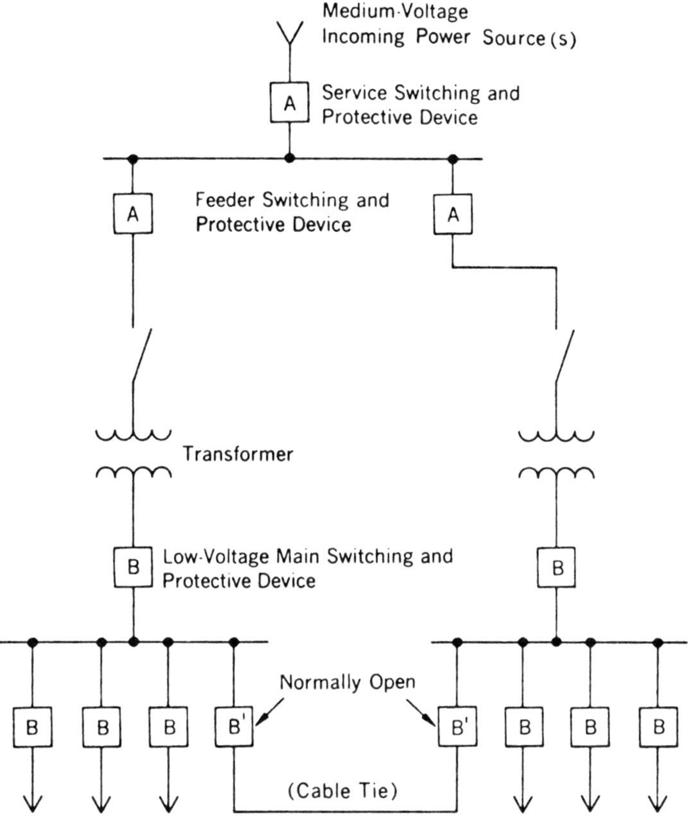

Figure 4.5 (b) Secondary selective system.

the time that a load is without power due to a fault in a unit substation transformer or its primary feeder. It provides flexibility in operation, particularly when equipment is being maintained or serviced. Any part of a primary feeder, or transformer and associated equipment, can be deenergized for inspection or maintenance without loss of power to the loads.

Under normal conditions the system is operated with the tie breaker open and each transformer supplies its own load. Should a fault occur in a transformer or its primary feeder, or if a transformer or primary feeder is deenergized for maintenance, the transformer secondary switching device or breaker is to be opened and the tie breaker is to be closed. The load connected to both buses is supplied by the energized transformer. Each transformer and its primary feeder must have sufficient capacity to carry the total load. Transformers used in this system are often equipped with cooling fans to provide additional capacity during emergency operations. Otherwise, nonessential load will have to be shed.

The tie breaker should be interlocked with the transformer secondary breakers to prevent the transformers from being operated in parallel. Parallel operation of the transformers would increase the available secondary short-circuit current and would risk the loss of power to both secondary buses in the case of a transformer fault or primary cable fault.

If electrically operated circuit breakers are used for the tie switching device and for the transformer secondary switching device, the control may be arranged to transfer the load automatically upon failure of one transformer or its primary feeder. This can reduce loss of power on either bus to a very short time.

The time-current characteristics of the several protective devices connected to the secondary bus should be arranged so as to provide selective operation between the transformer secondary protective device and the tie and the feeder or branch-circuit protective devices. This tends to minimize the amount of load being disconnected under fault conditions.

4.4 PRIMARY SELECTIVE SYSTEM

This system is another means of reducing the time required to restore voltage to a load in the event of loss of a primary feeder as compared with a radial system. In this system two or more primary feeders are provided. Two primary feeders are extended to each transformer, and selector switches are provided so that any transformer may be connected to either of the two primary feeders. Each primary feeder must have sufficient capacity to carry the maximum load that may be connected at one time.

Figure 4.6 shows a typical primary selective system. Under normal conditions, the system is operated with the load divided approximately equally on each primary feeder. When a fault occurs on one primary feeder, there will be an interruption of power to the load connected to that feeder. The interrupter switch connected to the faulted feeder will be opened and the respective transformer will be reconnected to the energized feeder. The two switches associated with one transformer should be mounted in separate individual metal enclosures in order that a deenergized feeder may be safely maintained while the other feeder is energized.

The primary selective system has a higher first cost than a radial system arranged to supply the same load. However, its first cost is usually lower than that of a comparable secondary selective system. The service reliability of this system lies between that of a radial system and that of a secondary selective system.

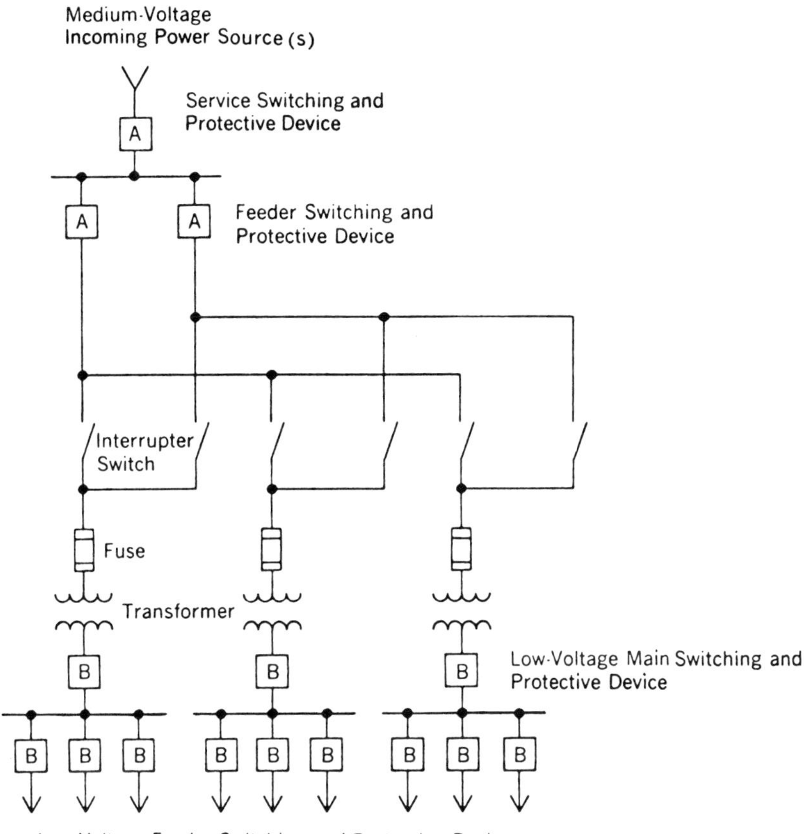

Figure 4.6 Primary selective system.

4.5 LOOPED PRIMARY SYSTEM

Two arrangements of the looped primary system are shown in Figure 4.7a & b. In Figure 4.7a, the primary loop is fed by a single medium-voltage circuit breaker. One loop-sectionalizing interrupter switch is located at the primary of each transformer. One section of the primary loop is connected directly to the primary of each transformer. In case of a fault in a transformer or on the primary loop, the primary circuit breaker will open and clear the fault. After the fault is located, the two interrupter switched at the ends of the faulted section are opened and the remainder of the loop is again energized. However, one transformer and one section of the primary loop remain out of service until the faulter equipment is put back into service.

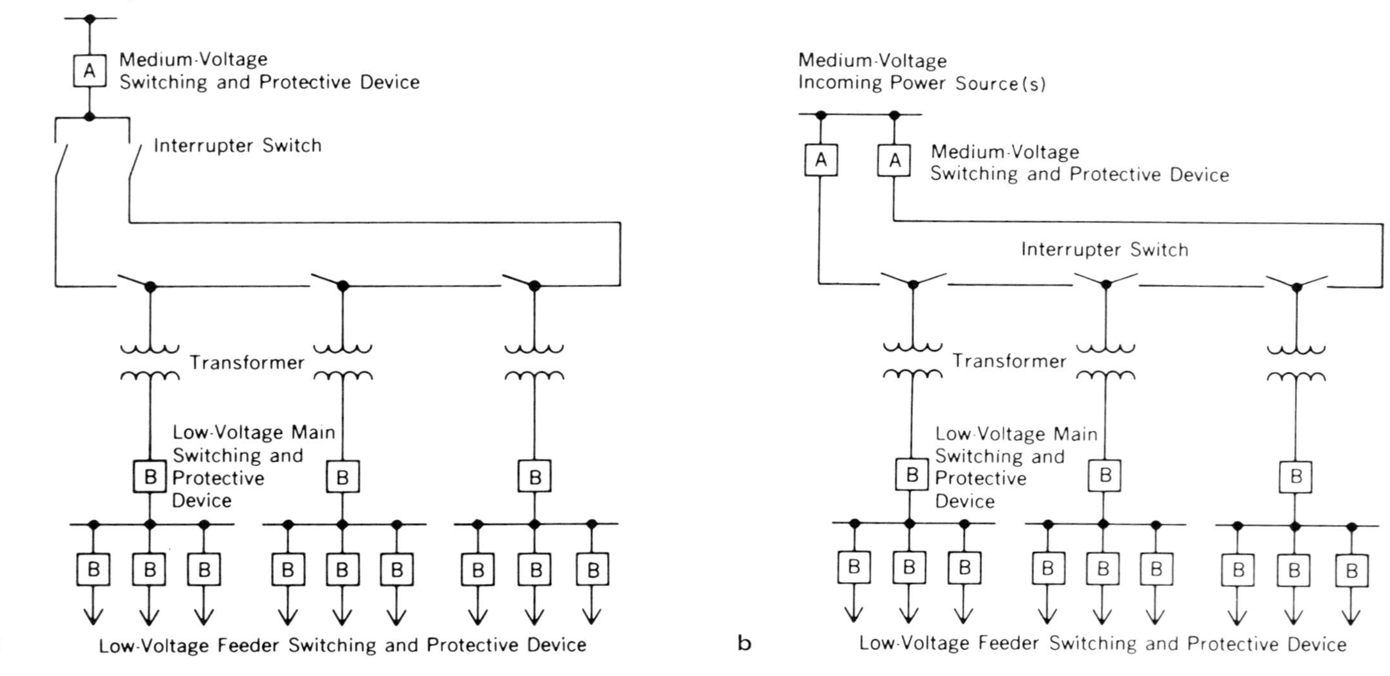

Figure 4.7 Looped primary system.

Two medium-voltage circuit breakers are used (Figure 4.7b), one on each end of the loop, and two interrupted switches are used at each transformer location. With this arrangement, the only time that any part of the load is without service for an extended period is when a fault takes place in a transformer or its secondary bus. Any fault on a section of the primary loop will not affect the transformers supplying their respective loads. The disadvantage of this arrangement lies in a complex and time-consuming operation in locating a fault on the primary loop or on any transformer, because the entire system must be deenergized for a time, and the time required to restore service may also be lengthy.

When the load centers are located relatively far apart, the looped primary system will cost little more than a comparable radial system. The initial cost can be reduced by using fused interrupter switched instead of medium-voltage circuit breakers. However, the interrupter switches must be capable of switching the entire load, and the limitations of the NEC, Section 450-3, on the ratings of the fuses may make coordination difficult.

4.6 SECONDARY NETWORK SYSTEM

The secondary network system can provide a very high degree of service reliability to all the loads. The arrangement of equipment in this system is similar to that of the secondary selective system. The difference is in the way the systems are operated. In the secondary selective system, the tie circuits between secondary buses are normally open and each transformer supplies its own load. In the secondary network system, the secondary buses are tied together and the transformers operate in parallel to supply the entire load. Figure 4.8 shows a typical secondary network system with primary selective system.

In the secondary network system, the transformer secondary switching and protective device is a special low-voltage power circuit breaker known as a netowrk protector. This is an electrically operated circuit breaker provided with relays arranged to trip the breaker on reverse power flow to the transformer and to reclose the breaker when normal voltage conditions return to the primary of the transformer. The relays are normally set so that the circuit breaker will open on magnetizing current to a transformer secondary winding from the low-voltage bus. Separate low-voltage fuses are usually installed on the load side of the network protector to mitigate damage to the protector and to protect the low-voltage system on the occurrence of high-magnitude fault current.

Under normal conditions, the total load is shared by all of the transformers operating in parallel. Should a fault occur in a primary

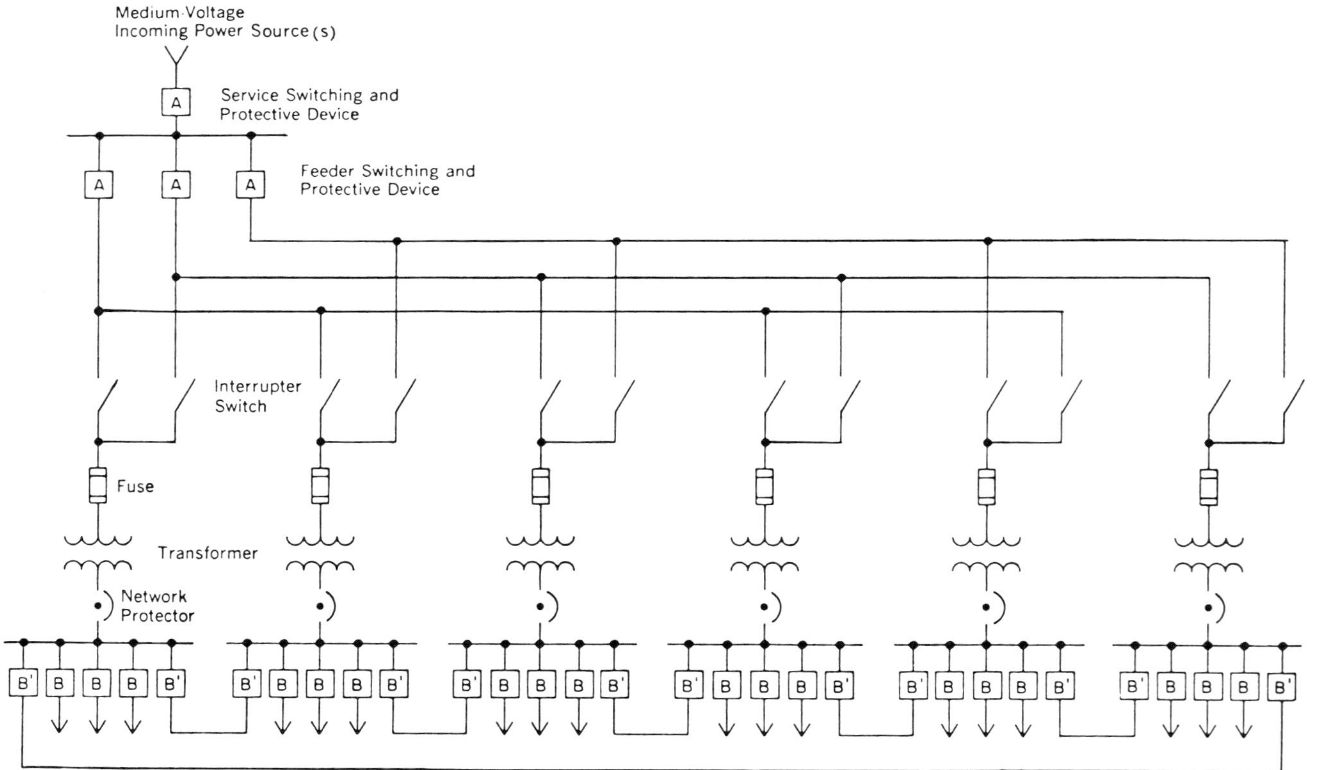

Figure 4.8 Secondary network system with primary selective system.

feeder or in a transformer or if voltage should fail on a primary feeder for any reason, the power flow from the secondary bus to the transformer will cause the network protector to open, thus disconnecting the transformer from the secondary bus. The remaining energized transformers will continue to supply power to the bus, and there will be no interruption of power to the loads. For an ideal operating condition, if total number of transformers is n, each transformer should normally be loaded only to (n − 1)/n percent, so that in an emergency, if one transformer is taken out of service, the remaining transformers can still carry full load of the plant. When normal voltage conditions are restored to a transformer that had been disconnected, the network protector will close automatically and the transformer will again carry its share of the load.

Figure 4.9 shows a spot network arrangement. This system may be used to advantage where there are concentrations of loads. In

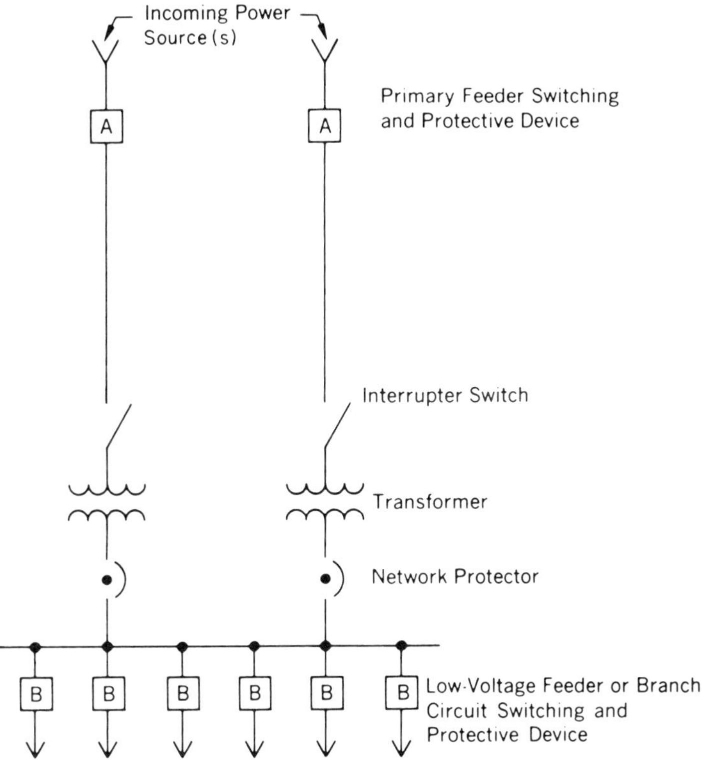

Figure 4.9 Spot network system.

this system, there are two or more transformers connected to a single bus through network protectors. The transformers and primary feeders must have sufficient capacity to carry all loads connected to the bus with one transformer out of service.

For larger installations, a system having a primary selective and secondary network arrangement can be used to advantage (Figure 4.8). Each of the primary feeders must have sufficient capacity to supply the entire load that is connected to it. If two primary feeders are used to supply the primary selective network system, half of the transformers would normally be connected to each feeder with adjacent transformers on different feeders. In case of a primary feeder fault, the fault is isolated from the system by automatic tripping of the primary feeder circuit breaker and all the network protectors associated with the faulted circuit.

The network protector is operated automatically by master and phasing relays. The network master relay trips the network protector for power flow from the network into the transformer. The relay is sensitive to power flow as small as the transformer core losses. On three-phase and line-to-line medium-voltage feeder faults, the master relay operates. On line-to-ground faults, the feeder circuit breaker trips first, then the master relay operated on fault current flow from the network or transformer exciting current. The master relay also recloses the network protector if the source voltage is higher than the network bus voltage; the phasing relay interprets proper voltage relationships to prevent network protector pumping. Following these tripping operations, the entire load will be supplied over the remaining feeder and half of the network transformers. All transformers can be restored to service by manually switching the deenergized units over to the remaining energized feeder.

Secondary networks are sometimes arranged in a closed loop. With this arrangement, if any one secondary tie circuit is out of service, the loads will all continue to operate in parallel. The tie circuits between the secondary buses are sometimes in the form of multiple parallel cables for each phase because a single conductor usually does not have sufficient current-carrying capacity. Each individual cable is protected by a special type of fuse known as a cable limiter, located at each end of the cable. Figure 4.10 shows an actual installation of the limiter box, which is a part of the secondary closed loop. If any cable is faulted, the cable limiters at each end will clear the faulted cable before its insulation is damaged. Each group of tie conductors is connected to the two secondary buses through nonautomatic circuit breakers or disconnecting switches. The tie circuit may also be made up with busways instead of parallel cables. For some tie circuits, a plug-in busway is used so that the load may be tapped off between substations.

Figure 4.10 Limiter box in a secondary closed loop system.

4.7 SELECTING A POWER DISTRIBUTION SYSTEM FOR AN INDUSTRIAL PLANT

4.7.1 Cost Analysis

The relative costs of the several systems discussed above will vary considerably between installations, depending on the total load connected, the areas over which the loads are distributed, and the demand factors of the loads. The first cost of a secondary network system will be higher than that of any of the other systems. However, the reliability of the service will be greatly increased. This is due to the increase in transformer capacity and the increase in the available short-circuit currents on the secondary system, which will in turn require higher-rated protective devices. An analysis should be made of the relative costs of the several systems for each proposed installation, and these costs balanced against probably unscheduled loss of power. The system providing the lowest overall annual operating cost consistent with safe operation should be the choice.

4.7.2 Utilizing Load Center Arrangements for System Formation

The most popular scheme is to have a main distribution system at higher voltages and local distribution centers for low voltages. This has proven to be economical in cutting line losses and in minimizing the effects of faults and power interruptions. Many considerations should go into the planning and specifying of load center substations. Secondary unit substations are those with a secondary voltage not greater than 1500 V. An articulated secondary load center substation should consist of:

1. *Incoming section:* provides for the connection of one or more incoming high-voltage circuits. Each of the circuits may or may not be provided with a power circuit breaker.
2. *Transformer section:* includes one or more transformers.
3. *Outgoing section:* provides for the connection of one or more outgoing feeders. Each of the feeders may or may not be provided with a low-voltage power circuit breaker as a component of the outgoing section.

In planning a load center substation, the primary and secondary voltages must be selected first. Standard secondary voltages in the United States are 575, 480, 240, and 208 V. The transformer secondaries may be wye connected. However, the 240Y/139 V system is rarely used. The 208Y/120 V system is to be used where a large percentage of the load is 120 V, single phase. If the 120 V load is less than half of the total power, a 480Y/277 V system with autotransformer will usually be more economical for supplying 120 V circuits. The 480 V system is the one most commonly used for industrial plants. The principal advantage over a lower-voltage system is the reduced conductor size required in the system.

The radial system is subject to relatively frequent power failures. One way to shorten an outage is to use a selective primary switch with a second power source. For still greater reliability, a selective-type substation may be used. In this setup, two transformers are employed; each is fed from separate power sources, and each is feeding an outgoing section. The two outgoing sections are connected through a tie breaker, which is normally kept open. If one of the power sources fails, the main transformer secondary breaker on the deenergized transformer will be tripped and the tie breaker closed. All loads will then be connected to one source. Tripping of the main breaker and closing of the tie may be accomplished manually or automatically on loss of voltage or when voltage falls below a fixed value.

In sizing transformers for the selective system, overload requirements must be considered. This will be needed when the entire substation is being fed from just one source. This may not mean that each transformer should be capable of supplying the full substation load. Some nonessential loads can be removed during the period. It should be remembered that transformers have an overload rating of 125% of rated capacity for a 2-hour period.

A variation of the selective-type substation has two main breakers but no tie breakers. Normally, the entire load is carried by one transformer. In case of power failure, the main secondary breaker is tripped and the main breaker from the standby transformer is closed. The transfer may be either automatic or manual. Usually, both transformers are of the same size so that load dumping is not necessary. The choice between these two types of selective substation is an economic one.

If power interruptions cannot be tolerated in some applications even for a few minutes, spot network substations may be the choice. Figure 4.9 shows a spot network substation in which two transformers are connected in parallel to a single outgoing section. Directional protective devices are employed so that in case of failure in one of the primary sources, the transformer will be disconnected from the outgoing section. As with the selective system, transformers may be sized to carry the complete substation load or provisions be made for dumping nonessential loads. In the spot network substation, maintaining service to the load depends on maintaining service to at least one of the two sources. Should both primary sources fail, a 480 V load will be without power. If this interruption is unacceptable, the distributed network type of substation may be preferred. The distributed network type of substation must be designed within the short-circuit capability of the network, which is not greater than 200,000 A. This becomes necessary because neither breaker nor fuses have been developed to interrupt more than 200,000 A.

BIBLIOGRAPHY

Everett, Max, Planning Secondary Unit Substations, *Plant Engineering*, Apr. 1967.

IEEE-JH 2112-1, Protection Fundamentals for Low-Voltage Electrical Distribution Systems in Commercial Buildings, 1974.

IEEE Standard 141-1986, Recommended Practice for Electric Power Distribution for Industrial Plants.

Smith, Robert L., Selecting a Reliable Power System Configuration, *Plant Engineering*, Jan. 7, 1982.

5
Fault Current Calculations

5.1 SOURCES OF FAULT CURRENT

Current that flows during a fault usually comes from two basic sources, synchronous and induction rotating machines. The rotating machines may be operating as generators, motors, or synchronous condensers. The current from each rotating machinery source is limited by the impedance of the machine and the impedance between the machine and the fault. These sources exhibit a variable reactance to the flow of fault current.

5.1.1 Generators

Fault current from a generator decreases exponentially from a relatively high initial value to a lower steady-state value sometime after initiation of a fault. Since a generator continues to be driven by its prime mover and to have its field energized from its separate exciter, the steady-state value of the fault current will persist unless interrupted by switching devices.

For fault current calculations, the variable reactance of a generator can be represented by three reactance values:

X_d'' = subtransient reactance, which determines current during first cycle after fault occurs

In about 0.1 s this value increases to

X_d' = transient reactance, which determines current after several cycles

In about 1/2 to 2 s, this reactance increases to

X_d = synchronous reactance, whch determines the current after a steady-state condition is reached

Most fault protective devices such as circuit breakers or fuses operate long before steady-state conditions are reached. Generator synchronous reactance is seldom used in calculating fault currents for selection of these devices.

5.1.2 Synchronous Motors and Condensers

Synchronous motors supply current to a fault in much the same manner as do synchronous generators. When system voltage drops due to a fault, the synchronous motor receives less power from the system for driving its load, and simultaneously the internal voltage will cause current to flow to the system fault. The inertia of the motor and its load acts as a prime mover, and with field excitation maintained, the motor acts as a generator as far as supply of fault current is concerned. The fault current diminishes as the magnetic field in the machine decays. The same designation is used to express the variable reactance of a synchronous motor as for a synchronous generator. However, numerical values of the three reactances X''_d, X'_d, and X_d will be different from that of the generators.

5.1.3 Induction Machines

The fault current contribution of an induction motor results from generator action produced by inertia driving the motor after the fault occurs. In contrast to the synchronous motor, the field flux of the induction motor is produced by induction from the stator rather than from a direct-current field winding. Since this flux decays on removal of source voltage resulting from a fault, the contribution of an induction motor drops off, disappearing after a few cycles. Since field excitation is not maintained, there is no steady-state value of fault current as for synchronous machines. Based on these facts, induction motors are assigned only a subtransient value of reactance (X''_d). This value is about equal to the locked-rotor reactance, hence the fault current contribution will be about equal to the full-voltage starting current of the machine. Large wound-rotor motors are operated with some external resistance maintained in their rotor circuits. They may then have sufficiently low short-circuit time constants that their contribution becomes insignificant. A specific investigation should be made to determine whether to neglect the contribution for a wound-rotor motor.

5.1.4 Electric Utility Systems

The utility generators are usually remote from the industrial plant. The current distributed to a fault in the remote plant appears to be merely a small increase in load current to the very large central station generators, and this current contribution tends to stay constant. Therefore, the utility system is usually represented at the plant by a single-valued equivalent impedance referred to the point of connection.

5.2 FUNDAMENTALS OF FAULT CURRENT CALCULATIONS

Ohm's law, I = E/Z, provides the relationship used in determining fault current, where I is the current to be determined, E the normal system voltage at point of fault, and Z the impedance from source to fault, including the impedance of the source. Rigorous calculations are generally tedious and time consuming. Simplifying assumptions can be made which detract little from accuracy and much from labor.

5.2.1 Type of Faults

In the usual procedures of fault current calculations, it is assumed that the fault is a zero-impedance, "bolted" fault with no current-limiting effect due to fault itself. Such calculations are used to determine the maximum short-circuit current value for the purpose of selecting devices of adequate interrupting rating, momentary rating, and to determne the maximum value of current at which time-current coordination need exist in the relay studies. The three-phase fault is usually the only one considered, since in an industrial system this type of fault generally results in maximum current.

5.2.2 Voltage and Impedance

The voltage that serves as a basis for fault current calculation is derived from the rated nameplate voltage of the generator or transformer supplying the faulted element of the system:

$$\text{line-to-neutral voltage} = \frac{\text{rated line-to-line voltage}}{\sqrt{3}}$$

In an ac circuit the impedance is the vector sum of resistance and reactance. The reactance of generators and transformers is usually at least five times the resistance. The fault current calculated by neglecting the resistance of such equipment will introduce only a few percent error. Because of this, the resistance of generators, transformers, motors, reactors, and large bus work is usually not considered regardless of the system voltage.

In systems below 600 V when calculatng faults on the branch feeder circuits, resistance should generally be included. The following procedure would normally be used for low-voltage feeder circuits. If the resistance of the feeder circuit is one-fourth or more of the total reactance from source to fault, resistance should be included in the calculations. An approximate total resistance may be obtained by adding to the resistance of the feeder a resistance equal to one-fourth of the total reactance of the system from source to the feeder. Assuming total equivalent resistance R and total equivalent reactance X, the impedance to the fault is expressed by the formula

$$Z = \sqrt{R^2 + X^2}$$

It is important to consider the reactance of all circuit elements in calculating fault currents in a low-voltage system.

5.2.3 Symmetry of Fault Current

In determining the maximum value of fault current that can occur at some point in a system, it must be considered that the fault current wave is likely not to be symmetrical about the zero current axis for several cycles after the fault occurs. System voltage and fault current are substantially sine wave in shape and are related in phase angle by the impedance angle of the system to the point of fault. Since the resistance will usually be negligible, the fault current will lag the source voltage by nearly 90 degrees. This means that when a fault occurs at or near the peak of the voltage wave, the fault current wave starts at zero and is symmetrical about the zero axis. When a fault occurs at or near the zero point of the voltage wave, the fault current again starts at zero on the original zero axis. However, most short-circuit currents are not symmetrical. They are offset from the normal-current axis for a period of several cycles. Figure 5.1 illustrates the case where a short circuit with low power produces the highest first peak of short-circuit current. The magnitude of current offset for a typical fault will be between the extremes of complete symmetry and complete asymmetry because the odds are against the fault occurring exactly at a voltage peak or a voltage zero.

An analysis of a typical asymmetrical current wave is made in Fig. 5.1. The offset of the asymmetrical current wave from a symmetrical wave having equal peak-to-peak displacement is a positive value of current that may be considered as a direct current. The asymmetrical current may therefore, be thought of as the sum of an alternating-current component b and a direct-current component a.

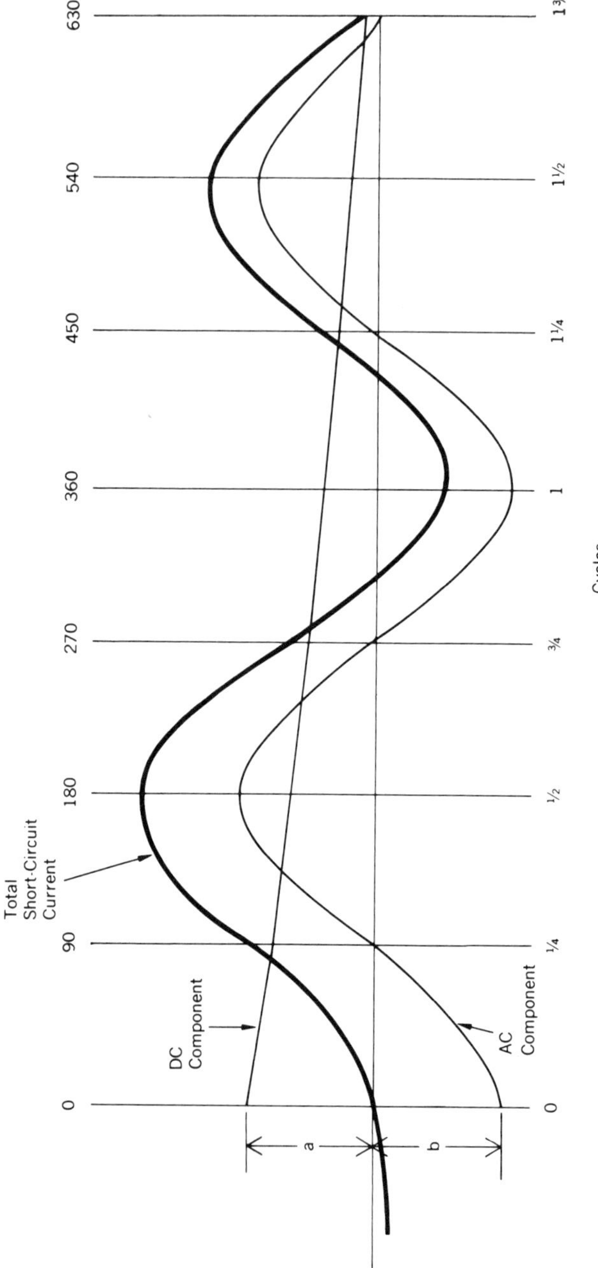

Figure 5.1 Analysis of asymmetrical current wave.

At the instant of fault occurrence, b is negative and a + b = 0. Just before the quarter cycle, the symmetrical alternating-current component is zero, and the total current is equal to the direct-current component. At near half-cycle, the total current is maximum, being equal to the sum of the maximum positive alternating-current component and the direct-current component. The initial rate of decay of the direct-current component is inversely proportional to the X/R ratio of the system from the source to the fault. The lower the X/R ratio, the more rapid is the decay, which is called "direct-current decrement." The total short-circuit current is thus affected by both an alternating-current decrement and a direct-current decrement before reaching its steady-state value.

5.3 PROCEDURES OF CALCULATIONS

5.3.1 Calculations Using Ohms, Percent Reactance, or Per Unit Reactance

The determination of short-circuit currents has been shown to be dependent on the reactance X from the source to the fault. The principal problem of short-circuit current determination is one of determining the reactance. To obtain it, the reactance of each element in the circuit must be determined and the elements combined in series or parallel. The reactance of the elements can be expressed either in ohms or in percent or in per unit on a chosen base value. It is often convenient to use the per unit system in calculations involving a system with several different voltage levels. When reactances are expressed as per unit quantities on a chosen kVA basis, they can be combined directly without regard for the transformer turns ratio in a multi-voltage-level system.

Formulas for converting these units from one system to another are as follows:

$$\text{per unit reactance} = \frac{\text{percent reactance}}{100} \tag{5.1}$$

$$\text{per unit reactance (on chosen kVA base)} = \frac{\text{ohms} \times \text{kVA base}}{1000 \times \text{kV}^2} \tag{5.2}$$

$$\text{per unit reactance (on chosen mVA base)} = \frac{\text{ohms} \times \text{mVA base}}{\text{kV}^2} \tag{5.3}$$

where ohms are line-to-neutral values (single conductor), kVA base is the three-phase base kVA, kV is line-to-line voltage, and mVA = kVA/1000.

5.3.2 Calculation Methods

Essentially, there are two methods of making short-circuit calculations: the direct method and the per unit method. Although they represent different calculation concepts, they produce results of the same degree of accuracy.

Direct Method

The direct method uses the system one-line diagram directly; uses system and equipment data such as volts, amperes, and ohms directly; and uses basic electrical equations and relationships directly without utilizing special diagrams, abstract units, or mathematical techniques. For those who do not specialize in short-circuit calculation work, the direct method is easier to comprehend since it uses the familiar system one-line diagram. It also instills confidence since system and equipment data are applied directly to familiar electrical equations, producing recognizable values that can be appraised. This method is particularly adapted to progressive analysis of an entire system or of a portion starting at the source, consisting each echelon step by step, and determining short-circuit values at each location out to the end of the various circuits. Such an analysis is very useful for planning the entire power system of a new building or facility where short-circuit values at all points must be determined before the appropriate equipment can be selected.

Per Unit Method

The per unit method involves converting the system one-line diagram into an equivalent impedance diagram and reducing this to a single impedance diagram. This can best be accomplished when several voltage levels are involved by using a special mathematical technique that establishes base values for volts, amperes, kVA, and ohms and then refers the actual parametrs to these bases in special equations as given in equations (5.2) and (5.3) to derive per unit values. Applying these values in special equations yields the short-circuit current values.

In this method of calculation, each point of fault is considered separately. A system equivalent impedance diagram is developed using those parameters that will have an effect on short-circuit current at that point. In establishing the impedance diagram, sometimes delta and wye conversion equations (5.4) and (5.5) will need to be applied. These are given in Figure 5.2. The conversion formulas are as follows:

FOR TRANSFORMING WYE TO DELTA

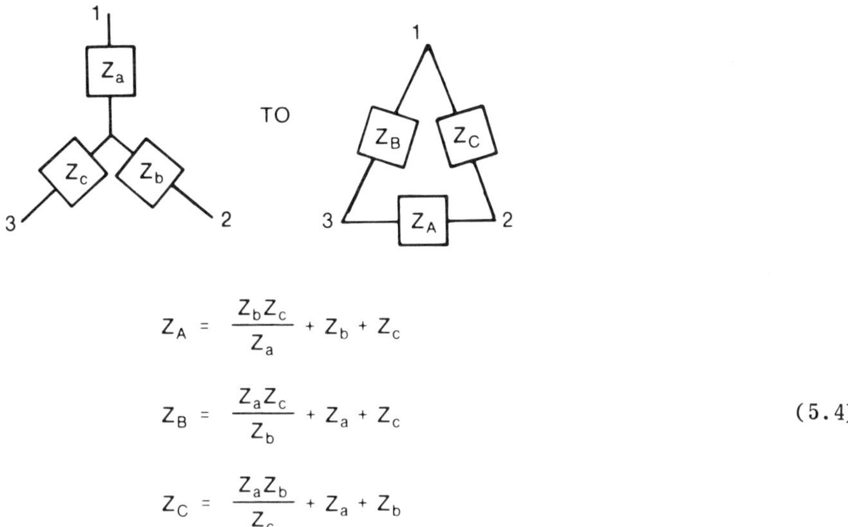

$$Z_A = \frac{Z_b Z_c}{Z_a} + Z_b + Z_c$$

$$Z_B = \frac{Z_a Z_c}{Z_b} + Z_a + Z_c \qquad (5.4)$$

$$Z_C = \frac{Z_a Z_b}{Z_c} + Z_a + Z_b$$

FOR TRANSFORMING DELTA TO WYE

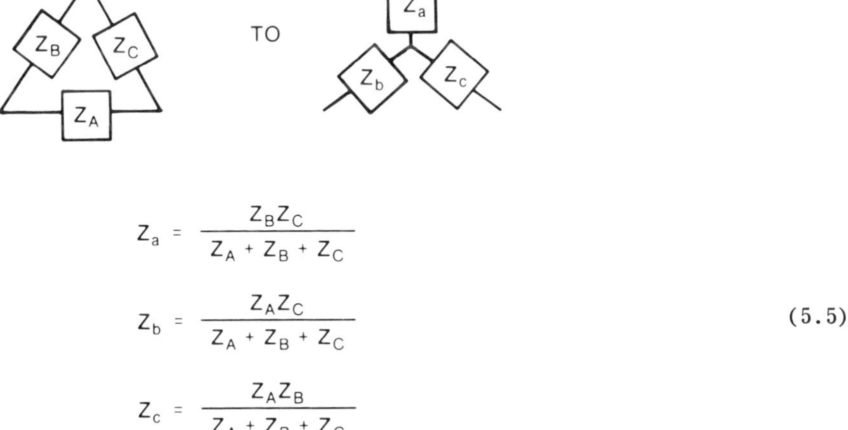

$$Z_a = \frac{Z_B Z_C}{Z_A + Z_B + Z_C}$$

$$Z_b = \frac{Z_A Z_C}{Z_A + Z_B + Z_C} \qquad (5.5)$$

$$Z_c = \frac{Z_A Z_B}{Z_A + Z_B + Z_C}$$

Figure 5.2 Transformer impedances connected in Delta or Wye.

Each fault location requires its own separate equivalent impedance diagram, subsequent reduction, and calculation. Since each point is considered separately, the per unit method tends to be more expedient when a single specific remote location is being analyzed. Most of the major short-circuit studies have pertained to systems involving many voltage levels. The per unit method is an advantage and therefore generally considered the official standard method of calculation as presented in many references, including ANSI standards and other IEEE publications.

5.3.3 Calculation Times

Some protective devices operate after a few cycles and others after a time delay. Short-circuit currents may need to be checked at the following suggested times:

1. First cycle. Maximum symmetrical values immediately after fault initiation are always required and are often the only values needed. These values are used in selecting proper short-circuit ratings for low-voltage equipment and when converted to asymmetrical values are the basis for selecting circuit medium-voltage switch and fuse ratings, and circuit breaker close and latch ratings. They often are used in selecting medium-voltage circuit breaker interrupting ratings. Even if a device does not interrupt until several cycles after fault initiation, thus allowing the fault current to decay, the protective devices and all series devices should withstand the maximum current as well as the total energy.

2. After 1.5 to 8 cycles (interrupting considerations). Maximum values after a few cycles are required for comparison with the interrupting ratings of medium-voltage circuit breakers.

3. About 30 cycles. These reduced fault currents are sometimes needed for estimating the performance of time-delay relays and fuses. Minimum values should be calculated to determine whether sufficient current is available to open the protective devices within a satisfactory time.

5.4 EXAMPLES OF SIMPLIFIED CALCULATIONS

5.4.1 Direct Method

One-Line Diagram

A one-line diagram is a graphical representation of the power system and should be prepared as the first step in making a short-circuit study. This diagram should show all sources of short-circuit current and all significant circuit elements. Reactance and resistance values of all elements should be included in the diagram. Reactance and resistance data can be obtained from Tables 5.1 to 5.7 or preferably from equipment suppliers.

Detailed Procedures

Most circuit component impedance values are given in ohms, except for utility and transformer impedances, which are found by the following procedure.

1. $$\text{Utility X (in ohms)} = \frac{1000(\text{secondary kV})^2}{\text{utility short-circuit kVA}}$$

2. $$\text{Transformer X (in ohms)} = \frac{10(\%X)(\text{secondary kV})^2}{\text{transformer kVA}}$$

$$\text{Transformer R (in ohms)} = \frac{10(\%R)(\text{secondary kV})^2}{\text{transformer kVA}}$$

Table 5.1 Transformer Impedance Data

High-Voltage Rating (volts)	Kilovolt Rating	Percent Impedance Voltage	
Secondary Unit Substation Transformers*			
2400 - 13 800	112.5 - 224	Not less than 2.0	
2400 - 13 800	300 - 500	Not less than 4.5	
2400 - 13 800	750 - 2500	6.75 ‡	
22 900	All	6.75 ‡	
34 400	All	7.25	
Liquid-Immersed Transformers, 501 - 30 000 kVA †			
		Low Voltage, 480 V	Low Voltage, 2400 V and Above
2400 - 22 900		6.75 ‡	6.5 **
26 400, 34 400		7.25	7.0
43 800		7.75	7.5
67 000			8.0
115 000			8.5
138 000			9.0

NOTES: (1) Ratings separated by hyphens indicate that all intervening standard ratings are included. Ratings separated by a comma indicate that only those listed are included.
(2) Percent impedance voltages are at self-cooled rating and as measured on rated voltage connection.
*From NEMA 210-1982
†From ANSI C57.12.10-1977
‡Three-phase transformers 5000 kVA and smaller with high-voltage windings rated 25 kV and below are commonly used in industrial applications and are normally built with impedance voltages of 5.75%.
**Three-phase transformers 5000 kVA and smaller with high-voltage windings rated 25 kV and below are commonly used in industrial applications and are normally built with impedance voltages of 5.5%.

Table 5.2 The reactance of Disconnecting Switches for Low–voltage Circuits (600 volts and below) is in the Order of Magnitude of 0.00008 Ohms per Pole to 0.00005 Ohms per Pole at 60 Cycles for Switches Rated 400–4000 Amperes Respectively.*

APPROXIMATE REACTANCES**

Switch Size (Amps)	X (Ohms)
200	0.0001
400	0.00008
600	0.00008
800	0.00007
1200	0.00007
1600	0.00005
2000	0.00005
3000	0.00004
4000	0.00004

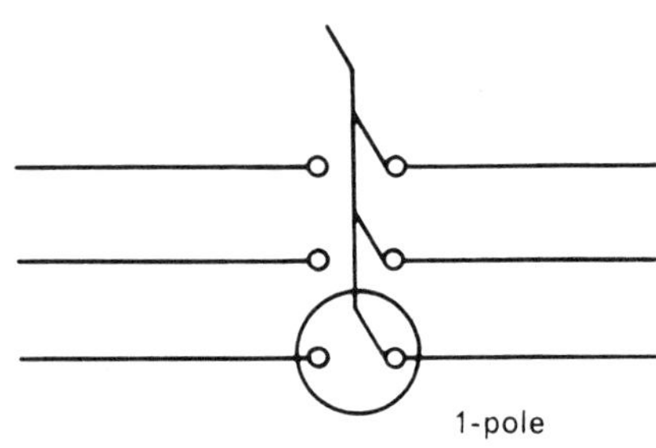

*Reprinted with permission of I.E.E.E., Ibid.

**For actual values, refer to manufacturer's data.

Table 5.3 Circuit Breaker Reactance Data

REACTANCE OF LOW-VOLTAGE POWER CIRCUIT BREAKERS†

Breaker Interrupting Rating — Amperes	Ampere Rating	Reactance in Ohms
15,000 and 25,000	15 to 35	0.04
	50 to 100	0.004
	125 to 225	0.001
	250 to 600	0.0002
50,000	200 to 800	0.0002
	1000 to 1600	0.00007
75,000	2000 to 3000	0.00008
100,000	4000	0.00008

TYPICAL MOLDED CASE CIRCUIT BREAKER IMPEDANCES

Molded Case Breaker Ampere Rating	Resistance in Ohms	Reactance in Ohms
20	.00700	Negligible
40	.00240	Negligible
100	.00200	.00070
225	.00035	.00020
400	.00031	.00039
600	.00007	.00017

NOTE: Due to the method of rating low-voltage power circuit breakers, the reactance of the breaker which is to interrupt the fault is not included in calculating fault current.†

†Reprinted with permission of I.E.E.E., Ibid.

Table 5.4 Resistance, Reactance, and Impedance of Copper Cable Circuits (ohms per 1000 feet)

Three Single Conductors

AWG or MCM	In Magnetic Duct						In Nonmagnetic Duct					
	600 V and 5 kv Nonshielded			5 kv Shielded and 15 kv			600 V and 5 kv Nonshielded			5 kv Shielded and 15 kv		
	R	X	Z	R	X	Z	R	X	Z	R	X	Z
8	.811	.0754	.814	.811	.0860	.816	.811	.0603	.813	.811	.0688	.814
8 (solid)	.786	.0754	.790	.786	.0860	.791	.786	.0603	.788	.786	.0688	.789
6	.510	.0685	.515	.510	.0796	.516	.510	.0548	.513	.510	.0636	.514
6 (solid)	.496	.0685	.501	.496	.0796	.502	.496	.0548	.499	.496	.0636	.500
4	.321	.0632	.327	.321	.0742	.329	.321	.0506	.325	.321	.0594	.326
4 (solid)	.312	.0632	.318	.312	.0742	.321	.312	.0506	.316	.312	.0594	.318
2	.202	.0585	.210	.202	.0685	.214	.202	.0467	.207	.202	.0547	.209
1	.160	.0570	.170	.160	.0675	.174	.160	.0456	.166	.160	.0540	.169
1/0	.128	.0540	.139	.128	.0635	.143	.127	.0432	.134	.128	.0507	.138
2/0	.102	.0533	.115	.103	.0630	.121	.101	.0426	.110	.102	.0504	.114
3/0	.0805	.0519	.0958	.0814	.0605	.101	.0766	.0415	.0871	.0805	.0484	.0939
4/0	.0640	.0497	.0810	.0650	.0583	.0929	.0633	.0398	.0748	.0640	.0466	.0792
250	.0552	.0495	.0742	.0557	.0570	.0797	.0541	.0396	.0670	.0547	.0456	.0712
300	.0464	.0493	.0677	.0473	.0564	.0736	.0451	.0394	.0599	.0460	.0451	.0644
350	.0378	.0491	.0617	.0386	.0562	.0681	.0368	.0393	.0536	.0375	.0450	.0586
400	.0356	.0490	.0606	.0362	.0548	.0657	.0342	.0392	.0520	.0348	.0438	.0559
450	.0322	.0480	.0578	.0328	.0538	.0630	.0304	.0384	.0490	.0312	.0430	.0531
500	.0294	.0466	.0551	.0300	.0526	.0505	.0276	.0373	.0464	.0284	.0421	.0508
600	.0257	.0463	.0530	.0264	.0516	.0580	.0237	.0371	.0440	.0246	.0412	.0479
750	.0216	.0445	.0495	.0223	.0497	.0545	.0194	.0356	.0405	.0203	.0396	.0445

Three-Conductor Cable

AWG or MCM	In Magnetic Duct and Steel Interlocked Armor						In Nonmagnetic Duct and Aluminum Interlocked Armor					
	600 v and 5 kv Nonshielded			5 kv Shielded and 15 kv			600 v and 5 kv Nonshielded			5 kv Shielded and 15 kv		
	R	X	Z	R	X	Z	R	X	Z	R	X	Z
8	.811	.0577	.813	.811	.0658	.814	.811	.0503	.812	.811	.0574	.813
8 (solid)	.786	.0577	.788	.786	.0658	.789	.786	.0503	.787	.786	.0574	.788
6	.510	.0525	.513	.510	.0610	.514	.510	.0457	.512	.510	.0531	.513
6 (solid)	.496	.0525	.499	.496	.0610	.500	.496	.0457	.498	.496	.0531	.499
4	.321	.0483	.325	.321	.0568	.326	.321	.0422	.324	.321	.0495	.325
4 (solid)	.312	.0483	.316	.312	.0508	.317	.312	.0422	.315	.312	.0495	.316
2	.202	.0448	.207	.202	.0524	.209	.202	.0390	.206	.202	.0457	.207
1	.160	.0436	.166	.160	.0516	.168	.160	.0380	.164	.160	.0450	.166
1/0	.128	.0414	.135	.128	.0486	.137	.127	.0360	.132	.128	.0423	.135
2/0	.102	.0407	.110	.103	.0482	.114	.101	.0355	.107	.102	.0420	.110
3/0	.0805	.0397	.0898	.0814	.0463	.0936	.0766	.0346	.0841	.0805	.0403	.090
4/0	.0640	.0381	.0745	.0650	.0446	.0788	.0633	.0332	.0715	.0640	.0389	.0749
250	.0552	.0379	.0670	.0557	.0436	.0707	.0541	.0330	.0634	.0547	.0380	.0666
300	.0464	.0377	.0598	.0473	.0431	.0640	.0451	.0329	.0559	.0460	.0376	.0596
350	.0378	.0373	.0539	.0386	.0427	.0576	.0368	.0328	.0492	.0375	.0375	.0530
400	.0356	.0371	.0514	.0362	.0415	.0551	.0342	.0327	.0475	.0348	.0366	.0505
450	.0322	.0361	.0484	.0328	.0404	.0520	.0304	.0320	.0441	.0312	.0359	.0476
500	.0294	.0349	.0456	.0300	.0394	.0495	.0276	.0311	.0416	.0284	.0351	.0453
600	.0257	.0343	.0429	.0264	.0382	.0464	.0237	.0309	.0389	.0246	.0344	.0422
750	.0216	.0326	.0391	.0223	.0364	.0427	.0197	.0297	.0355	.0203	.0332	.0389

Resistance based on tinned copper at 60 cycles. 600 volt and 5 kv unshielded based on varnished cambric insulation. 5 kv shielded and 15 kv cable based on Neoprene insulation. Values shown are for 1000 feet of cable at 75°C.

Table 5.5 Cross-Linked Polyethylene Insulated Cable Resistance, Reactance, and Impedance of Aluminum Conductor Cable (approximate ohms per 1000 feet)

Three Single Conductors.

AWG or MCM	In Magnetic Duct						In Nonmagnetic Duct					
	600 V and 5kv Nonshielded			5 kv Shielded and 15 kv			600 V and 5 kv Nonshielded			5 kv Shielded and 15 kv		
	R	X	Z	R	X	Z	R	X	Z	R	X	Z
6	.847	.053	.849	—	—	—	.847	.042	.848	—	—	—
4	.532	.050	.534	.532	.068	.536	.532	.040	.534	.532	.054	.535
2	.335	.046	.338	.335	.063	.341	.335	.037	.337	.335	.050	.339
1	.265	.048	.269	.265	.059	.271	.265	.035	.267	.265	.047	.269
1/0	.210	.043	.214	.210	.056	.217	.210	.034	.213	.210	.045	.215
2/0	.167	.041	.172	.167	.055	.176	.167	.033	.170	.167	.044	.173
3/0	.133	.040	.139	.132	.053	.142	.133	.037	.137	.132	.042	.139
4/0	.106	.039	.113	.105	.051	.117	.105	.031	.109	.105	.041	.113
250	.0896	.0384	.0975	.0892	.0495	.102	.0894	.0307	.0945	.0891	.0396	.0975
300	.0750	.0375	.0839	.0746	.0479	.0887	.0746	.0300	.0804	.0744	.0383	.0837
350	.0644	.0369	.0742	.0640	.0468	.0793	.0640	.0245	.0705	.0638	.0374	.0740
400	.0568	.0364	.0675	.0563	.0459	.0726	.0563	.0291	.0634	.0560	.0367	.0700
500	.0459	.0355	.0580	.0453	.0444	.0634	.0453	.0284	.0535	.0450	.0355	.0573
600	.0388	.0359	.0529	.0381	.0431	.0575	.0381	.0287	.0477	.0377	.0345	.0511
700	.0338	.0350	.0487	.0332	.0423	.0538	.0330	.0280	.0433	.0326	.0338	.0470
750	.0318	.0341	.0466	.0310	.0419	.0521	.0309	.0273	.0412	.0304	.0335	.0452
1000	.0252	.0341	.0424	.0243	.0414	.0480	.0239	.0273	.0363	.0234	.0331	.0405

Three-Conductor Cables

AWG or MCM	In Magnetic Duct						In Nonmagnetic Duct					
	600 V and 5kv Nonshielded			5 kv Shielded and 15 kv			600 V and 5 kv Nonshielded			5 kv Shielded and 15 kv		
	R	X	Z	R	X	Z	R	X	Z	R	X	Z
6	.847	.053	.849	—	—	—	.847	.042	.848	—	—	—
4	.532	.050	.534	—	—	—	.532	.040	.534	—	—	—
2	.335	.046	.338	.335	.056	.340	.335	.037	.337	.335	.045	.338
1	.265	.048	.269	.265	.053	.270	.265	.035	.267	.265	.042	.268
1/0	.210	.043	.214	.210	.050	.216	.210	.034	.213	.210	.040	.214
2/0	.167	.041	.172	.167	.049	.174	.167	.033	.170	.167	.039	.171
3/0	.133	.040	.139	.133	.048	.141	.133	.037	.137	.133	.038	.138
4/0	.106	.039	.113	.105	.045	.114	.105	.031	.109	.105	.036	.111
250	.0896	.0384	.0975	.0895	.0436	.100	.0894	.0307	.0945	.0893	.0349	.0959
300	.0750	.0375	.0839	.0748	.0424	.0860	.0746	.0300	.0804	.0745	.0340	.0819
350	.0644	.0369	.0742	.0643	.0418	.0767	.0640	.0245	.0705	.0640	.0334	.0722
400	.0568	.0364	.0675	.0564	.0411	.0700	.0563	.0291	.0634	.0561	.0329	.0650
500	.0459	.0355	.0580	.0457	.0399	.0607	.0453	.0284	.0535	.0452	.0319	.0553
600	.0388	.0359	.0529	.0386	.0390	.0549	.0381	.0287	.0477	.0380	.0312	.0492
700	.0338	.0350	.0487	.0335	.0381	.0507	.0330	.0280	.0433	.0328	.0305	.0448
750	.0318	.0341	.0466	.0315	.0379	.0493	.0309	.0273	.0412	.0307	.0303	.0431
1000	.0252	.0341	.0424	.0248	.0368	.0444	.0239	.0273	.0363	.0237	.0294	.0378

Values Are for 1000 Circuit Feet at 90°C Conductor

Table 5.6 Busway Impedance Data (ohms per 1000 feet)

Plug-In Busway

Ampere Rating	Ohms per 1000 feet line to neutral 60 cycles		
	Resistance	Reactance	Impedance
Copper Bus Bars			
225	0.0836	0.0800	0.1157
400	0.0437	0.0232	0.0495
600	0.0350	0.0179	0.0393
800	0.0218	0.0136	0.0257
1000	0.0145	0.0135	0.0198

Ampere Rating	Ohms per 1000 feet line to neutral 60 cycles		
	Resistance	Reactance	Impedance
Aluminum Bus Bars			
225	0.1090	0.0720	0.1313
400	0.0550	0.0222	0.0592
600	0.0304	0.0121	0.0327
800	0.0243	0.0154	0.0288

Low-Impedance Feeder Busway

Ampere Rating	Ohms per 1000 feet line to neutral 60 cycles		
	Resistance	Reactance	Impedance
800	0.0219	0.0085	0.0235
1000	0.0190	0.0050	0.0196
1350	0.0126	0.0044	0.0134
1600	0.0116	0.0035	0.0121
2000	0.0075	0.0031	0.0081
2500	0.0057	0.0025	0.0062
3000	0.0055	0.0017	0.0058
4000	0.0037	0.0016	0.0040

Current Limiting Busway

Ampere Rating	Ohms per 1000 feet line to neutral 60 cycles			X/R Ratio
	Resistance	Reactance	Impedance	
1000	0.013	0.063	0.064	4.85
1350	0.012	0.061	0.062	5.08
1600	0.009	0.056	0.057	6.22
2000	0.007	0.052	0.052	7.45
2500	0.006	0.049	0.049	8.15
3000	0.005	0.046	0.046	9.20
4000	0.004	0.042	0.042	10.50

Table 5.7 Asymmetrical Factors

Short-circuit power factor, percent	Short circuit X/R ratio	Ratio to symmetrical Rms amperes		
		Maximum 1-phase instantaneous peak amperes Mp	Maximum 1-phase Rms amperes at 1/2 cycle M_m (asymmetrical factor)	Average 3-phase Rms amperes at 1/2 cycle M_a
0	∞	2.828	1.732	1.394
1	100.00	2.75	1.696	1.374
2	49.993	2.743	1.665	1.355
3	33.322	2.702	1.630	1.336
4	24.979	2.663	1.598	1.318
5	19.974	2.625	1.568	1.301
6	16.623	2.589	1.540	1.285
7	14.251	2.554	1.511	1.270
8	13.460	2.520	1.485	1.256
9	11.066	2.487	1.460	1.241
10	9.9301	2.455	1.436	1.229
11	9.0354	2.424	1.413	1.216
12	8.2733	2.394	1.391	1.204
13	7.6271	2.364	1.372	1.193
14	7.0721	2.336	1.350	1.182
15	6.5912	2.309	1.330	1.171
16	6.1695	2.282	1.312	1.161
17	5.7947	2.256	1.294	1.152
18	5.4649	2.231	1.277	1.143
19	5.1672	2.207	1.262	1.135

20	4.8990	2.183	1.247	1.127
21	4.6557	2.160	1.232	1.119
22	4.4341	2.138	1.218	1.112
23	4.2313	2.11	1.205	1.105
24	4.0450	2.095	1.192	1.099
25	3.8730	2.074	1.181	1.093
26	3.7138	2.054	1.170	1.087
27	3.5661	2.034	1.159	1.081
28	3.4286	2.015	1.149	1.075
29	3.3001	1.996	1.139	1.070
30	3.1798	1.978	1.130	1.066
31	3.0669	1.960	1.121	1.062
32	2.9608	1.943	1.113	1.057
33	2.8606	1.926	1.105	1.053
34	2.7660	1.910	1.098	1.049
35	2.6764	1.894	1.091	1.046
36	2.5916	1.878	1.084	1.043
37	2.5109	1.863	1.078	1.039
38	2.4341	1.848	1.073	1.036
39	2.3611	1.833	1.068	1.033
40	2.2913	1.819	1.062	1.031
41	2.2246	1.805	1.057	1.028
42	2.1608	1.791	1.053	1.026
43	2.0996	1.778	1.049	1.024
44	2.0409	1.765	1.045	1.022
45	1.9845	1.753	1.041	1.020
46	1.9303	1.740	1.038	1.019

3. The impedance (in ohms) given for current transformers, large switches, and large circuit breakers is essentially all X.
4. Determine (a) cable and bus X (in ohms) and (b) cable and bus R (in ohms).
5. Total all X and all R in the system to the point of fault.
6. Determine the impedance (in ohms) of the system by

$$Z_T = \sqrt{(R_T)^2 + (X_T)^2}$$

7. Calculate short-circuit symmetrical root-mean-square (rms) amperes at the point of fault.

$$I_{sc\ sym\ rms} = \frac{\text{secondary line voltage}}{\sqrt{3}\ (Z_T)}$$

8. Determine the motor loads and add up the full-load motor currents.
9. The short-circuit current that the motor load can contribute is an asymmetrical current usually approximated as being equal to the locked rotor current of the motors. As a close approximation with a margin of safety, use

$$\text{asymmetrical motor contribution}^* = 5 \times (\text{full-load motor current})$$

10. The symmetrical motor contribution can be approximated by using the average asymmetry factor associated with the motor in the system. This asymmetry factor varies according to motor design and 1.25 is used for approximate calculation purposes. To solve for the symmetrical motor contribution:*

$$\text{symmetrical motor contribution}^* = \frac{\text{asymmetrical motor contribution}}{1.25}$$

11. The total symmetrical short-circuit rms current is calculated as

$$^\dagger\text{total } I_{sc\ sym\ rms} = I_{sc\ sym\ rms} + \text{symmetrical motor contribution}$$

---

*A more exact determination depends on the subtransient reactances of the motors in question and associated circuit impedances. A less conservative method would involve the total motor circuit impedance to a common bus and proceed therefrom.

†Arithmetical addition results in conservative values of fault current. More finite values involve vectoral addition of the currents.

12. Determine the X/R ratio of the system to the point of the fault.

$$X/R \text{ ratio} = \frac{\text{total X (ohms)}}{\text{total R (ohms)}}$$

13. The asymmetrical factor corresponding to the X/R ratio in step 12 is found in Table 5.7, column M_m. This multiplier will provide the worst-case asymmetry occurring in the first half-cycle. Where the average three-phase multiplier is desired, use column M_a.
14. Calculate the asymmetrical rms short-circuit current:

$$I_{sc\ asym\ rms} = I_{sc\ sym\ rms} \times (\text{asymmetrical factor})$$

15. The total asymmetrical short-circuit rms current is calculated as:

$$^{\dagger}\text{total } I_{sc\ asym\ rms} = I_{sc\ asym\ rms} + \text{asymmetrical motor contribution}$$

Examples for Fault Current Calculations

Figures 5.3 shows a one-line diagram for a simple radial system with all components information given in details. To start short-circuit calculations, we must obtain the available short-circuit kVA from the local utility company. The utility estimates that their system can deliver a short-circuit of 100,000 kVA at the transformer. Since the X/R ratio of the utility system is usually very high, only the reactance need be considered. With this short-circuit fault information available, we can begin to make the necessary calculations to determine the fault current at X_1 and X_2 in the system.

Figure 5.4 shows a simplified impedance diagram of the system for calculating fault current at X_1 with X and R values as tabulated alongside, using the formulas given previously in the detailed procedures. From Figure 5.4 we have

$$\text{total R} = 0.000674\ \Omega$$

$$\text{total X} = 0.003855\ \Omega$$

Hence

$$\text{total Z} = \sqrt{(0.000674)^2 + (0.003855)^2} = 0.00392\ \Omega/\text{phase}$$

$$I_{sc\ sym\ rms}\ (\text{amperes}) = \frac{240}{\sqrt{3}\ (0.00392)} = 35{,}300\ \text{A}$$

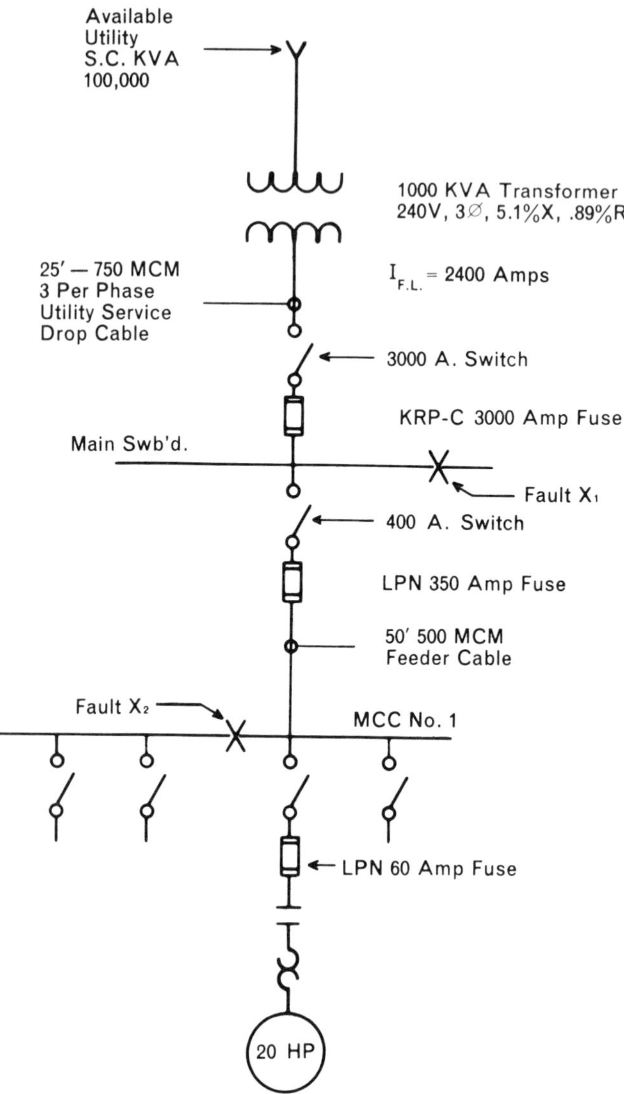

Figure 5.3 One-line diagram for a radial system.

asymmetrical motor contribution (100% motor load) $= 5 \times 2400$
$= 12{,}000$ A

symmetrical motor contribution $= \dfrac{12{,}000}{1.25} = 9600$ A

total $I_{sc\ sym\ rms}$ (amperes) (fault X_1) $= 35{,}300 + 9600$
$= 44{,}900$ A

X/R ratio $= \dfrac{0.003855}{0.000674} = 5.72$

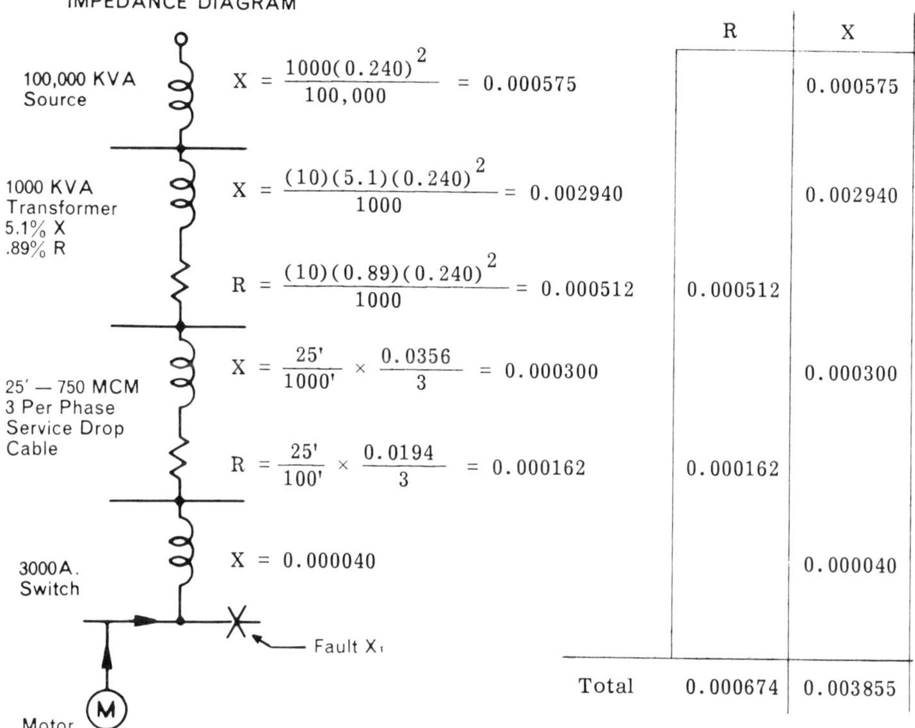

	R	X
		0.000575
		0.002940
	0.000512	
		0.000300
	0.000162	
		0.000040
Total	0.000674	0.003855

Figure 5.4 Direct method—simplified impedance diagram and short-circuit calculation at fault X_1.

asymmetrical factor* = 1.290 (Table 5.7)

$I_{sc\ asym\ rms}$ (amperes) = 1.290 × 35,300 = 45,500

total $I_{sc\ asym\ rms}$ (amperes) (fault X_1) = 45,500 + 12,000
= 57,500 A

Figure 5.5 shows a simplified impedance diagram of the system for calculating fault current at X_2 with X and R values as tabulated alongside, using the previous formulas as given in the detailed proceudres. From Figure 5.5 we have

total R = 0.002144 Ω

total X = 0.006265 Ω

Hence

$$\text{total } Z = \sqrt{(0.002144)^2 + (0.006265)^2} = 0.00662\ \Omega/\text{phase}$$

$$I_{sc\ sym\ rms}\ (\text{amperes}) = \frac{240}{\sqrt{3}\ (0.00662)} = 20{,}930\ \text{A}$$

asymmetrical motor contribution (100% motor load) = 5 × 2400
= 12,000 A

$$\text{symmetrical motor contribution} = \frac{12{,}000}{1.25} = 9600\ \text{A}$$

total $I_{sc\ sym\ rms}$ (amperes) (fault X_2) = 20,930 + 9600
= 30,530 A

$$X/R \text{ ratio} = \frac{0.006265}{0.002144} = 2.92$$

asymmetry factor* = 1.112

$I_{sc\ asym\ rms}$ (amperes) = 1.112 × 20,930 = 23,310

total $I_{sc\ asym\ rms}$ (amperes) (fault X_2) = 23,310 + 12,000
= 35,310 A

---

*Multiplier for maximum one-phase rms amperes at half-cycle.

IMPEDANCE DIAGRAM

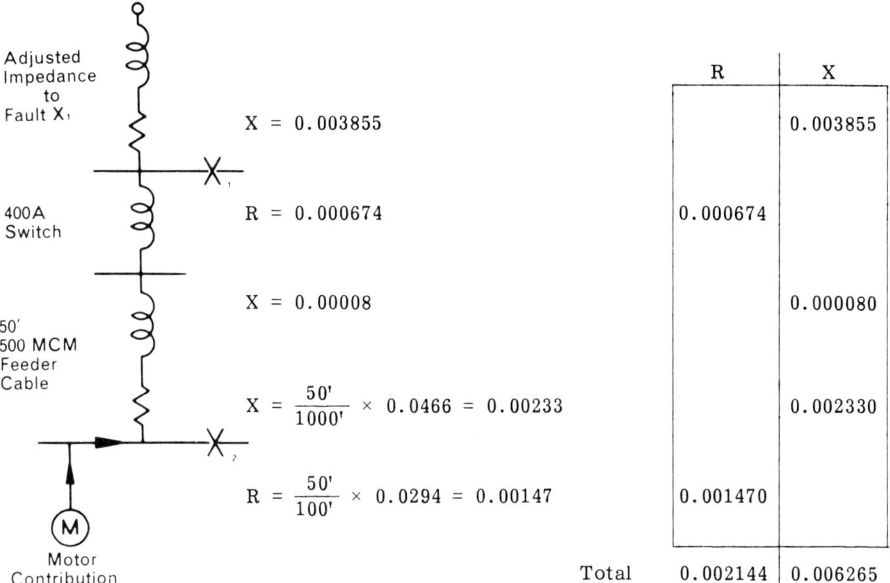

Figure 5.5 Direct method—simplified impedance diagram and short-circuit calculation at fault X_2.

5.4.2 Per Unit Method

One-Line Diagram

The need for a one-line diagram is the same as that for the direct method. For simplicity, Figure 5.3 will again be used as the system for the per unit method of calculation.

Detailed Procedures

After establishing a one-line diagram of the system, we proceed with the calculations as follows.

1. Utility per unit $X = \dfrac{\text{base kVA}}{\text{utility sc kVA}}$

2. Transformer per unit $X = \dfrac{(\%X)(\text{base kVA})}{(100)(\text{transformer kVA})}$

$$\text{Transformer per unit R} = \frac{(\%\text{R})(\text{base kVA})}{(100)(\text{transformer kVA})}$$

3. Component per unit X (cable, switches, CT, bus) $= \dfrac{(\text{ohms X})(\text{base kVA})}{(1000)(\text{kV})^2}$

4. Component per unit R (cable, switches, CT, bus) $= \dfrac{(\text{ohms R})(\text{base kVA})}{(1000)(\text{kV})^2}$

5. Total all per unit X and all per unit R in system to point of fault.
6. Determine the per unit impedance of the system by

$$\text{per unit } Z_T = \sqrt{(\text{per unit } R_T)^2 + (\text{per unit } X_T)^2}$$

7. Calculate the symmetrical rms short-circuit current at the point of fault.

$$I_{\text{sc sym rms}} = \frac{\text{base kVA}}{\sqrt{3}\ \text{kV (per unit } Z_T)}$$

8. Determine the motor load. Add up the full-load motor currents.
9. The short-circuit current that the motor load can contribute is an asymmetrical current usually approximated as being equal to the locked-rotor current of the motors. As a close approxmation with a margin of safety, use:

$$\text{asymmetrical motor contribution}^* = 5 \times (\text{full-load motor current})$$

10. The symmetrical motor contribution can be approximated by using the average asymmetry factor associated with the motor in the system. This asymmetry factor varies according to motor design and 1.25 is used for approximate calculation purposes. To solve for the symmetrical motor contribution:

$$\text{symmetrical motor contribution}^* = \frac{\text{asymmetric motor contribution}}{1.25}$$

11. The total symmetrical short-circuit rms current is calculated as

$$^{\dagger}\text{total } I_{\text{sc sym rms}} = I_{\text{sc sym rms}} + \text{symmetric motor contribution}$$

---

* † See Footnotes on p. 68.

12. Determine the X/R ratio of the system to the point of fault.

$$X/R \text{ ratio} = \frac{\text{per unit } X_T}{\text{per unit } R_T}$$

13. From Table 5.7, column M_m, obtain the asymmetrical factor corresponding to the X/R ratio determined in step 12. This multiplier will provide the worst-case asymmetry occurring in the first half-cycle. Where the average three-phase multiplier is desired, use column M_a.
14. The asymmetrical rms short-circuit current can be calculated as

$$I_{sc\ asym\ rms} = I_{sc\ sym\ rms} \times (\text{asymmetrical factor})$$

15. The total asymmetrical short-circuit rms current is calculated as:

$$^{\dagger}\text{total } I_{sc\ asym\ rms} = I_{sc\ asym\ rms} + \text{asymmetrical motor contribution}$$

Examples for Fault Current Calculations

For easy comparison we use the same one-line diagram (Figure 5.3) for a simple radial system as for the direct method. Figure 5.6 shows a simplified impedance diagram of the system for calculating the fault current at X_1 with per unit R and per unit X values as tabulated alongside, using the formulas as given in the detailed procedures (Section 5.4.2). From Figure 5.6 we have

total per unit R = 0.1171

total per unit X = 0.6689

Hence

$$\text{total per unit } Z = \sqrt{(0.1171)^2 + (0.6689)^2} = 0.6800$$

$$I_{sc\ sym\ rms} \text{ (amperes)} = \frac{10,000}{\sqrt{3}\ (0.240)(0.6800)} = 35,300 \text{ A}$$

$$\text{asymmetrical motor contribution (100\%)} = 5 \times 2400 = 12,000 \text{ A}$$

$$\text{symmetrical motor contribution} = \frac{12,000}{1.25} = 9600 \text{ A}$$

---

* † See Footnotes on p. 68.

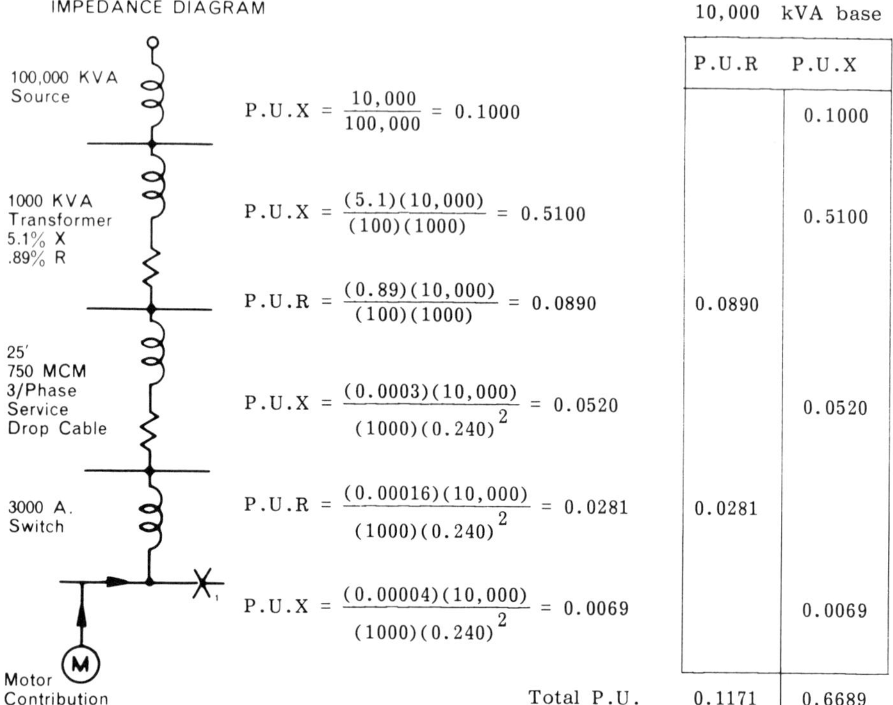

Figure 5.6 Per unit method—simplified impedance diagram and short-circuit calculation at fault X_1.

$$\text{total } I_{sc\ sym\ rms} \text{ (amperes) (fault } X_1) = 35,300 + 9600 = 44,900 \text{ A}$$

$$X/R \text{ ratio} = \frac{0.6689}{0.1171} = 5.72$$

asymmetrical factor* = 1.290 (Table 5.7)

*Multiplier for maximum one-phase rms amperes at half-cycle.

$I_{sc\ asym\ rms}$ (amperes) = 1.290 × 35,300 = 45,500

total $I_{sc\ asym\ rms}$ (amperes) (fault X_1) = 45,500 + 12,000
= 57,500 A

Figure 5.7 shows a simplified impedance diagram of the system for calculating fault current at X_2 with per unit R and per unit X values as tabulated alongside, using the formulas given in the detailed procedures. From Figure 5.7 we have

total per unit R = 0.3722

total per unit X = 1.0878

Hence

total per unit Z = 0.3722 + 1.0878 = 1.150

$$I_{sc\ sym\ rms}\ (\text{amperes}) = \frac{10{,}000}{\sqrt{3}(0.240)(1.150)} = 20{,}930\ \text{A}$$

asymmetrical motor contribution (100%) = 5 × 2400
= 12,000 A

$$\text{symmetry motor contribution} = \frac{12{,}000}{1.25} = 9600\ \text{A}$$

total $I_{sc\ sym\ rms}$ (amperes) (fault X_2) = 20,930 + 9600
= 30,530 A

$$X/R\ \text{ratio} = \frac{1.0878}{0.3722} = 2.92$$

asymmetrical factor* = 1.112 (Table 5.7)

$I_{sc\ asym\ rms}$ (amperes) = 1.112 × 20,930 = 23,310

total $I_{sc\ asym\ rms}$ (amperes) (fault X_2) = 23,310 + 12,000
= 35,310 A

---

*Multiplier for maximum one-phase rms amperes at half-cycle.

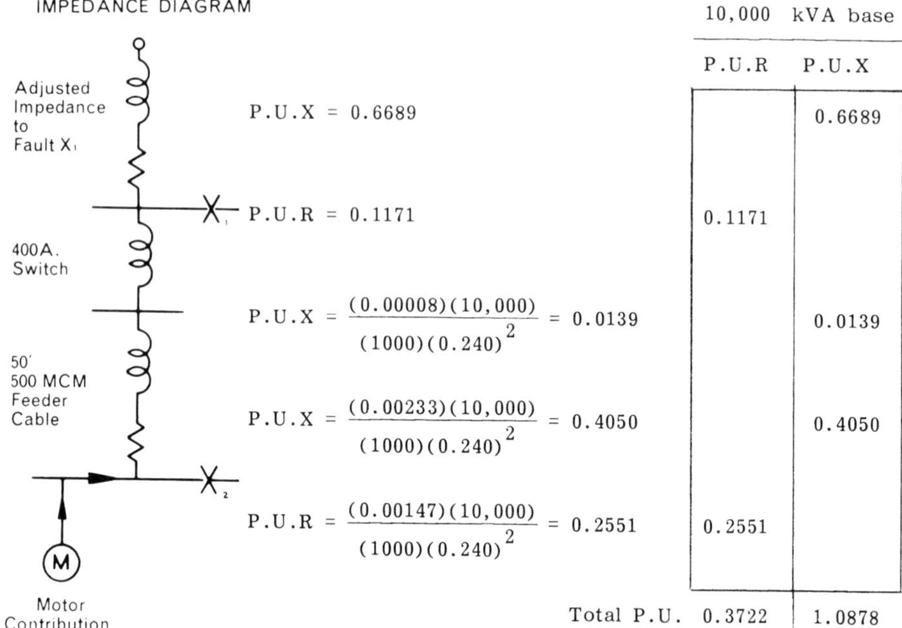

Figure 5.7 Per unit method—simplified impedance diagram and short-circuit calculation at fault X_2.

5.5 SHORT-CIRCUIT RATINGS OF PROTECTIVE EQUIPMENT

5.5.1 Reasons for Making a Short-Circuit Study

A power system short-circuit study as discussed in the preceding sections is necessarily used to determine:

1. The calculated system fault current duties, which can be compared with the short-circuit current rating of circuit-interrupting devices, such as circuit breakers and fuses
2. The selection and rating or setting of short-circuit protective devices, sch as direct-acting trips, fuses, and relays

In the previous sections, basic formulas are given for calculating system resistances and reactances in either ohms or per unit values. For a simple radial system, calculations of short-circuit currents can easily be made with a simple calculator. However, for a complex electric power system, calculations can best be carried out with the aid of a computer. Each of the power system components (utility sources, generators, motors, transformers, cables, etc.) is represented by a resistance value and a reactance value.

The computer program places an assumed three-phase fault on each bus location in the system, and a set of short-circuit currents is calculated which can then be compared with the published short-circuit ratings of the power system equipments. The computer short-circuit program used to calculate three-phase short-circuit duties automatically simulates a fault on each bus and records:

1. The total and symmetrical short-circuit current duty at the faulted bus
2. The short-circuit contribution from all buses connected to the faulted bus
3. The system X/R ratio at the fault point for high-voltage circuit breaker interrupting duty with its associated multiplying factor
4. The symmetrical and asymmetrical current for the breaker momentary duty

5.5.2 Switchgear Ratings

The short-circuit rating assigned to a high-voltage power circuit breaker by the manufacturer is significant in two ways. First, the rating represents a conservative statement of the actual capability of the breaker to close against, to withstand, and to interrupt short-circuit currents. Thus the rating is the maximum condition under which the breaker may safely be applied. Second, the rating is the maximum condition of application for which the manufacturer guarantees that the breaker will perform satisfactorily. Therefore, it is important that a circuit breaker be applied within the rating assigned to its design if the installation is to be safe. One purpose of a short-circuit study is to determine the conditions under which breaker will be applied in a specific system. The short-circuit rating of a circuit breaker is its capability at the maximum voltage at which the breaker may be applied. There is a distinction that must be made between the rating of the breaker and its capability in a specific application.

Prior to 1964, breakers were assigned a short-circuit interrupting capacity in asymmetrical mVA, and it was stated that the interrupting capacity was a constant over a defined range of voltages. An equivalent interrupting capacity in amperes could be calculated at each voltage level. This is called a total current basis of rating breakers. Since 1964, however, breakers have been assigned an interrupting capacty in symmetrical rms amperes at a specified maximum voltage, and the capability is said to increase in inverse proportion to voltage up to a specified maximum current. This is the so-called symmetrical basis of rating. Under the new rating structure, a mVA rating is still assigned to breakers for class distinction, but it is not the interrupting capacity of the device in most cases.

Under the symmetrical current basis of rating switchgear, the factor K defines the permissible range of voltage and fault current. The interrupting capability of the breaker then falls into one of three bands:

1. Voltage greater than maximum rated voltage. The breaker may not be applied.
2. Voltage between maximum rated voltage and 1/K times rated voltage. The interrupting capacity is equal to

$$\frac{\text{(interrupting capacity at rated voltage)(rated voltage)}}{\text{actual voltage}}$$

3. Voltage less than 1/K times rated voltage. The interrupting capacity is K times the interrupting capacity at rated voltage.

The momentary current capacity, defined as the fully offset rms fault current against which the breaker must be able to close and latch its contacts, is 1.6K times the symmetrical rms interrupting capacity of the breaker at maximum rated voltage and is not a function of the actual voltage of application. Low-voltage breakers usually have direct-acting trips that operate in the first cycle of fault currents, and thus ratings are based on the total symmetrical current after the first cycle. Low-voltage breakers are tested and applied in accordance with ANSI/IEEE C37.13-1981. Older low-voltage breakers (prior to 1957) were rated on a total asymmetrical current after the first half-cycle and had a slightly different test procedure. Fuses are fast-acting, protective devices that operate in the first cycle of fault and are rated on a total asymmetrical fault current.

5.5.3 Momentary Duty Versus Interrupting Duty

Momentary duty calculated by following ANSI C37.010-1979 is compared with the closing and latching capability of medium- and high-voltage circuit breakers. Total impedances or reactance portions of line, transformer, motor, generator, and utility source impedances are used for the momentary current calculations. The reactance used for the utility source, generator, and synchronous machines are subtransient reactances. The circuit E/X current at the fault point is the symmetrical momentary duty for the breaker. The close and latch duty is found by multiplying the symmetrical duty by 1.6.

Present-day low-voltage breaker ratings are compared to the symmetrical current obtained by E/Z at the fault point, while some older low-voltage ratings are compared to an average asymmetrical current (1.25 times the symmetrical current) on symmetrically rated low-voltage breakers. Where the X/R ratio is greater than 6.6, the calculated duty is multiplied by 1.15 for comparison with the breaker rating. Fuse ratings are compared to an asymmetrical current equal to 1.6 times the symmetrical currents in some cases.

The interrupting duty, calculated by following ANSI C37.010-1979 for a symmetrical current rated breaker, is compared with the medium- and high-voltage breaker interrupting ratings. The interrupting current is lower than the momentary current because it takes into account the short-circuit decrement with respect to time while the power circuit breaker is opening. Data in Table 5.8 are useful for these calculations as well as the momentary duty calculations. The interrupting duty is found by calculating the short-circuit current (E/X) from the reactance network only and then finding a resistance network reduction. The breaker interrupting time, electrical distance away from the generators, and X/R ratio at the fault are used to determine a multiplying factor to be applied to the symmetrical current to take into account the appropriate direct-current decrements for breakers rated from two to eight cycles interrupting time. The multipliers can be found from curves given in ANSI C37.010-1979 for symmetrically rated breakers.

5.5.4 Equipment Evaluation and Coordination Using Calculated Short-Circuit Current Values

After the calculated values of first cycle and interrupting rms symmetrical short-circuit current are determined, each piece of equipment should be evaluated to be sure that its short-circuit rating is adequate. Active equipment, devices that interrupt fault currents, should have first-cycle and interrupting ratings equal to or higher than the calculated values. Passive equipment, such as busway and cable, should have a short-circuit withstand rating equal to or higher than the calculated values. For devices rated for asymmetrical values of momentary fault current, such as medium-voltage fuses, multiply the calculated symmetrical values by appropriate factors to determine the asymmetrical values.

The calculated values of short-circuit current are also needed for checking time-current coordination between protected devices. Calculated values are also necessary to determine how some devices are sized and set to protect other equipment or themselves. Overcurrent relays have thermal damage characteristics that may require them to be set for fast operation at locations where available short circuits are high. Transformers also have thermal damage characteristics that require their protective devices to clear secondary faults

Table 5.8 Multipliers for Source Short-Circuit Current Contributions

	Calculation					
	First cycle		Interrupting		Medium-voltage circuit breaker close and latch	
Source	Multiply SCA by	Multiply X''_d by	Multiply SCA by	Multiply X''_d by	Multiply SCA by	Multiply X''_d by
Utility	1.0	1.0	1.0	1.0	1.0	1.0
Generators[a]	1.0	1.0	1.0	1.0	1.0	1.0
Synchronous motors	1.0	1.0	0.667	1.5	1.0	1.0
Induction motors						
Above 1000 hp at 1800 rpm	1.0	1.0	0.667	1.5	1.0	1.0
Above 250 hp at 3600 rpm	1.0	1.0	0.667	1.5	1.0	1.0
All others 50 hp and above	1.0	1.0	0.333	3.0	0.833	1.2
All smaller than 50 hp	1.0	1.0	Neglect	Neglect	Neglect	Neglect

[a]Use 0.75 X''_d for hydrogenerators without amortisseur windings.

before a specified length of time. These and other aspects of coordination are covered more fully in Chapter 7.

BIBLIOGRAPHY

ANSI C57.12.10-1977, American National Standard Requirements for Transformer 230,000 Volts and Below, 833/958 Through 8333/10,417 kVA, Single-Phase, and 750/862 Through 60,000/80,000/100,000 kVA, Three-Phase.

ANSI C57.12.22-1980, American National Standard Requirements for Pad-Mounted, Compartmental-Type, Self-Cooled, Three-Phase Distribution Transformers with High-Voltage Bushings; High-Voltage, 34500 Grd Y/19920 Volts and Below; 2500 kVA and Smaller.

ANSI/IEEE C37.010-1979, IEEE Application Guide for AC High-Voltage Circuit Breakers Rated on a Symmetrical Basis.
ANSI/IEEE C37.13-1981, IEEE Standard for Low-Voltage AC Power Circuit Breakers Used in Enclosures.
ANSI/NFPA 70-1987, National Electrical Code.
Engineering Dependable Protection for an Electrical Distribution System, Part I: A Simple Approach to Short-Circuit Calculations, Bussmann Mfg. Division, McGraw-Edison Co., St. Louis, Mo.
IEEE Standard 141-1986, Recommended Practice for Electrical Power Distribution for Industrial Plants.
IEEE Standard 242-1986, Recommended Practice for Protection and Coordination of Industrial and Commercial Power Systems.

6 Grounding

6.1 SYSTEM GROUNDING

Whether or not to ground a power system is a question that must be answered sometime by most design engineers charged with planning power distribution systems. This chapter presents basic reasons for grounding or not grounding and reviews general practices and methods of system grounding.

When an industrial power system consists of power generating equipment and distribution circuits, the reasons for grounding these components are often the same as those for grounding similar components of an utility system and other large power systems, and the methods of grounding would generally be similar under similar conditions of service. However, in some cases the reasons for grounding and the methods for grounding certain components of an industrial power system may differ according to the requirements of manufacturing or process operations.

The National Electrical Code, ANSI/NFPA 70-1987, contains regulations pertaining to system and equipment grounding applicable to industrial, commercial, and health-care facilities. These rules are considered minimum requirements for the protection of life and property, and should be reviewed during the course of system design.

6.1.1 Ungrounded Versus Grounded System

The following discussions are intended to highlight the merits or disadvantages of both ungrounded and grounded systems.

Service Continuity

For many years, many industrial plant distribution systems have been operated ungrounded at one or more voltage levels. In most cases this has been done with the hope of gaining an additional degree of service continuity. Any contact between one phase of the system and ground is unlikely to cause an immediate outage to any load that may represent an advantage to many industrial processes. On the other hand, grounded systems are designed to isolate the faulted circuit immediately, thus resulting in an attendant outage of the loads on that circuit.

Although a ground fault on one phase of an ungrounded system generally does not cause a service interruption, the occurrence of a second ground fault on a different phase will result in an outage. If the second fault is on a different feeder, both feeders may be deenergized. Therefore, an adequate detection system is considered important for operation of an ungrounded system. Experience has indicated that multiple ground faults are rare on a grounded neutral system.

Arcing Fault Hazards

In recent years, many cases of arcing fault burndowns have been reported. In most cases, severe damage or complete destruction of electrical equipment was caused by the energy of arcing fault currents. It is generally recognized that prevention of arcing fault burndowns must rely on fast and sensitive detection of the arcing fault current, accompanied by an interruption of the faulty current within 10 to 20 cycles. In a solidly grounded-neutral system, this fast sensitive detection is possible since an arcing fault will produce a current in the ground path. Under normal conditions, there is no significant current in the ground return path. Monitoring the solidly grounded-neutral system for currents in the ground circuit provides an easy means for detecting and removing destructive arcing faults to ground.

The inherent problem of ground-fault protection device typically installed in low-voltage main circuit breakers, service breakers, and so on, is that a ground fault in a cable or equipment of a small sub-feeder is likely to trip the large main circuit breaker, thus shutting down the whole system. Coordinated ground-fault protective devices can be installed at each feeder, subfeeder, and at each local branch of the system to avoid the tripping of a main circuit breaker.

Safety

Most of the hazards to personnel and property existing in some industrial electrical systems are the result of poor or nonexistent grounding of electrical equipment and metallic structures. It is therefore important to note here that regardless of whether or not the system is grounded, safety considerations require thorough grounding of equipment and structures.

It is erroneous to believe that on an ungrounded system a person may contact an energized phase conductor without personal hazard. As shown in Figure 6.1, an ungrounded system with balanced phase-to-ground capacitance has normal line-to-neutral voltage between any phase conductor and ground. To contact such a conductor accidentally or intentionally may present a serious or even lethal shock hazard.

Other hazards of shock and fire may result from inadequate grounding of equipment in either grounded or ungrounded systems. Accidental ground faults are inevitable. A high-impedance ground circuit may not permit enough current flow to operate protective devices, with the result that a potential fire or safety hazard may exist. On the other hand, the relatively high ground-fault currents associated with solidly grounded systems may present a hazard to exposed workers from hot arc products and flying molten metal. However, this problem has become less serious because of the universal application of metal-enclosed equipment in recent years.

Power System Overvoltage

Some of the more common sources of overvoltage on a power system are the following:

1. Lightning
2. Switching surges
3. Static
4. Contact with a high-voltage system
5. Line-to-ground fault
6. Resonant conditions
7. Restriking ground faults

Generally speaking, a neutral grounding can effectively help reduce the hazard that will otherwise develop from system overvoltage.

Cost

The cost differential between a grounded and an ungrounded neutral system will vary, depending on the method of grounding, the degree of protection desired, and whether a new or an existing system is to be grounded. Power transformers with wye-connected secondaries and wye-connected generators are available as standard options, and there is no cost factor for establishing the system neutral. The additional cost items are the neutral grounding resistor or reactor if chosen, and the cost of ground-fault relaying. To ground an existing ungrounded delta-connected system requires an additional

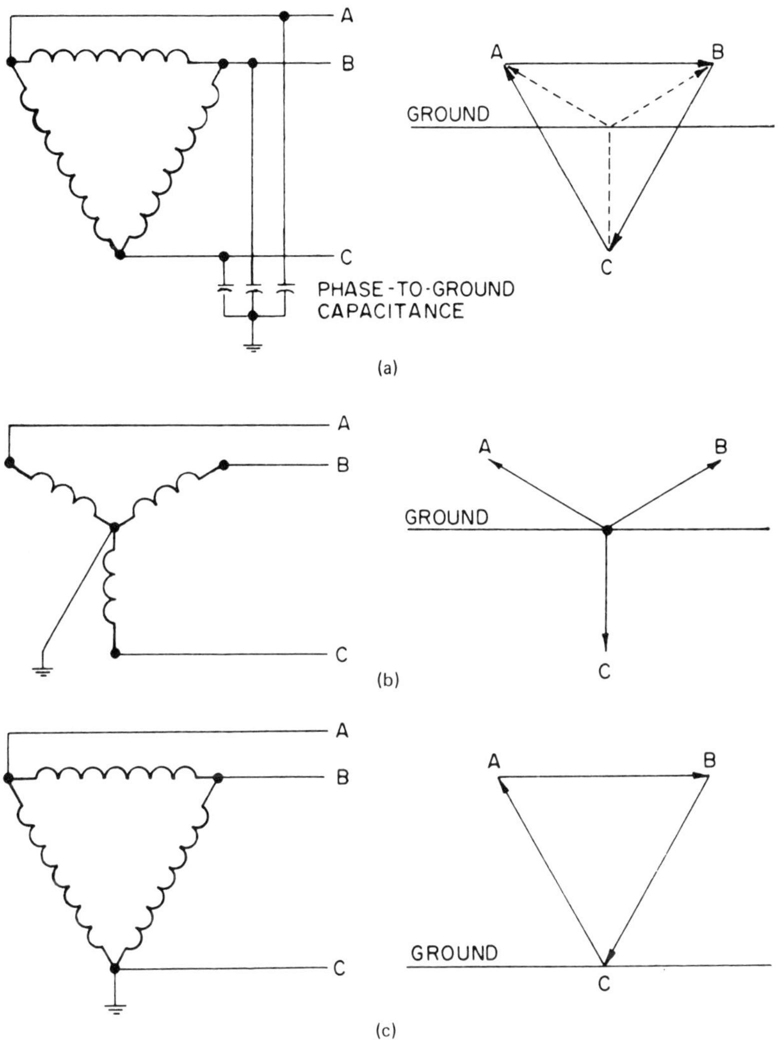

Figure 6.1 Diagram showing voltages to ground under steady-state conditions for (a) ungrounded system, (b) grounded Wye system, and (c) grounded Delta system.

cost item: the grounding transformers for establishing the system neutral. Also, the existing relay schemes may have to be modified to obtain sensitive ground-fault detection.

The purpose of converting an existing ungrounded system to grounded operation is usually to limit transient overvoltages. Older systems with degraded insultion levels due to aging are most vulnerable to failure due to transient overvoltages. Therefore, the cost of conversion would be small compared with the cost of replacing cables and/or transformers. Many industrial power system operators believe that an ungrounded system offers greater service continuity than a grounded system does provided that good-quality electrical maintenance is available. However, many users whose maintenance practices are not quite as extensive feel that a grounded-neutral system gives them more continuous service than does an ungrounded system.

6.1.2 Methods of System Grounding

Grounding methods are very dependent on system voltages. For industrial plant power systems, low-voltage distribution systems are defined as 600 V or less; medium-voltage systems, 2.4 to 69 kV; and high-voltage systems, 115 kV and above. Most grounded systems employ some method of grounding the system neutral at one or more points. Figure 6.2 shows the equivalent diagrams for ungrounded and various types of grounded-neutral systems.

Ungrounded Systems

An ungrounded system is defined as a system of conductors with no intentional connection to ground. A simple one is diagrammed in Figure 6.3. Since all circuit conductors are separated from each other and ground (earth) by an insulating medium (normally rubber and/or air), the conductors are capacitively coupled to each other (phase to phase) and to ground. The phase-to-phase capacitive coupling has little influence on the grounding characteristics of the system, so it is excluded from this discussion.

The voltage from each phase to neutral equals the voltage from each phase to ground, and the neutral is at ground potential, as long as none of the conductors is grounded. However, if one conductor develops a bolted fault to ground, the system neutral is in effect shifted from ground by a voltage value equal to the system line-to-neutral voltage, thus increasing the voltage stress on the insulation of the unfaulted conductors. Moreover, the ungrounded (or capacitive-coupled) system is very likely to produce transient overvoltages due to arcing line-to-ground fault and to series resonance. (Arcing ground faults are also called "restriking" or "intermittent" ground faults.) Because the overvoltages due to arcing ground faults and series resonance are caused by excessive neutral

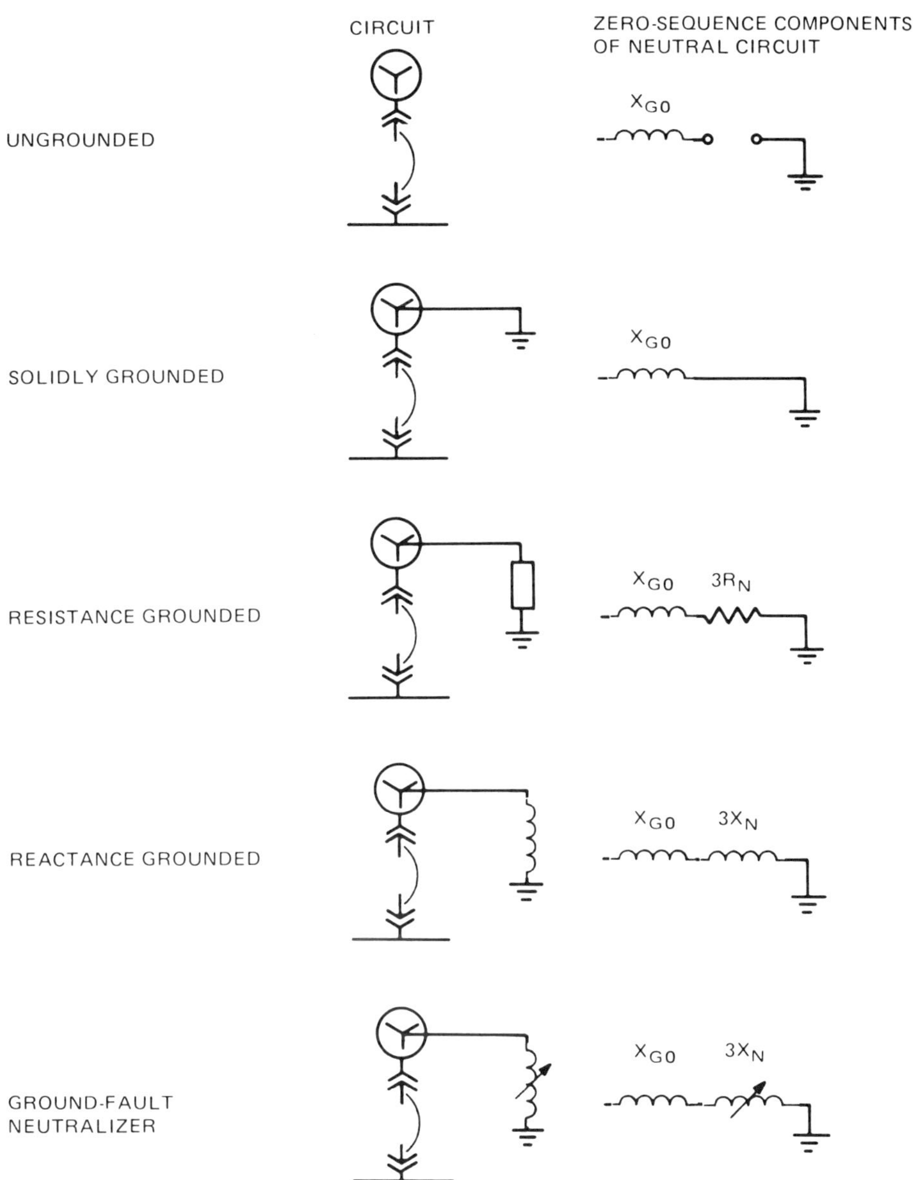

X_{GO} = Zero-sequence reactance of generator or transformer
X_N = Reactance of grounding reactor
R_N = Resistance of grounding resistor

Figure 6.2 System neutral circuit and equivalent diagrams.

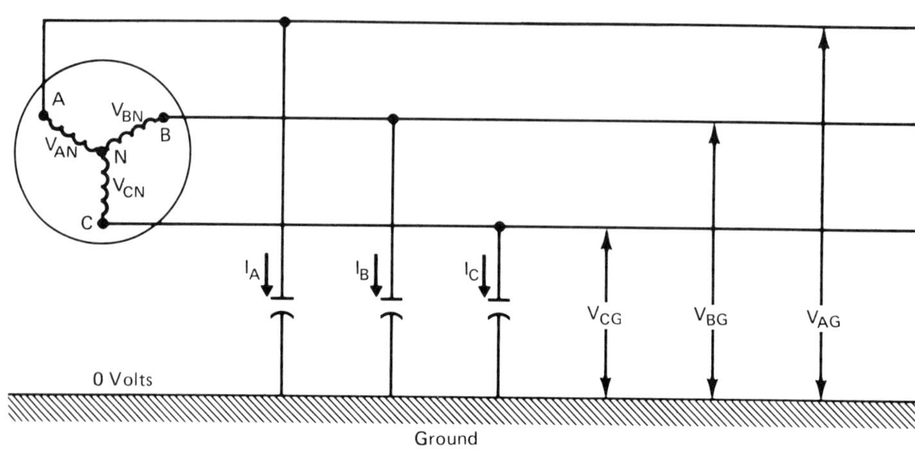

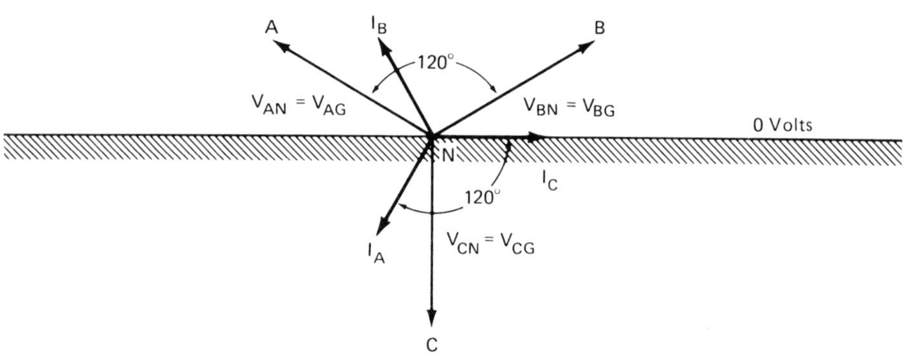

Figure 6.3 Simplified diagram for an ungrounded system.

displacement with reference to ground, the amount of overvoltage can be held to tolerable limits if the neutral voltages can be stabilized. Stabilization is accomplished by grounding the system.

Grounded Systems

A grounded system is defined as a system of conductors in which at least one conductor, usually the system neutral, is intentionally grounded, either solidly or through a resistor or other current-limiting device. Both solid and resistance grounding can limit transient overvoltage to a safe level (250% of normal voltage); therefore, other system parameters determine the choice. In general, there are several methods of grounding, discussed below.

Solid Grounding. This refers to the connection of the neutral of a generator, power transformer, or grounding transformer directly to the station ground or to the earth. If the reactance of the generator or transformer is too great with respect to the total system reactance, the objective sought in grounding, principally freedom from transient overvoltages, may not be achieved. In terms of resistance and reactance, effective grounding of a system is accomplished only when $R_0 \leqslant X_1$ and $X_0 \leqslant 3X_1$, and such relationships exist at any point in the system. X_1 is the positive-sequence reactance of the complete system, including the subtransient reactance of all rotating machines.

In most generators solid grounding without external impedance may permit the maximum ground-fault current from the generator to exceed the maximum three-phase fault current that the generator can deliver and for which its windings are braced. Neutral-grounded generators should be grounded through an impedance that will limit the ground-fault current to a value no greater than the generator three-phase fault current.

Resistance Grounding. In resistance grounding the neutral is connected to ground through one or more resistors. With the exception of transient overvoltages, the line-to-ground voltages that exist during a line-to-ground fault are nearly the same as those for an ungrounded system. A system properly grounded by resistance will not be subject to destructive transient overvoltages. For resistance grounded systems at 15 kV and below, such overvoltages will not be of a serious nature if the resistance value lies within the following boundary limits: $R_0 \leqslant (1/3)X_{C0}$, $R_0 \geqslant 2X_0$, where R_0 is zero-sequence resistance, X_{C0} is system phase capacitance reactance to ground, and X_0 is the zero-sequence reactance of the system.

Resistance grounding may be either of two classes, high resistance or low resistance, distinguished by the magnitude of ground-fault current permitted to flow. Both types are designed to limit transient overvoltage to a safe level (within 250% of neutral). The low-resistance method has the advantage of immediate and selective clearing of the grounded circuit but requires that the minimum ground-fault current be large enough to actuate the ground-fault relay positively. High-resistance grounding is a method that can be applied to an existing medium-voltage ungrounded system to obtain transient overvoltage protection without the modification expense of adding ground relays to each circuit.

Reactance Grounding. A reactor is connected between the system neutral and ground. Since the ground-fault current that flows in a reactance-grounded system is a function of the neutral reactance, the magnitude of the ground-fault current is often used as a criterion for describing the degree of grounding. In a reactance-grounded system, the available ground-fault current should be at least 25%, and preferably 60% of the three-phase fault, to prevent serious transient overvoltage. In reactance-grounded systems the ratio of X_0/X_1 should be 10 or less, where X_0 is the zero-sequence inductive reactance of the system, including the neutral reactance, and X_1 is the positive-sequence inductive reactance of the system, including the subtransient reactance of all rotating machines. This is significantly higher than the level desirable in a resistance-grounded system. Therefore, reactance grounding is usually not considered an alternative to resistance grounding.

Ground-Fault Neutralizer (Resonant Grounding). A ground-fault neutralizer is a reactor connected between the neutral of a system and ground and having a specially selected relatively high value of reactance. A line-to-ground fault causes line-to-neutral voltage to be impressed across the neutralizer, which passes an inductive current. This current is approximately equal in magnitude to the resultant of the system charging current of the two unfaulted phases. When these currents neutralize each other, the only remaining current in the fault is due to resistance, insulator leakage, and corona. The current is relatively small and in phase with the line-to-neutral voltage. The arc is extinguished without restriking, and the flashovers are quenched without removing the faulted line section from service. This method of grounding has been used primarily on systems above 15 kV, consisting largely of overhead distribution lines.

6.1.3 Selection of System Grounding Arrangements

For Systems 600 V and Below

Low-voltage systems are frequently operated solidly grounded. The principal reason for this is the extensive use of 480/277-V systems with line-to-neutral connected loads, and the requirements in the

National Electrical Code, ANSI/NFPA 70-1987, for solidly grounding the neutral of such systems. For the low-voltage system without line-to-neutral loading, solid grounding or high-resistance grounding are the practical devices. Low-resistance grounding is usually not considered. For reasons of economy, the solidly grounded system depends on adequate ground-fault current at locations remote from the source to operate phase fault devices. It is important that the equipment grounding network provide a very low impedance return path for the ground-fault current.

For Systems 2.4 to 15 kV

Modern power systems in this range are usually low-resistance grounded to limit the damage due to ground faults in the windings of rotating machines and yet permit sufficient fault current for the detection and selective isolation of individual faulted circuits. Ground faults are detected by an overcurrent relay connected in the residual circuit of the three-phase current transformers (Fig. 7.12) or by a relay connected to a window or doughnut-type current transformer which encloses all the phase conductors (Fig. 7.13). Using either method, positive tripping can be accomplished with low magnitude of ground-fault current. However, greater sensitivity is available with the zero-sequence current transformer method. For this reason low-resistance grounding is commonly used. Alternatively, high-resistance grounding can also be used. This method contributes to better service continuity by permitting continued operation with one ground fault, and is the good choice provided that a good maintenance crew is available.

For Systems Above 15 kV

Systems above 15 kV are almost always effectively grounded, because these are usually circuits with open lines in which surge arresters rated for grounded neutral service are desirable for overvoltage protection and lower cost. The cost of resistors for resistor grounding at these voltages is prohibitive.

6.1.4 Calculation of Ground-Fault Current

The magnitude of the current that will flow in the event of a line-to-ground fault on a grounded system is determined by the impedance of the ground return path.

Resistance Grounding

When a simple line-to-ground fault occurs on a resistance-grounded system, a voltage appears across the resistor nearly equal to normal line-to-neutral voltage of the system. In low-resistance grounded systems the resistor current is approximately equal to the fault current.

Standard grounding resistors have a voltage rating equal to the line-to-neutral voltage, and a current rating equal to the current that flows when this voltage is applied to the resistor.

The method just mentioned applies to faults on lines or buses or at the terminals of machines or transformers. If the fault is internal to a rotating machine or transformer, the ground-fault current will be less. In the case of wye-connected equipment at intermediate points in the winding between the neutral and a terminal, the fault current will be intermediate between zero and the current due to a terminal fault. In the case of the delta-connected machines, the internal voltage to neutral may be considered to be 100% at the terminal and 50% at the midpoint of the windings. The midpoints have the lowest potential with respect to the electric neutral of any part of the winding.

Reactance Grounding

In this system with a single line-to-ground fault, the ground-fault current may be calculated from the following formula by neglecting resistance:

$$I_g = \frac{3E}{X_1 + X_2 + X_0 + 3(X_N + X_{Gp})} \tag{6.1}$$

where

I_g = ground-fault current, amperes
X_1 = system positive-sequence reactance, ohms per phase, including the subtransient reactance of rotating machines
X_2 = system negative-sequence reactance, ohms per phase, including the subtransient reactance of rotating machines
X_0 = system zero-sequence reactance, ohms per phase
X_{Gp} = reactance of ground-return circuits, ohms
X_N = reactance of neutral grounding reactor, ohms
E = line-to-neutral voltage, volts

In most industrial distribution systems without in-plant generation, X_2 can be considered equal to X_1.

Solid Grounding

In this system the ground-fault current for a single line-to-ground fault may be calculated from the equation

$$I_g = \frac{3E}{X_1 + X_2 + X_0 + 3X_{Gp}} \tag{6.2}$$

6.1.5 Specifying Grounding Equipment Ratings

Resistor Ratings

For low-resistance grounded-neutral systems the determination of the resistor value, in ohms, and thus the magnitude of the ground-fault current is based on the following conditions:

1. Providing sufficient current for satisfactory performance of the system relaying requirements
2. Limiting ground-fault current to a value which will minimize damage at the point of fault without resulting in system overvoltages

Reactor Ratings

To minimize transient overvoltages, the ground-fault current must not be less than 25% of the three-phase fault current. This corresponds to a ratio of $X_0/X_1 = 10$, where X_0 and X_1 represent the total electrical system values for any possible ground-fault condition on the system. If the neutral reactance is selected in accordance with the following relationship, the current in the winding of the faulted phase will not exceed the three-phase fault current of the machine, ragardless of system reactance:

$$X_N = \frac{X_1 - X_{G0}}{3} \tag{6.3}$$

where

X_N = reactance of neutral reactor
X_1 = generator positive-sequence subtransient reactance
X_{G0} = generator zero-sequence reactance

The current rating can be calculated from equation (6.1). The neutral-grounding reactor should be selected to carry the available current under all practical operating conditions.

Grounding-Transformer Rating

The electrical specifications of a grounding transformer are typically as follows:

1. *Voltage:* line-to-line voltage of the system.
2. *Current:* maximum neutral current.
3. *Time:* amount of time for which the transformer is designed to carry rated current; usually for a short time, such as 10 or 60 s; for high-resistance grounding, the rating should be continuous

4. *Reactance:* a function of the positive-sequence short-circuit reactance of the system, or X_1.

The grounding transformer reactance, when used to effect reactance-type grounding, is based on the following criterion: The X_0/X_1 ratio must not exceed 10, and preferably, not exceed 3, in order to eliminate the possibility of transient overvoltage from a forced current zero interruption. The maximum limitation of 3 for the X_0/X_1 ratio together with a maximum limitation of 1 for the R_0/X_1 ratio will also satisfy the criteria for an effectively grounded system and will permit the use of line-to-ground voltage rated surge arresters for greater economy and protection.

6.2 EQUIPMENT GROUNDING

6.2.1 Basic Objectives

The main objective of equipment grounding is to provide safety for operators, electricians, repairpersons, and the general public. This can be accomplished by ensuring that all parts of steel structures, motor and generator frames, control equipment enclosures, switchgear, conduit, portable electric equipment, and any metallic body enclosing or near an electric cirucit that is accessible to a person must be at ground potential. This is clearly stated in the National Electrical Code. The actual grounding of the conducting body allows sufficient current flow in the event of an accident to insulation to ensure positive operation of ground-detecting devices and the operation of fuses and circuit breakers. It also tends to prevent the potential between a conducting body and ground from rising to a dangerous value. Failure to provide a continuous path in the enclosure will result in arcing and heating at breaks and joints.

6.2.2 Types of Equipment to Be Grounded

Structures

The steel framework of buildings should be grounded at the base of every corner column and intermediate columns at a distance not greater than 60 ft. The connection from the ground or grid to the structure should not be less than 2/0 copper, and should be made by brazing, thermit welding, equivalent process, or a suitable solderless terminal. The metal structures for switchgear, lightning arresters, disconnect switches, transformers, and so on, should be individually connected to the ground bus, similar to the building framework.

Outdoor Stations

The ideal size for the ground bus will depend on the magnitude of the available ground-fault currents and operating time of protective equipment, but for practical purposes it should have an equivalent conductivity of not less than 4/0 AWG copper for small substations or 500,000-circular mil (cmil) copper for medium-sized and large installations. The grid type of ground is often used. The ground bus should be connected to any metallic water pipe, metallic drain, or sewer pipe located in the station area or within reasonable distance. Connection should consist of a conductor of not less than the size of the ground bus. A metal fence surrounding a outdoor station should be grounded. A ground conductor of not less than 1/0 AWG copper should be installed around the fence, approximately 12 to 24 in. away from it and about 12 in. deep. This should be brazed or welded to ground rods installed at 10- to 40-ft intervals.

Large Generator and Motor Rooms

In large station rooms, a ground bus of adequate size should be run around the periphery of the building. Conductor material should be soft-drawn or medium-hard-drawn copper wire, copper bar, or equivalent. The ground bus should be connected at two or more points to grounding electrode, the building structure, and to water mains, metallic drain, and sewer pipe in order to keep the resistance of the ground bus to earth as low as possible. A typical grounding system for a large generator and motor room is shown in Figure 6.4.

Conductor Enclosures

Conductor enclosures include cabinets, junction boxes, outlet boxes, controllers, service raceway, conduit, couplings, fittings, cable armor, lead sheath, and grillwork. Metal boxes, cabinets and fittings, or non-current-carrying metal parts of other fixed equipment, if metallically connected to grounded cable armor or metal raceway, are considered to be grounded by such connection. The lead sheath, shield, and armor of large single-conductor cables (500,000 cmils and above) should be grounded at one end only to prevent circulating currents. The sheath, shield, and armor of such a cable should be insulated from ground throughout the remainder of its length unless the cable is too long, in which case insulating joints must be provided to permit grounding at a number of points to keep sheath voltage down to desirable limits.

Portable Electric Equipment

1. Portable equipment operating at line voltages above 600 V should be supplied through a suitable portable cable permanently

connected at both ends. The complete equipment, including any associated housing or structure, should be grounded through a grounding wire or wires in this cable equal in current-carrying capacity to the largest line conductor. It is desirable that such equipment be operated from a wye-connected system with its neutral grounded through a resistor that limits ground-fault current to 50 A or less. Suitable ground-fault relaying should be provided.

2. Protable equipment operating at line voltages of 600 V or less should be grounded through a separate grounding wire or wires in the connecting cable equal in current-carrying capacity to the largest line conductor.

3. Portable equipment operating from a single-phase circuit at 150 V and below and not rated over 15 A should be grounded through a separate grounding wire in the connecting cable equal in current-carrying capacity to the line conductors. Grounding should be through separate grounding contacts in the plug and receptacle. Figure 6.5 shows the desired grounding-conductor connection arrange-

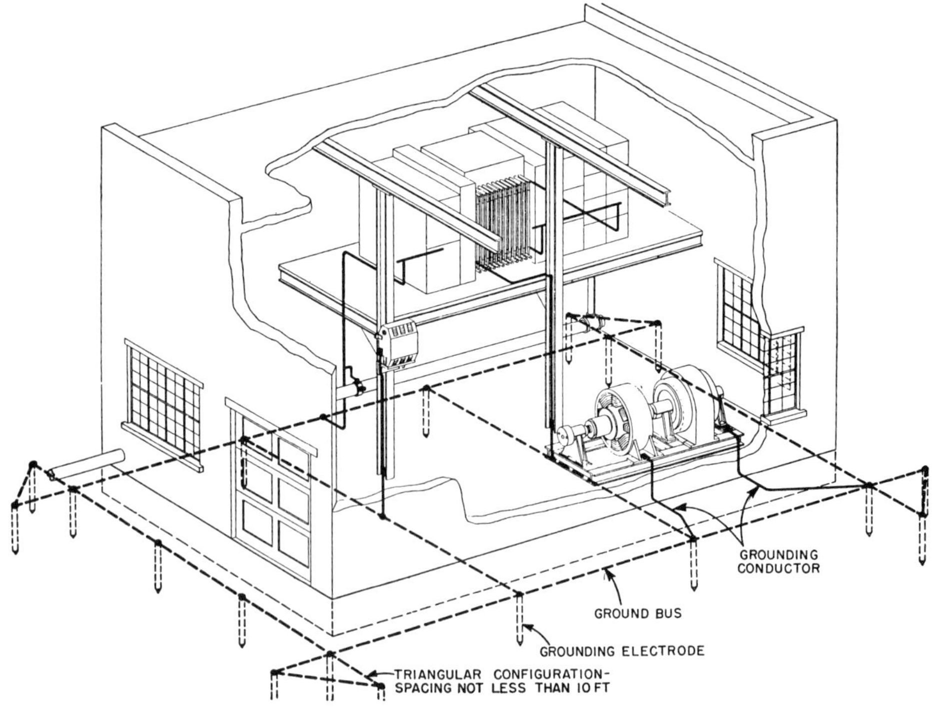

Figure 6.4 Typical grounding system for a large motor room.

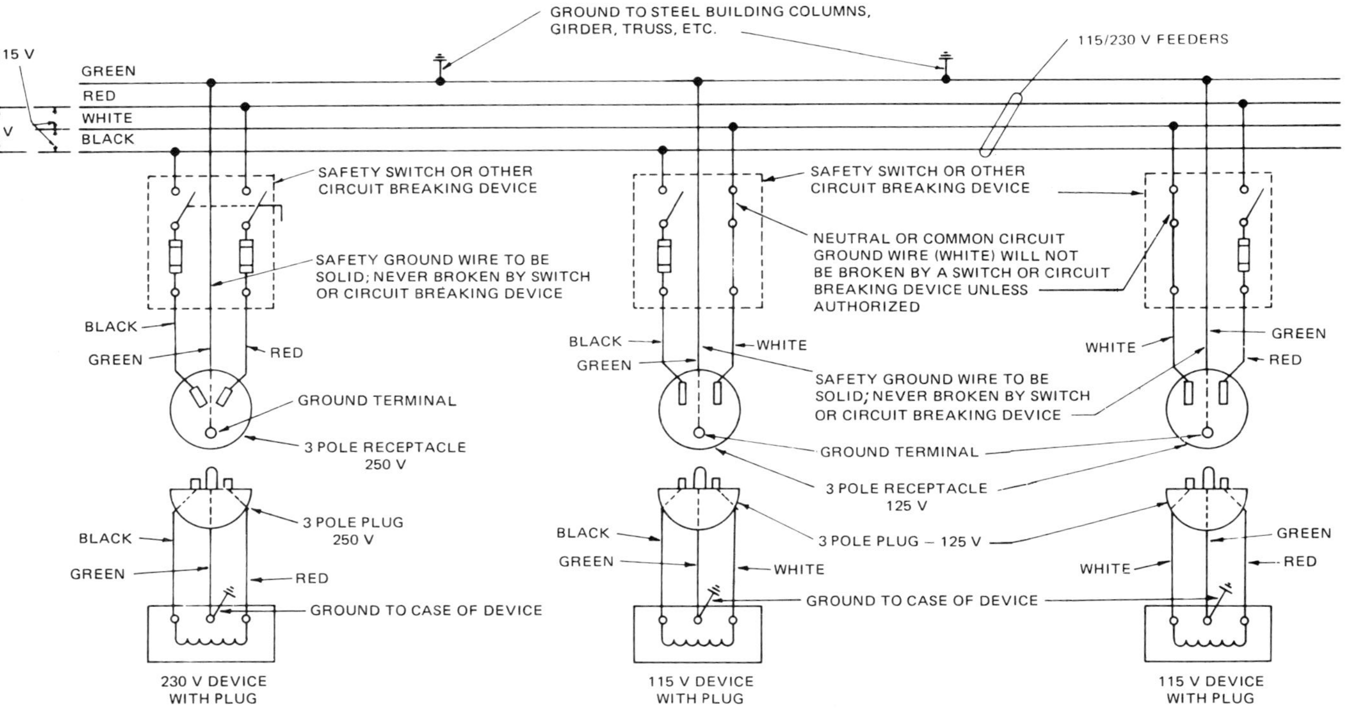

NOTES: (1) When looking at front face of switch, circuit breaking device, transformer, etc. the left-side hot wire will be black, the right-side hot wire will be red.
(2) The circuit neutral or common circuit wire (grounded) will be white.
(3) The safety ground wire will be green. (This wire ties on to the case or shell of the device.)
(4) The standard two-pole plug will fit the receptacle.

Figure 6.5 Typical power circuit and equipment grounding with receptacles.

ment for a variety of power circuit patterns, and clearly shows the distinction between the grounding and the grounded conductors.

6.3 STATIC AND LIGHTNING PROTECTION GROUNDING

6.3.1 Basic Objectives of Static Grounding

Industrial plants handling solvent, dusty materials, or other flammable products often have a potential hazardous-operation condition because of static accumulating on equipment, on materials being handled or even on operating personnel. The discharge of a static charge to ground or to other equipment in the presence of flammable or explosive materials is often the cause of fires and explosions which may result in the loss of lives and huge financial losses. Protection of human life is the first objective in attempting to control static charges. The other objectives are to prevent the loss of (1) capital investment in buildings and equipment, (2) operating funds in stored materials, and (3) profits because of the loss of production.

6.3.2 Fundamental Causes of Static

A difference in potential will exist between two substances when one holds a positive charge and the other holds a negative charge. These charges accumulate when dissimilar substances are brought into contact with each other. Under this condition the negative electrons will migrate from the surface of one substance to that of the second substance, leaving the positive ions on the first substance. This means that one substance has a positive charge and the second has a negative charge. Upon separation of two such oppositely charged materials, a static discharge may take place. Voltages that have been observed in a few industries are shown in Table 6.1.

Table 6.1 Observed Static Voltage Ranges of Selected Industrial Equipment

Equipment	Voltage range observed
Belted drives	60,000–100,000
Fabric handling	15,000–80,000
Paper machines	5,000–100,000
Tank trucks	Up to 25,000
Grain belt conveyors	Up to 45,000

For a static spark to produce ignition in a combustible vapor and air mixture, sufficient energy must be sotred in the charge body. The amount of energy that is stored and available from a capacitive-type discharge can be calculated from the following formula:

$$E = 1/2 \times CV^2 \times 10^{-9}$$

where

C = capacitance, picofarads
V = potential, volts
E = energy, millijoules

Approximate values of capacitance in picofarads of some objects are shown in Table 6.2. For the static electricity to be able to cause ignition, in addition to the requirement of sufficient energy in the spark discharge, it must take place in an ignitable mixture. If the mixture is too thin, ignition may not occur.

6.3.3 Methods of Static Control

Grounding and Bonding

Many static problems can be solved by bonding the various parts of the equipment together and grounding the entire system. Reference should be made to the National Fire Protection Association's and other bulletins on static electricity for some detailed methods. Therefore, no attempt will be made here to describe in detail how and where such bonding and grounding should be made.

Table 6.2 Capacitance Values of Selected Objects

Objects	Capacitance value (pF)
Human being	100–400
Automobile	500
Tank truck (2000 gal)	1000
12-ft-diameter tank with insulated lining	100,000

Humidity Control

Many insulating materials contain a certain amount of moisture in equilibrium with the surrounding air. This moisture or relative humidity controls the surface conductivity of these insulating materials. At normal humidity (30% or more) an invisible film of water provides an electrical leakage path over most solid insulating bodies that drains away static charges. When relative humidity is 30% or less, the same materials dry out and become good insulators. Static manifestations become noticeable. Humidifying the entire atmosphere near the point of static electricity has proved to be a solution to the static problem.

Ionization

When a charged object is brought in contact with ionized air, the static charge is dissipated. The charge is either conducted to the ground through the ionized air, or the charged object attracts a sufficient number of positively or negatively charged ions from the air to neutralize it. Ionization of air can be obtained by flame, alternating electric fields generated by high-voltage ultraviolet light, or radioactivity.

Conductive Floors

The use of conductive floors or floor coverings may be required to prevent the accumulation of static charge by grounding personnel and conductive objects together, since the human body in dry locations can also accumulate a dangerous static charge. Such flooring must be of nonsparking materials, such as conductive rubber, lead, or other conductive compounds.

Conductive Footwear and Casters

When conductive flooring is used, operators must wear conductive nonsparking footwear. The resistance between the wearer and ground must not exceed 1 megohm, which is the total resistance of the conductive footwear plus the resistance of the floor.

Special Precautions

In addition to the use of conductive floors and shoes, other controls may be considered:

1. Providing wearing apparel with low static-producing qualities
2. Establishing rigid operating procedures
3. Using conductive rubber mats where conductive flooring is not used throughout an area

The subject of dissipation of static electricity is well covered in ANSI/NFPA 77-1977, "Static Electricity." In industrial areas with extremely hazardous conditions, it may be well to consider these recommendations.

6.3.4 Lightning—Nature and Its Hazard

Lightning is an electric discharge between clouds or between clouds and earth. Charges of one polarity are accumulated in the clouds and of the opposite polarity in the earth. When the charge increases to the point that the insulation between can no longer contain it, a discharge takes place. The discharge is evidenced by a flow of current, usually great in magnitude, but extremely short in time. Damage to buildings and structures is the result of heat and mechanical forces produced by the passage of current through resistance in the path of discharge. Although the discharge takes place at the low-resistance path, it is not uncommon for the current to follow a path of high resistance. This may be a tree, a masonary structure, or a porcelain insulator. Lightning can cause damage to structures by direct stroke and to equipment by surges coming in over exposed power lines.

6.3.5 Need for Protection Against Lightning

Damage to structures and equipment due to surge effect is a subject in itself, and protection against this type of damage is not within the scope except as grounding is involved. A lightning protection system consists of terminals projecting into the air above the uppermost parts of the structure with interconnecting and grounding conductors. Terminals should be placed so as to project above all points that are likely to be struck. Conductors should present the least possible impedance to earth. Each projecting terminal above the structure should have at least two connecting paths to earth and more if practicable. Each conductor running down from the terminals on top of the structure should have an earth connection. Properly made connections to earth are an essential feature of a lightning rod system for protection of buildings. Electrodes should be at least 2 ft away from and should extend below building foundations. Experiments have indicated that a vertical conductor will divert to itself direct hits that might otherwise fall within a cone-shaped space, of which the apex is the point and the base is a circle whose radius is approximately equal to the height of the point.

6.3.6 Protection of Power Stations and Substations

Station protection against direct stroke should include effective shielding of the station structure itself and at least the first 2000 to 2500 ft of exposed lines adjacent to the stations. This shielding may take the form of masts or extensions of the steel structure to provide a proper cone of protection to apparatus and circuits within the station area. The height of a single mast or single ground wire shielding is usually based on a shielding angle of 30 degrees. When two or more masts or ground wires are used, that part of the shielding angle that lies between masts or ground wires is increased to 60 degrees.

Lightning protection of power stations and substations includes the protection of station equipment by means of lightning arresters. For lightning arresters a local grounding connection should be made by driving electrodes into the earth near the arresters as shown in Figure 6.6. In addition, the lightning arrester grounding conductor should be connected into the common station ground bus. For the average case an arbitrary upper limit of 5 Ω resistance to ground has been established. For connection from arrester to ground should be as short and as straight as possible. The National Electrical Code, ANSI/NFPA 70-1987, states that a lightning arrester ground connection should not be smaller than No. 6 AWG, but larger sizes may be desirable with large systems.

6.4 CONNECTION TO EARTH

6.4.1 Resistance to Earth

The grounding resistance of an electrode is made up of:

1. Resistance of the electrode
2. Contact resistance between the electrode and the soil
3. Resistance of the soil from the electrode surface outward

The first two resistances are very small fractions of an ohm and for all practical purposes can be neglected. The third element is the one discussed here. Around a rod this resistance is the sum of the series resistance of virtual shells of earth, located progressively outward from the rod. The shell nearest the rod has the smallest circumferential cross section, so it has the highest resistance. Successive shells outside this one have progressively lower resistance. As the radius outward from the rod increases to about 20 ft, the incremental resistance per unit of radius decreases to nearly zero. The first few inches away from the rod are the most important one as far as reducing the resistance is concerned. In high-soil-resistivity locations, chemical treatment or the use of concrete will be most useful in improving the effectiveness of grounding electrode system. Adding more electrodes to the first one does not affect the resistance close to the electrode. So the resistance will be higher than the value obtained by dividing the resistance of a single rod by the number of rods in the grounding system unless the rods are separated by impractically great distance.

Figure 6.6 Typical method of grounding a lightning arrester.

The connection to earth or the electrode system needs to have a sufficiently low resistance to permit prompt operation of circuit protective devices in the event of a ground fault. System ground resistances of less than 1 Ω may be obtained by use of a number of individual electrodes connected together. Such a low resistance may only be required for large substations. Resistances in the range 2 to 5 Ω are generally suitable for industrial plant substations and buildings. The 25-Ω value noted in the National Electrical Code, ANSI/NFPA 70-1987, applies to the maximum resistance for a single electrode. There is no implication that 25 Ω per se is a satisfactory level for a grounding system.

6.4.2 Grounding Electrodes

Basically all ground electrodes may be divided into two groups. The first group comprises underground metallic piping systems, metal building frameworks, well casings, steel piling, and other underground metal structures installed for other purposes. The second group comprises made electrodes specifically designed for grounding purposes.

Existing Electrodes

The metal building frames are normally attached to their concrete foundation footings by long anchor bolts. The anchor bolts in concrete serve as electrodes, while the metal building frame is simply a grounding conductor. The National Electrical Code states that continuous underground water or gas piping systems in general have a resistance to earth of less than 3 Ω and that metal building frames, local metallic underground piping systems, metal well casings, and the like have a resistance to earth of substantially less than 25 Ω. For safety grounding and for small distribution systems where the ground currents are of low magnitude, such electrode are usually satisfactory. However, care should be exercised to ensure that all parts are effectively bonded together.

Made Electrodes

Made electrodes may be subdivided into driven electrodes, steel reinforcing bars in below-ground concrete, buried strips or cables, grids, buried plates, and counterpoises. The type selected will depend on the type of soil encountered and the available depth. Driven electrodes are generally more satisfactory and economical where bedrock is 10 ft or more below the surface, while grids, buried strips, or cables are preferred for lesser depth. Grids are frequently used for substations or generating stations to provide equipotential areas throughout the entire station for greater safety to personnel. Buried plates have not been used extensively in recent years because of the high cost. Also when used in small numbers they are the least effective type of made electrodes. The counterpoise is a form of buried cable electrode used to ground transmission-line towers and structures. In selecting the number and size of grounding terminals, their current discharge limitations must be recognized. If these are exceeded, the earth around the electrode may be exploded by steam generation or may be dried out to the extent of becoming nonconductive.

6.4.3 Choices of Rods

Grounding rods are manufactured in diameters of 3/8, 1/2, 5/8, 3/4, and 1 in. (9.53, 12.7, 15.88, 19.05, and 25.4 mm) and in lengths of 5 to 40 ft (1.5 to 12.2 m). For most applications, the diameters of 1/2, 5/8, and 3/4 in., in lengths of 8, 10, 12, and 16 ft, are satisfactory. The NEC, ANSI/NFPA 70-1987, specifies that rods of steel or iron shall be at least 5/8 in. in diameter, and that rods of nonferrous materials shall not be less than 1/2 in. in diameter. Copper-clad steel, one of the most common type of rods, permits driving to considerable depth without destruction of the rod itself, while the copper coat permits direct copper-to-copper connection between the ground wire and the rod. For ordinary soil conditions, the 10-ft (3-m) length of rod has become fairly well established as a minimum standard length to meet the code requirement of a minimum of 8 buried feet (2.44 buried meters).

BIBLIOGRAPHY

ANSI/NFPA 77-1977, Static Electricity.
ANSI/NFPA 78-1980, Lightning Protection Code.
ANSI/NFPA 70-1987, National Electrical Code.
Hedlund, C. F., Lightning Protection for Buildings, *IEEE Transactions on Industry and General Applications*, vol. IGA-3, Jan./Feb. 1967, pp. 26–30.
IEEE Standard 32-1972, Terminology and Test Procedure for Neutral Grounding Devices.
IEEE Standard 80-1976, Guide for Safety in AC Substation Grounding.
IEEE Standard 142-1982, Recommended Practice for Grounding of Industrial and Commercial Power Systems.
Kaufmann, R. H., Some Fundamentals of Equipment-Grounding Circuit Design, *AIEE Transactions (Applications and Industry)*, vol. 73, Nov. 1954, pp. 227–232.

Regotti, A. A., and Wargo, H. W., Grounding for Industrial and Commercial Distribution Systems, *Westinghouse Engineer*, Apr. 1974, pp. 41–45.

Static Electricity, Circular C-438, National Bureau of Standards, Boulder, Colo., U.S. Government Printing Office, Washington, D.C.

7
System Protection

7.1 SYSTEM BEHAVIOR AND PROTECTION NEEDS

7.1.1 System Behavior

The principal abnormalities in a power distribution system are short circuits and overloads. Short circuits may be caused in many ways, including failure of insulation due to excessive moisture, mechanical damage to electrical distribution equipment, and failure of utilization equipment as a result of overloading or other abuses. Circuit may become overloaded simply by connecting additional utilization equipment to the circuit, or by improper installation and maintenance, or improper operating procedures, such as too frequent starting or obstructed ventilation.

Short circuits may occur between two phase conductors, between all phases of a polyphase system, or between one or more phases and ground. It may be "solid" or welded, in which case the short circuit is permanent and has a relatively low impedance. The short circuit may include an arc having relatively high impedance. It may or may not extinguish itself. The flashover may lead to ionization and more extensive short circuit. Other sources of abnormality, such as lightning, load surges, and loss of synchronism, usually have little or no effect on system overcurrent selectivity, but should not be ignored. Protection can best be handled on an individual basis for the specific equipment involved such as transformers, motors, and generators.

Design engineers of industrial power systems have available techniques to minimize the effects of abnormalities occurring on the system or on the utilization equipment that it supplies. They must design into the system features that will and can:

1. Quickly isolate the affected portion of the system
2. Minimize the magnitude of the available short-circuit current
3. Provide alternate circuits and automatic throwovers to minimize the durection or the extend of outages

7.1.2 Protection Needs

System protection is one of the most essential features of an electrical system and must be considered concurrently with all other features. It is important to the safety of personnel and the reliability of electrical supply. The need for higher production from industrial plants has created demands for greater reliability of the system. Trends to network systems and parallel operation with the utilities have resulted in sources having very high overcurrent during fault conditions. The high costs of power distribution equipment and the time required to repair or replace damaged equipment such as transformers, cable, high-voltage circuit breakers, and so on, make it imperative that a system be given the best protection within economic constraints.

A service interruption in a chemical plant can cause loss of product and create major cleanup and restart problems. It may be desirable to tolerate a short-time overload condition and associated reduction in life expectancy of the electric apparatus affected. Other industries, such as refineries, paper mills, automotive plants, steel mills, and food-processing plants, are similarly affected.

7.1.3 Protection Needs for Grounded Versus Ungrounded System

Grounding of industrial power systems is treated in Chapter 6. It is necessary here only to observe the effect on basic relaying methods of the choice between a grounded and an ungrounded system. In a grounded system, phase-to-ground faults produce current of sufficient magnitude to operate ground-fault responsive overcurrent relays, which automatically detect the fault, determine which feeder has failed, and initiate the tripping of the correct circuit breaker to deenergize the faulted portion of the system without interrupting service to healthy circuits.

In an ungrounded system as shown in Figure 6.3, phase-to-ground faults produce a relatively insignificant value of fault current. In a small isolated neutral industrial installation, the ground-fault current for a single line-to-ground fault may be under 1 A, while the large plant containing miles of cable to provide electrostatic capacitance to ground may produce not more than 20 A of ground-fault current. These currents usually are not of significant magnitude for the operation of overcurrent relaying to locate and remove such faults. It is possible to provide phase-to-neutral relays that will operate an alarm on the occurrence of a ground fault.

7.2 PRINCIPLES OF RELAYING FOR INDUSTRIAL PLANTS

Fault protection relaying can be classified into two groups, primary relaying, which should operate first to remove faulted equipment from the system, and backup relaying, which operates only when primary relaying fails. Figure 7.1 shows a one-line diagram illustrating zones of protection. It illustrates the basic principles of primary relaying, with separate areas of protection established around each system element which can be isolated by a separate interrupting device. Any

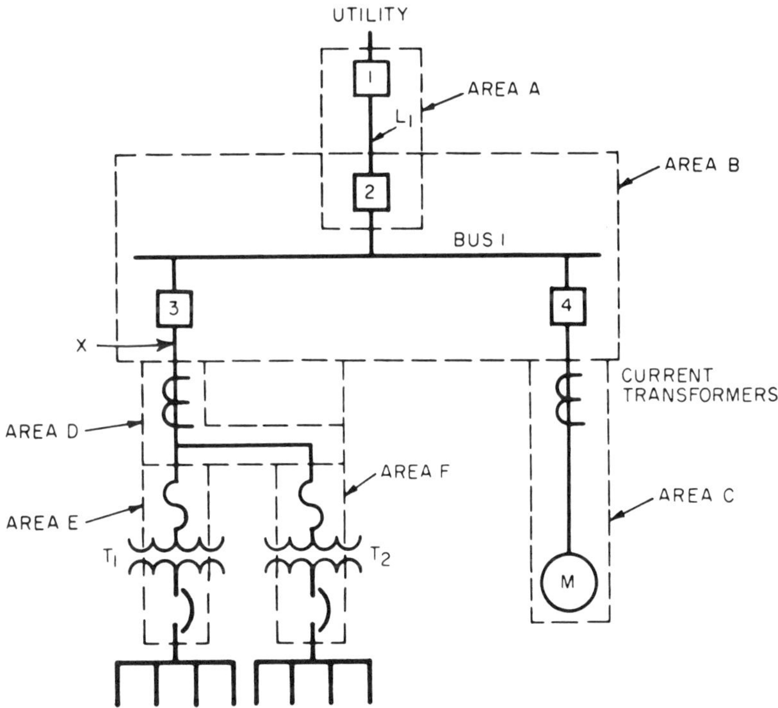

Figure 7.1 One-line diagram showing zones of protection.

equipment failure occurring with a given area will cause tripping of all circuit breakers applying power to that area.

To assure that all faults within a given zone will operate the relays of that zone, the current transformers associated with that zone should be placed on the line side of each circuit breaker so that the circuit breaker itself is a part of two adjacent zones. This is known as "overlapping." In radial circuits the consequences of this lack of overlap are not usually very serious. When the current transformers are located immediately at the load bushings of the circuit breaker, the amount of circuit exposed to this problem is minimized. The consequences of lack of overlap become more serious in the case of tie circuit breakers between differentially protected buses and bus feeds protected by differential or pilot-wire relaying. In applying protective relays to industrial power systems, safety, simplicity, reliability, maintenance, and the degree of selectivity required must be considered. Before attempting to design a protective relaying plan, the elements that make up the distribution system together with the operating requirements must be examined.

7.2.1 Typical Small-Plant Relay System

One of the simplest industrial power systems consists of a single service entrance circuit breaker and one distribution transformer which steps the utility primary distribution voltage down to utilization voltage, as illustrated in Figure 4.1. There would be several circuits on the secondary side of the transformer protected by either fuses or circuit breakers. Protection for the circuit between the incoming line and the circuit breaker on the transformer secondary would normally consist of conventional overcurrent relays. Preferably the relays should have the same time-current characteristics as the relays on the utility system, so that the service entrance breaker can be set to trip before the utility supply-line circuit breaker. The phase relays should have instantaneous elements to be able to clear higher current faults promptly.

This simple system provides both primary and backup relay protection. It can be expanded by tapping the primary feeder and providing fuse protection on the primary of each distribution transformer as shown in Figure 4.3. An additional step or area of protection is included over the simple system. All secondary circuit breakers as before, while faults within the transformer should now be cleared by the transformer primary fuses. The transformer secondary breakers or primary fuses may act as backup protection for the faults that are not cleared by the secondary feeder protective devices. The primary feeder faults will be cleared by circuit breaker (A). This will, in turn, act as backup protection for the transformer primary fuses.

7.2.2 Protective Relaying for a Large Industrial Power System

For a large industrial plant power system, which may incorporate an intricate network of medium- and low-voltage distribution substations, uninterruptible power sources, and in-plant generation required to operate in parallel with or isolated from local utility networks, the number of sequential steps of relaying will increase to provide an inherently selective scheme within each zone of protection.

Depending on the degree of complexity of the system, various types of relays may be required for protective schemes for the system and equipment in the system, such as transformers, motors, and/or generators. Generally speaking, the following relays are probably employed to varying degrees:

1. Directionally controlled overcurrent relays (67) are provided to trip on current flow toward the high side of the transformer. Their sensitivity is not limited by the flow of load current in the normal or nontrip direction. They have inverse-time characteristics.
2. Differential protective relays (87B) are instantaneous in operation and inherently selective within themselves. Without such relays, high current bus faults must be cleared by proper operation of overcurrent devices on the several sources, resulting in long-time clearing.
3. Time-delay ground relays (51N) are used to realize maximum sensitivity. They are connected to the output of current transformers measuring the current in the neutral connection to ground.
4. Pilot-wire differential relays (87L) are used to provide superior protection for the cable tie between buses. In addition to being instantaneous in operation, pilot-wire Schemes are inherently selective within themselves and require only two pilot wires. For backup protection provided by overcurrent relays, separate current transformers are required to provide reliability and flexibility in the application of other protective devices.
5. Current balance relays (46) protect motors against damage from excessive rotor heating caused by single phasing or unbalanced voltage conditions.
6. Surge protection is provided by the surge arrester and surge capacitor combination, which should be located as close as possible to the motor terminals.
7. Many other protective relays may be used for protection of generators connected to the incoming bus of the system, such as percentage differential relays (87G); loss of excitation protection is provided by device (40), external unbalanced fault current (negative-sequence) protection by device (46), antimotoring protection

by device (32), and backup overcurrent protection by device (51V/50).

All device numbers cited above are listed and defined in ANSI C37.2-1979, Electrical Power System Device Function Numbers.

7.3 PROTECTIVE RELAYS AND THEIR APPLICATIONS

7.3.1 Overcurrent Relays (50, 51, 50/51)

Instantaneous Overcurrent Relays (50)

There are two types of instantaneous relays using the principle of electromagnetic attraction: the solenoid or plunger type, and the clapper or hinged armature type. The basic elements of the solenoid type are a solenoid and a movable plunger of soft iron. The pickup current is determined by the position of the plunger in the solenoid. A calibration screw may be provided to adjust the position of the plunger. In the clapper type, a hinged armature that is held open by a restraining spring is attracted to the pole face of an electromagnet. The magnetic pull of the electromagnet is proportional to the coil current. The pickup current may be calibrated over a specified range. This type of relay is nromally found in an induction relay case when a "50/51" (time-overcurrent with instantaneous) function is specified.

Induction-Type Time-Delay Overcurrent Relays (51)

The most commonly used time-delay relays for system protection are the induction disk type. The principal components of an induction-type overcurrent relay are an overcurrent unit, an indicator contactor switch, and an indicating instantaneous-trip unit. The principal components and their location are shown in Figure 7.2.

Solid-State Overcurrent Relay

Some new overcurrent relay design utilizes solid-state technology. Time-current curves are obtained through the use of RC digital timing circuits. The instantaneous circuit can be set to the desired setting on the instantaneous tap multiplier potentiometer on the front panel. The setting is in multiples of the tap setting. The relay will respond to current above the setting per typical time curve (Figure 7.3). This unit is responsive to total current, including the dc component, and an allowance must be made in the setting to avoid overtripping. In general, the time-current characteristic curves and tap ranges are similar to those provided in induction relays. However, solid-state overcurrent relays can provide faster reset times and have no significant overtravel.

Characteristic Curves of Overcurrent Relays

Time-overcurrent relays are available with many different current ranges and tap settings. Typical available range of tap settings are as follows:

Per-unit amperes			
Time	Instantaneous	Taps	Time dial
0.5–2.5 (or 0.5–2.0)	20–80	0.5, 0.6, 0.8, 1.0 1.5, 2.0, 2.5	1/2–11
1.5–6 (or 2–6)	20–80	1.5, 2.0, 2.5, 3.0 3.5, 4, 5, 6	1/2–11
4–16 (or 4–12)	20–80	4, 5, 6, 7, 8, 10 12, 16	1/2–11

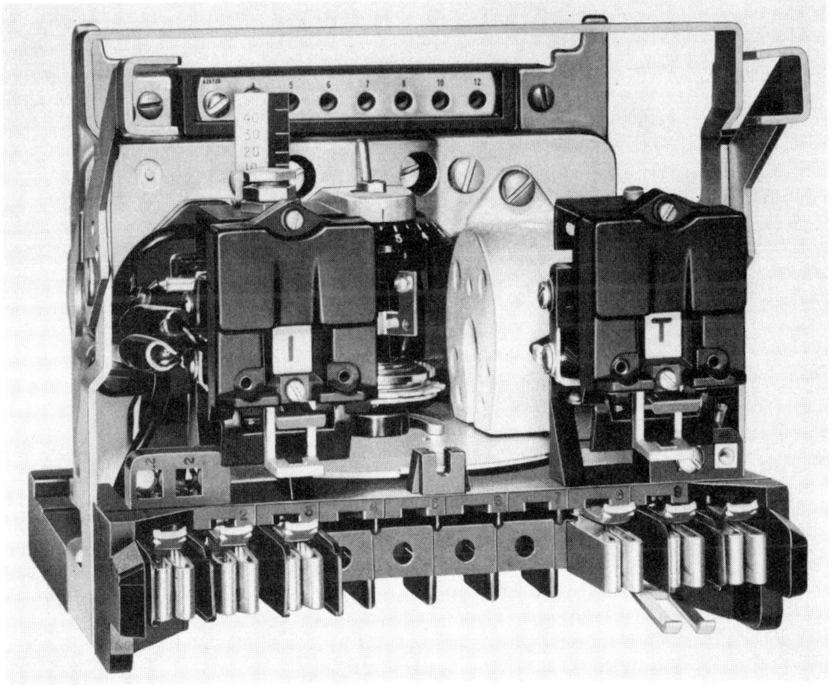

Figure 7.2 Induction-disc overcurrent relay (relay removed from drawout case).

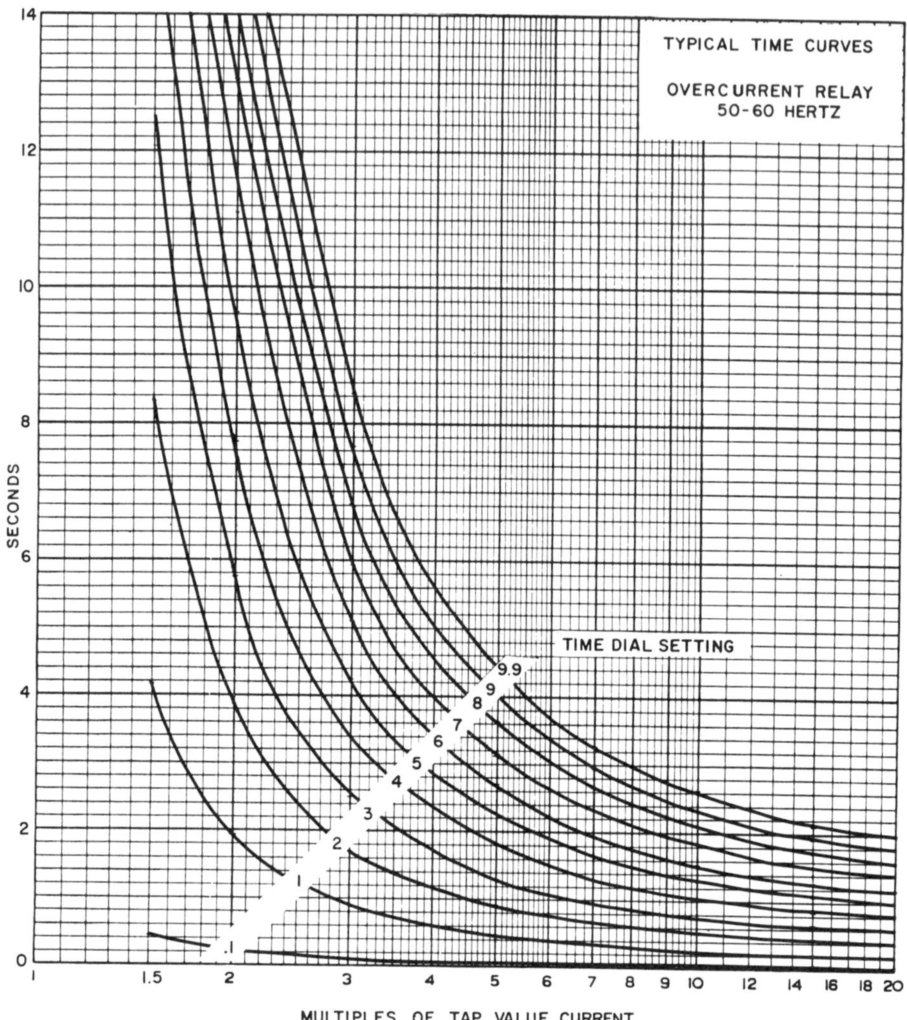

Figure 7.3 Time-current curves of a solid-state inverse time overcurrent relay.

Solid-state overcurrent relays have the following current ratings:

Per-unit amperes			
Time	Instantaneous	Taps	Time dial
0.5–12	10–40	0.5, 1.0, 1.5, 2.0, 2.5, 3, 3.5, 4, 4.5, 5, 6, 7, 8, 10, 12	1/2–11

Most overcurrent relays are equipped with a time delay that permits a current several times in excess of the relay setting to persist for a limited period of time without closing the contacts. If a relay operates faster as current increases, it is said to have inverse-time characteristics.

Overcurrent relays are available with many different time-current characteristics. Induction disk overcurrent relays have a provision for variation of the time adjustment and permit change of operating time for a given current. This adjustment is called the time lever or time dial setting. Figure 7.4 shows a family of time-current operating curves available with a typical inverse-time overcurrent relay. Similar curves are available for other overcurrent relays having different time-delay characteristics with increasing current values. The relay operating time will decrease in an inverse manner down to a certain minimum value. Figure 7.5 shows the characteristic curves of inverse (A), very inverse (B), and extremely inverse (C) time relays when set on their minimum and maximum time dial posotions. It also shows the characteristics of the instantaneous element (D) that is normally supplied in these relays. Table 7.1 shows the different types of overcurrent relays and their common field of application.

Figure 7.5 illustrates the comparative slopes of three induction-type relay curves at 10 TD and again at 1/2 TD. Solid-state overcurrent relays have very similar time-current characteristics. For same tap and time dial settings, the time required for responding to an overcurrent condition is usually shorter for a solid-state relay.

Special Types of Overcurrent Relays

Voltage-Controlled and Voltage-Restrained Overcurrent Relays. In a voltage-controlled overcurrent relay, an auxiliary undervoltage element controls operation of the induction disk element. When the applied voltage drops below a predetermined level, an undervoltage contact is closed in a shaded-pole circuit, permitting the relay to develop torque and operate as a conventional overcurrent relay. The

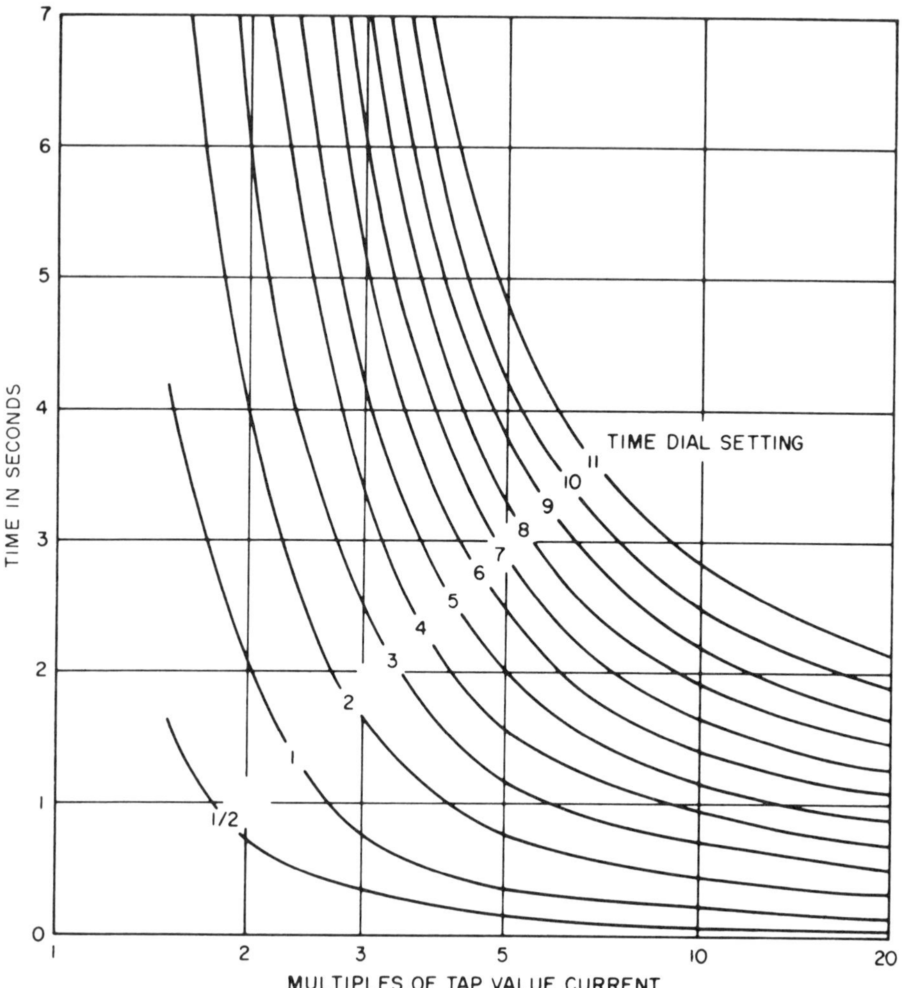

Figure 7.4 Time-current curves of a typical inverse time overcurrent relay.

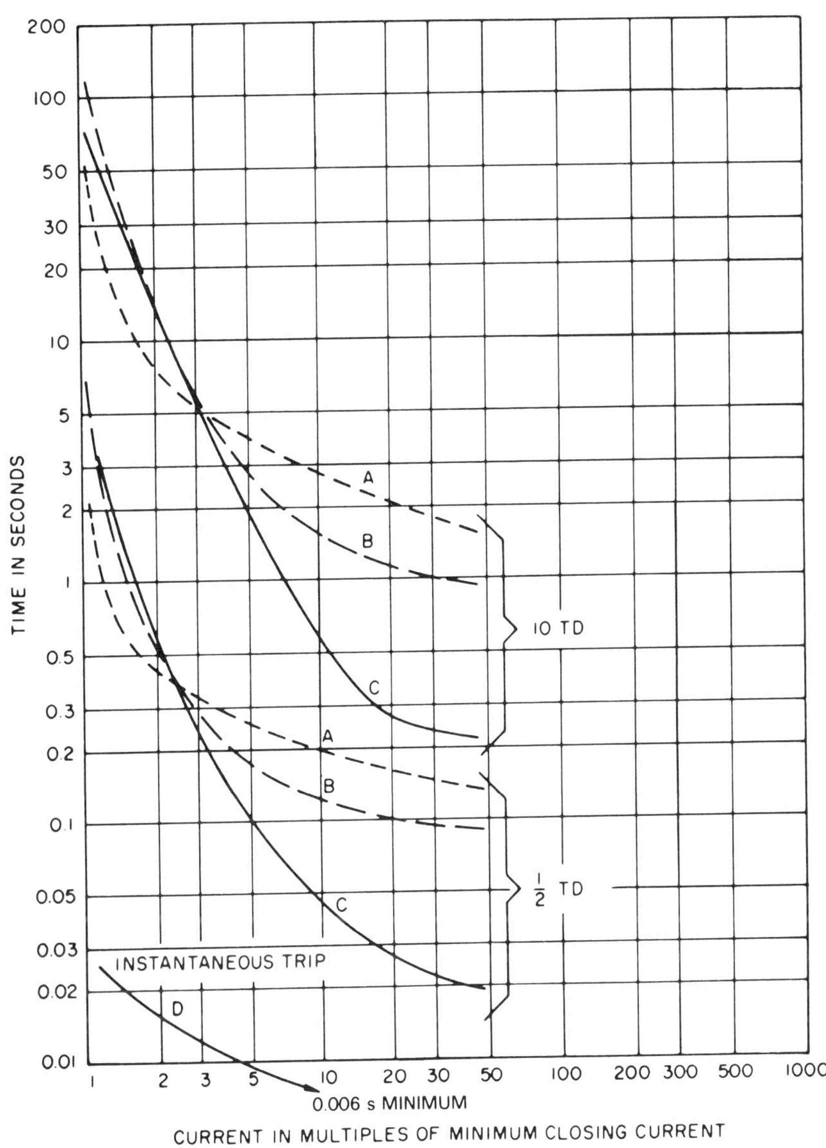

A Inverse
B Very Inverse
C Extremely Inverse
D Instantaneous
TD Relay Time Dial Setting

Figure 7.5 Comparative time-current characteristics of three induction type overcurrent relays.

voltage-restrained relay has a voltage element that provides restraining torque proportional to voltage and thus actually shifts the relay pickup current. Hence the relay becomes more sensitive the larger the voltage drop, but is relatively insensitive at normal voltage. The relay is set to ride through permissible power swings at nominal voltage. Since the voltage and overcurrent units of the voltage-controlled relay are independently adjustable, it is preferrable to a voltage-restrained type. The timing of a voltage-restrained relay is affected by the bus voltage variation, making it difficult to coordinate with primary relays. Voltage-controlled overcurrent relay is often recommended for generator backup protection.

Harmonic Restraint Overcurrent Relays. These relays consist of a clapper-type instantaneous unit operated by second harmonics filtered through an air-gap transformer, a block filter, two full-wave rectifiers, a varistor, and a polar unit. The harmonic restraint unit provides high-speed supervision of overcurrent relay, which protects

Table 7.1 Common Fields of Application for Overcurrent Relays

Time Inverse	Operating time[a] (s)	Protection
Instantaneous	0.016	Overcurrent instantaneous trip
Short-time	0.47	Bus or generator differential
Long-time	25.00	Motor overcurrent
Definite-time	2.00	Overcurrent on high-low short circuit
Moderate-inverse	2.48	Overcurrent on high-low short circuit
Inverse	2.52	Overcurrent on lines/feeders
Very inverse	1.53	Overcurrent on lines/feeders
Extremely inverse	0.80	Overcurrent relay and fuse coordination

[a]At 10 times dial and 10 times overcurrent.

transformers. The taps and characteristics of these relays are as follows:

Tap amperes	Continuous amperes	One-second amperes
0.87	10	300
2.00	18	300
4.00	12	300

High-Low Current Relays. These relays have the same basic construction as that of overcurrent relay, except that high-low contacts are provided. They are generally available in the following current ranges:

Range	Taps
0.5–2.5	0.5, 0.6, 0.8, 1.0, 1.5, 2.0, 2.5
1–12	1.0, 1.2, 1.5, 2.0, 2.5, 3.0, 3.5, 4.0, 5.0, 6.0, 7.0, 8.0, 10.0, 12.0

These relays may have either single- or double-circuit closing contacts for tripping either one or two circuit breakers. Their time-current characteristics curves are identical to those of overcurrent relays.

7.3.2 Overvoltage (59) and Undervoltage Relay (27)

Time Inverse Over- and/or Undervoltage Relays

These relays consist of a voltage unit employing the electromagnetic induction disk construction and an indicating contact switch unit. The voltage unit consists of an E-shaped laminated core that has a main tapped coil and a shading on one of the outer legs to shift the flux out of phase. The out-of-phase flux interacts with the main coil flux to create a torque on the disk. Rotation of the disk is opposed by a spiral spring which resets the contacts and disk to their normal position when the applied voltage falls below the tap setting value.

On undervoltage relays, the contacts open at tap value or above. On overvoltage relays the contacts close at tap value or above. On both types, the tap value is the voltage at which the relay's front contact closes. The back contact will close within 5% of this value.

Induction disk voltage relays are generally available in the following ratings:

Taps	ICS-A	Time dial
120 (55–140)	0.2/2.0	1/2–11
240 (110–280)	0.2/2.0	1/2–11
480 (220–560)	0.2/2.0	1/2–11

The time-current characteristics for under- and overvoltage relays are shown in Figures 7.6 and 7.7. The following table shows the different types of voltage relays and their field of application:

Voltage relay	Operating time (s)	Protection
Undervoltage	0–140 0–28	Induction motor transitory faults
Overvoltage	0–140 0–14	Generator ground protection

Special Voltage Relays

Frequency-Compensated Over- or Undervoltage Relays. These relays employ the same basic induction disk voltage unit and indicating contactor switch and operate similarly to under- or overvoltage relays except that they include a frequency-compensating resistor connected in the outer leg coil circuit of the E-type electromagnet. The compensating resistor enables them to maintain their 60-Hz pickup voltage and/or drop out values within 5% over a variable input frequency from 30 to 90 Hz.

Third-Harmonic Filtered Overvoltage Relays. These relays consist of the same basic induction disk voltage unit and indicating contactor switch as other types of under- or overvoltage relays, but in addition include a built-in capacitor connected in series with the electromagnet coil to filter out third-harmonic voltage. They are designed for low pickup values (8% continuous voltage rating) in generator ground-fault detection schemes. The taps and time dials for these relays are as follows:

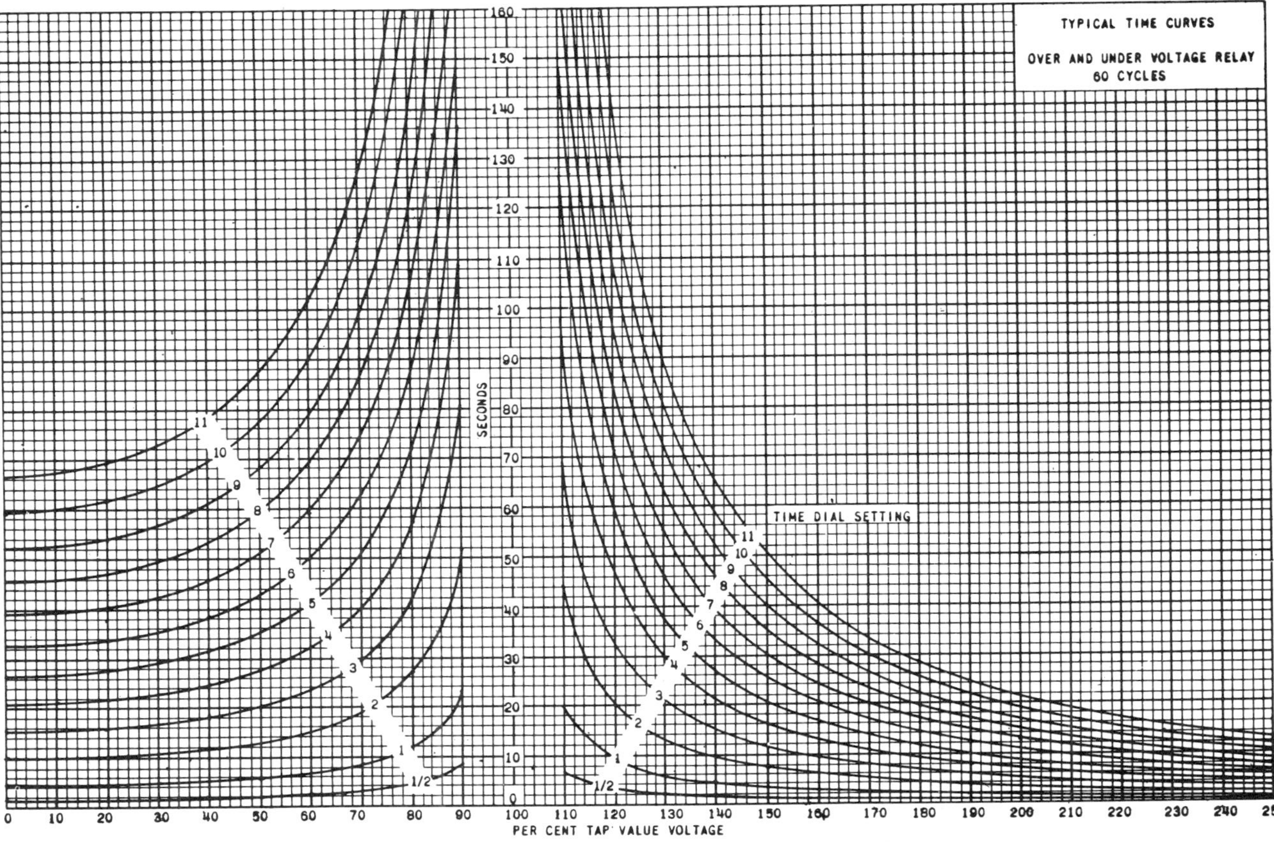

Figure 7.6 Typical time-current characteristic curves of long time under- and overvoltage relay.

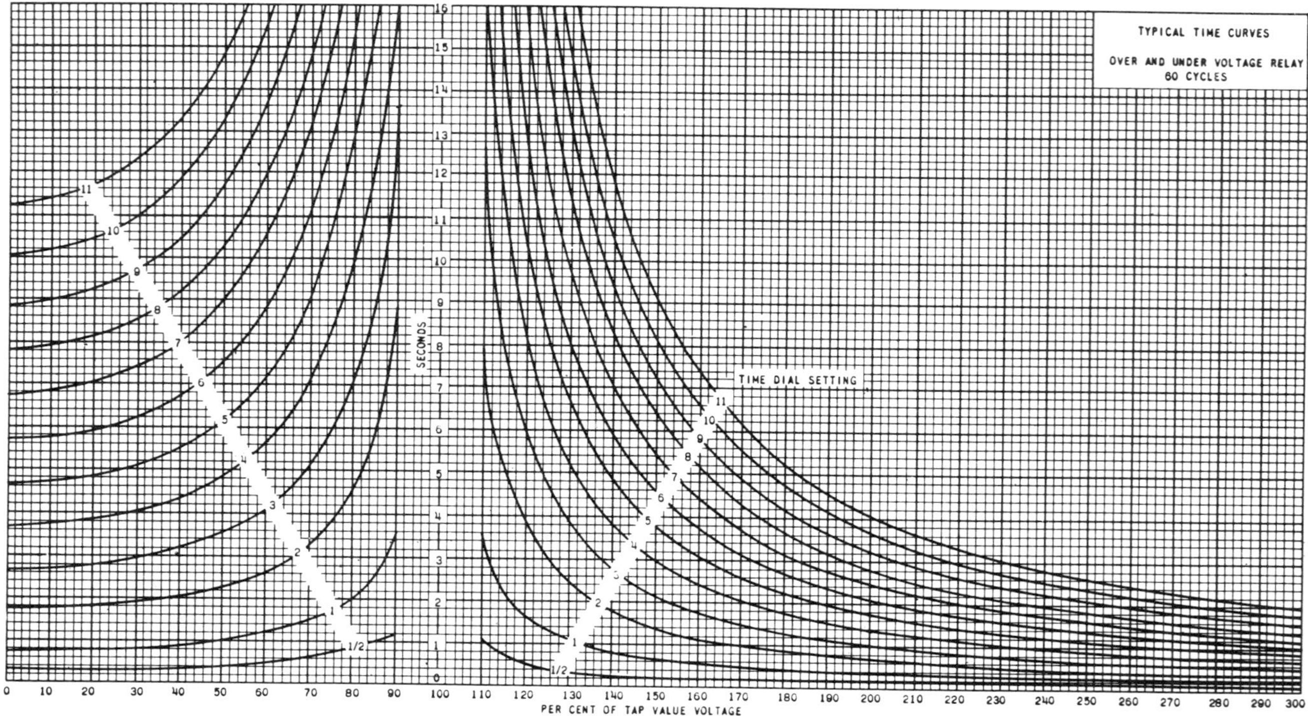

Figure 7.7 Typical time-current characteristic curves of short time under- and overvoltage relay.

Tap V	Time dial
67	1/2–11
199	1/2–11
290	1/2–11

Negative-Sequence Under- and Overvoltage Relays. These relays are basically the same as time inverse under- and overvoltage relays except that a negative-sequence filter is added to provide negative-sequence voltage to the voltage unit of the relay. They provide instantaneous and time-delay detection of negative-sequence overvoltage as well as responding with time delay to phase-to-phase undervoltage. When used in motor protection, the relay protects against system undervoltage, single phasing of the supply, and reversal of phase rotation of the supply.

When the relay is used for overvoltage protection, the back contacts are made at normal voltage and the negative-sequence element is committed to an instantaneous function. The negative-sequence overvoltage pickup is adjustable from 5 to 10% of rated line-to-neutral voltage. The setting of the under- and overvoltage unit can be defined either by tap setting and time dial position or by tap value voltage.

Solid-State Under- and Overvoltage Relays

The single-phase solid-state under- and overvoltage relay consists of a solid-state voltage unit, a reactor, a series resistor, and a full-wave rectifier. The resistor and reactor are proportioned to maintain a constant effective ampere turn to the voltage unit for 20 to 60 Hz.

7.3.3 Directional Relays

Directional Overcurrent Relays (67)

Directional overcurrent relays are used to provide sensitive tripping for fault currents in one direction and nontripping for load or fault currents in the other direction. These relays consist of two units, an overcurrent element and a directional element, as shown in Figure 7.8. Operation of this element is controlled by the directional element. When current is flowing in the tripping direction, the directional contacts that are in the lag coil circuit close, thus enabling the overcurrent element to operate when the current exceeds its tap setting. The directional element has an operating coil. The latter is energized by either voltage or current in order to determine the direction of current flow. The maximum torque may be produced when the operating current leads the voltage by 45 degrees.

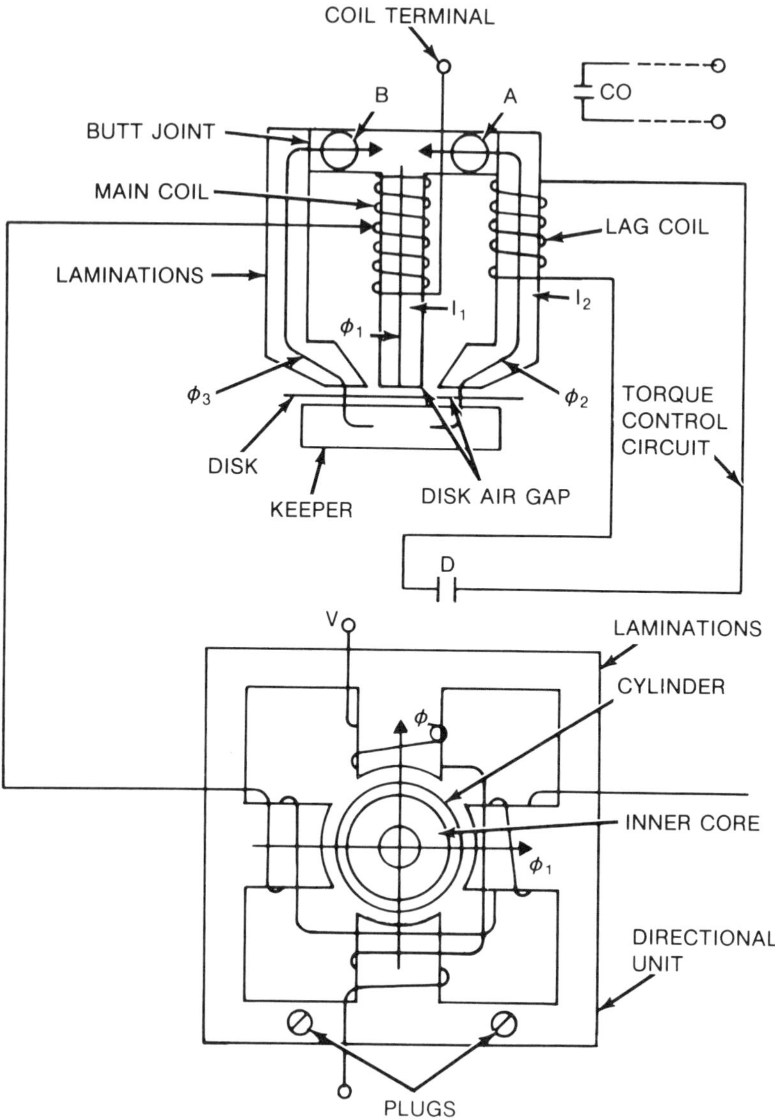

Figure 7.8 Directionally controlled overcurrent relay.

Instantaneous Directional Overcurrent Relays

An electromagnetic relay has an instantaneous induction cup element that is controlled by an instantaneous power directional element. The operating current is adjustable over a selected range, and the directional characteristics must be identified and applied in the same manner as described in the preceding paragraph.

High-Speed Directional Overcurrent Relays

A high-speed directional element and an overcurrent element are coordinated to operate on the occurrence of a short circuit that reverses the normal flow of current. Proper coordination of elements requires that a contact on one high-speed element opens before the contact on another can close when the relay is suddenly deenergized or a quick reversal in power flow occurs. This type of relay can be used to protect in-plant generation operating in parallel with a utility system from a fault on the utility system.

Directional Ground Relays

These relays operate on the product of the current in the operating coil and the voltage/current in the polarizing coil. It provides very sensitive protection in the desired direction of current flow. The operating element of an electromagnetic relay may be either an induction disk having an adjustable time delay for selectivity, or an induction cup element for high-speed operation. The directional characteristics of the relay should be determined, to assure correct application. This type of relay is normally reserved for use in ground-fault protection of generators and transformers.

7.3.4 Differential Relays (8T)

Application of Differential Relays

Differential relays operate on summing the current flowing into and out of a protected circuit zone. Normally, the current flowing into a circuit zone equals the current flowing out in which no differential current flows in the relay. If a fault occurs in the circuit zone, part of the current flowing in will be deflected into the fault, and the current flowing out will be less than the current flowing in. If the differential current is above the preset value, the relay will trip.

Differential relays provide high-speed, sensitive, and inherently selective protection. The types of relays often used are:

1. Overcurrent differential
2. Percentage differential
   a. Fixed percentage differential

   b. Variable percentage differential
   c. Harmonic restraint percentage differential
3. High-impedance differential
4. Pilot wire differential

For differential protection, proper matching of relay and current transformer characteristics is a prime design requirement.

Overcurrent Differential Relays

Overcurrent differential protection is used extensively to protect motors, generators, and transformers against internal faults. Figure 7.9 shows differential protection applied to a generator. Both ends of the windings must be available for installation of the current transformers, which have their secondary windings connected in series. Under normal conditions the current flowing in each current transformer secondary winding will be the same and the differential current flowing through the relay operating winding will be zero. In case of an internal fault in the generator, the secondary currents will no longer be the same and the differential current will flow through the relay operating coil and cause the trip circuit to be energized. In this application, the overcurrent relays have to be set so that they

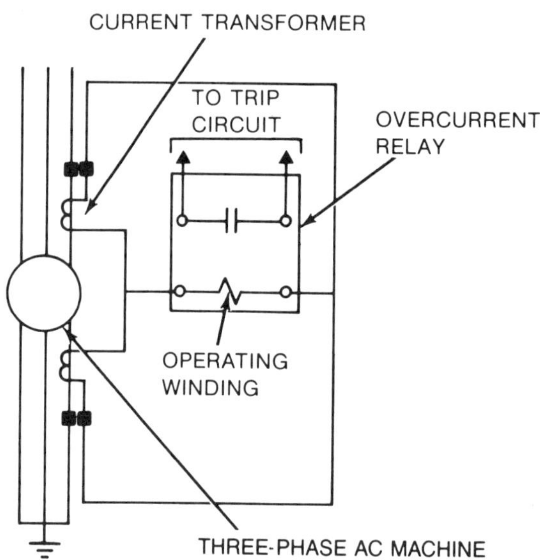

Figure 7.9 Overcurrent relay used for differential protection of an AC generator.

do not operate on the maximum error current that can flow in the relay during an external fault. To solve this problem without sacrificing sensitivity, the percentage differential relay is usually the answer.

Percentage Differential Relays (87T, 87B, 87M, and 87G)

As discussed in the preceding section, percentage differential relays are generally used in transformer, bus, motor, or generator applications. There are three types: fixed percentage, variable percentage, and harmonic restraint percentage. The harmonic restraint percentage differential relay is used only for transformer applications. Figure 7.10 shows a basic relay connection (one phase) for a fixed percentage restraint differential relay. Under normal conditions, current circulates through the current transformer and relay restraining coils R_1 and R_2; no current in coil O. The amount of differential or operating current required to overcome the restraining torque and close the relay is a fixed percentage of the restraining current.

The solid-state percentage differential relay consists of various solid-state functional circuit elements, connected together to provide a three-phase relay. The elements consist of a restraint circuit, operating circuit, sensing circuit, amplifier circuit, trip circuit, and

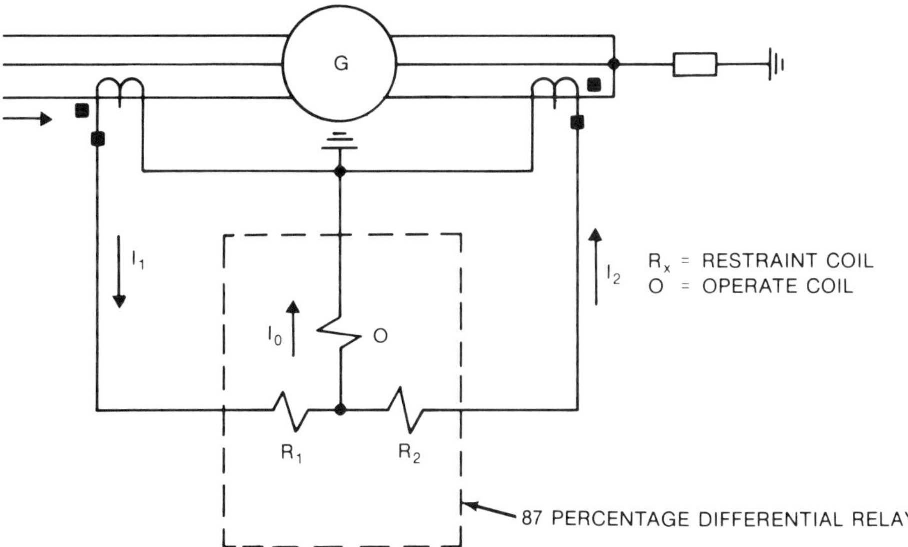

Figure 7.10 Basic relay connections (one phase only) for fixed percentage restraint differential relay.

indicating circuit. The restraint circuit senses each phase current and produces an output voltage proportional to the phase current with the largest magnitude. The operating circuit senses the differential current for each phase and produces an output voltage proportional to the differential current with the largest magnitude. The relay receives outputs of the restraint circuit and the operating circuit, and combines them into an output that reflects the difference between the two. The output is fed into the amplifier circuit. If the magnitude is sufficient to trigger the amplifier, its signal causes the trip circuit to operate, thus tripping the circuit breaker.

7.3.5 Current Balance Relays (60)

Electromagnetic Type

An electromagnetic consists of two or three induction disk elements, each having two current coils as shown in Figure 7.11. These coils are connected to different phases so that a closing torque is produced on the disk that is proportional to the difference or unbalance between the currents in the two phases. The amount of unbalance current required to close the contacts may be a fixed percentage, typically 25%.

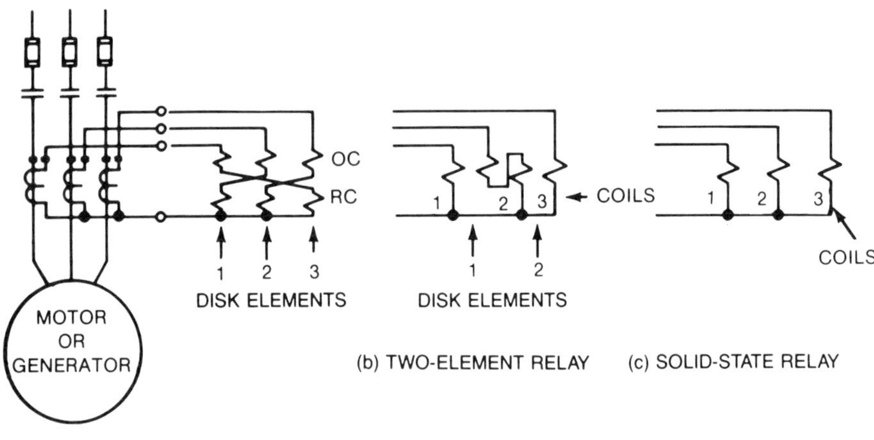

Figure 7.11 Current balance relay connection diagram.

Solid-State Type

The relay determines the difference between the line currents and trips when the difference exceeds a preset ampere value. Trip time is either inversely proportional to the phase unbalance current or a definite time.

7.3.6 Ground-Fault Relaying

Common methods employed to provide ground-fault protection are discussed below.

Residual Connection

A residually connected ground relay is widely used to protect medium-voltage systems. The scheme, which uses individual relays and current transformers, is not often applied to low-voltage circuit breakers. However, there are available low-voltage circuit breakers with three current transformers built into them and connected residually with the solid-state trip devices of the circuit breakers to provide ground-fault protection. The basic residual scheme is shown in Figure 7.12. Each phase relay is connected to the output circuit of its respective current transformer while a ground relay connected in the common or

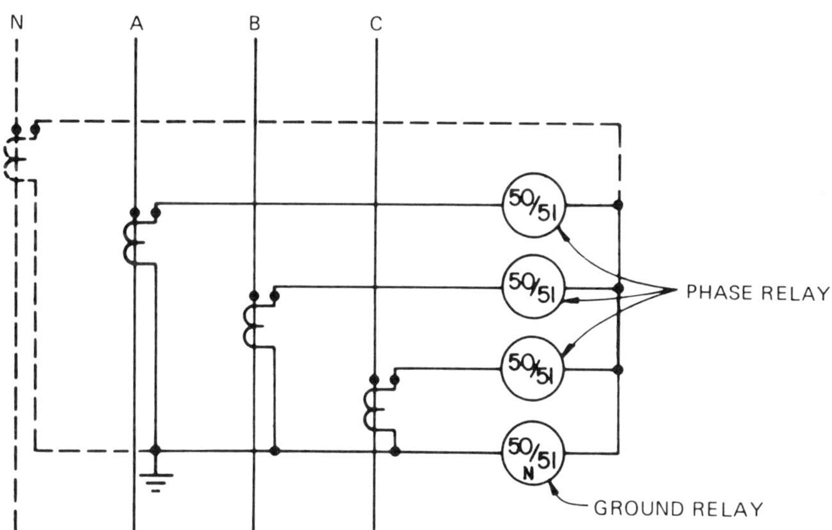

Figure 7.12 Standard connections for residually connected ground relay.

residual circuit will measure the ground fault current. In Figure 7.12, no current flows in the residual leg under normal conditions, since the net effect of the three current transformers is zero. When a ground fault occurs, current bypasses the phase conductors and the current transformers, the net flux is no longer zero, and current flows in the residual leg, resulting in operation of the relay. Residually connected relays cannot have sensitive settings because of unequal saturation of the current transformers. If sensitive ground-fault protection is required, use the core balance method.

Core-Balance Method

The core-balance current transformer is the basis of several low-voltage ground-fault protective systems introduced in recent years. Figure 7.13 shows a core-balance current transformer circuit. Under normal conditions, all current flows out and returns through the current transformer. The net flux produced in the current transformer core will be zero and no current will flow in the ground relay. When a ground fault occurs, the ground-fault current returns through the equipment grounding circuit conductor, bypassing the current transformer. The flux produced in the current transformer core is proportional to the ground-fault current, and a proportional current flows in the relay circuit. Relays connected to the core-balance current transformers can be made very sensitive, detecting even current in milliamperes. Many ground protective systems now have solid-state relays specially designed to operate with core-balance current transformers. The relays, in turn, trip the circuit protective device.

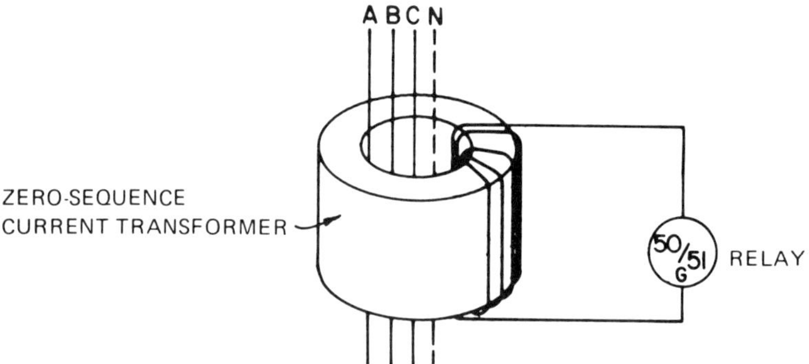

Figure 7.13 Core-balance current transformer application diagram.

Neutral Relaying

Another method that provides a convenient, low-cost scheme of detecting ground faults is to employ a time overcurrent relay connected to a current transformer located in the grounded neutral of a transformer or generator. The relay (51G) can be set to operate on very low current. This scheme is commonly used on 5- and 15-kV systems. It can be set to minimum values of current pickup and time delay which will be selective with the feeder ground-fault relays. This scheme is also used on solidly grounded 480-V three-phase three- or four-wire systems. Another form of neutral relaying is used when the neutral resistor is sized to limit the ground-fault current to a few amperes (i.e., 1 to 10 A). This method, known as high-resistance grounding, limits the damage at the fault site such that the fault will be detected and an alarm initiated.

7.3.3 Other Types of Relays

Pilot-Wire Relays (87L)

The relaying of tie lines between an industrial system and a utility system or between major load centers within a industrial system often presents a special problem. Such lines should be capable of carrying maximum emergency load currents for any length of time, and they should be removable from service quickly if a fault should occur. Pilot-wire relaying can respond very quickly to faults in the protected line. Faults are promptly cleared, resulting in minimum damage and/ or disturbance. Various types of pilot-wire relaying schemes all operate on the principle of comparing the conditions at the terminals of the protected line. The information for indicating a fault in the line is transmitted between terminals over a pilot-wire circuit. However, this scheme does not provide protection for faults of the adjacent station bus or beyond it.

Distance Relays (21)

These relays comprise a family of relays that measure voltage and current, and the ratio is expressed in terms of impedance. In general, impedance is an electric measure of the distance along a transmission line from the relay location to a fault. These relays are considered more expensive than overcurrent relays and can be obtained with one to three zones of operation. The first zone provides instantaneous protection for up to about 90% of protected line. The second and third zones, if used, are time delayed and extend protection into the area protected by the service entrance relays. Distance relays are needed to get selective tripping over a wide variation of fault current magnitudes. They may also be useful for low fault currents, which are difficult to distinguish from load currents.

Distance-controlled overcurrent relays (51 and 21) can be used in combination to provide fast tripping for faults on the primary of supply transformers, plus backup time delay for low-side faults, with some limitations. This combination is useful where overcurrent alone cannot be set to respond to transformer low-side faults. The three principal types of distance relay and their applications are as follows:

1. *Impedance type:* phase-fault relaying for moderate-length lines
2. *Mho type:* phase-fault relaying for long lines or where severe synchronizing power surges may take place; used for large synchronous machine loss-of-field protection
3. *Reactance type:* ground-fault relaying and phase-fault relaying for very short lines

Frequency Relays (81)

Frequency relays sense under- or overfrequency conditions during system disturbances. The speed of operation depends on the deviation of the actual frequency from the relay setting. The usual application of this type of relay is to drop system load selectively, based on the frequency decrement, in order to restore normal system stability.

Temperature-Sensitive Relays

Temperature-sensitive relays usually operate in conjunction with temperature-detecting devices such as thermocouples and are used for protection against overheating of large motors (above 1500 hp), generator stator windings, and large transformer windings. For generators and large motors, several temperature detectors are embedded in the stator windings, and the hottest reading detector is connected into the temperature relay bridge circuit. The bridge circuit is balanced at this temperature, and an increase in winding temperature will increase the resistance of the detector, unbalance the bridge circuit, and cause the relay to operate.

Pressure-Sensitive Relays (63)

Pressure-sensitive relays are used in power systems to respond either to the rate of rise of gas pressure (sudden pressure relay) or to a slow accumulation of gas (gas detector relay), or a combination of both. A sudden rise in the gas pressure above the liquid insulating medium in a liquid-filled transformer indicates that a major internal fault has occurred. The sudden pressure relay will respond quickly to this condition and isolate the faulted transformer. Slow accumulation of gas indicates the presence of a minor fault, such as loose contacts, grounded parts, short-circuited turns, and so on. The gas detector relay will respond to this condition and either sound an alarm or isolate the faulted transformer.

Auxiliary Relays

Auxiliary relays are used in protection schemes whenever a protective device cannot itself provide all the functions necessary for satisfactory fault isolation. Some of the most common applications of auxiliary relays are circuit breaker lockout (86), targeting, multiplication of contacts, timing, and alarming.

7.4 PROTECTIVE DEVICES

This section is intended to cover some of the most commonly used protective devices in an industrial power system. Some of the devices, such as circuit breakers, are used in conjunction with protective relays. Others, such as fuses, are used to protect the power system and/or equipment independently, or in combination with circuit breakers. A more detailed discussion of these devices appears in Chapter 8. Protective relays cannot work alone. They must work in conjunction with a circuit breaker or other switching device. Protective devices are required for opening and closing or for changing the circuit connections. In general, they consist of switches, fuses, contactors, or circuit breakers.

7.4.1 Circuit Breakers

Protective relays are applied in conjunction with circuit breakers to form a complete protection system in most circuits above 600 V. Circuit breakers rated 600 V and below have traditionally been divided into two types: power circuit breakers, sometimes known as metal frame breakers, and molded-case circuit breakers. In these breakers, tripping units are field adjustable over a wide range and are interchangeable within their frame sizes. The tripping units are mostly of the electromagnetic overcurrent direct-acting type; however, solid-state tripping units are now available from most manufacturers.

Power circuit breakers 600 V and below can be used with integral current-limiting fuses in drawout construction to meet interrupting current requirements up to 200,000 A rms symmetrical. A molded-case circuit breaker is a switching device and an automatic protective device assembled in an integral housing of insulating material. These breakers are generally capable of clearing a fault more rapidly than are power circuit breakers.

7.4.2 Switches

A disconnecting switch is used to isolate a circuit or equipment from the source of power. It is intended to be operated only after the circuit has been opened by other means. With a fused load/break switch combination, fast fault clearing and circuit isolating can be achieved. This application, if properly coordinated to interrupt load currents within the switching rating, may be more economical than a circuit breaker.

7.4.3 Fuses

A fuse is an overcurrent protective device with a circuit-opening fusible part that is heated and severed by the passage of overcurrent through it. Fuses are available in a wide range of voltage, current, and interrupting ratings, current-limiting types, and for indoor and outdoor applications. Current-limiting fuses 600 V and below are extremely fast in operation at very high values of fault current. Although their published ratings are expressed in symmetrical amperes, current-limiting fuses interrupt a short circuit within the first half-cycle, and their equivalent asymmetrical rating includes a 1.6 multiplier to provide for the maximum expected current asymmetry. Non-current-limiting fuses are widely applied above 600 V. They are available in higher current ratings, but lower interrupting ratings, than those of current-limiting fuses.

7.4.4 Contactors

Motor starters are equipped with overload relays. These relays may be of the magnetic or thermal type. In the magnetic type, a dashpot provides the necessary delay time for the flow of starting current, whereas in the thermal type, the time delay is derived from the behavior of certain subcomponents of the overload in response to internally generated heat. Both the magnetic and thermal types are sized according to the motor ratings.

7.5 SURGE PROTECTION

7.5.1 Nature of the Surge Voltage

Surge voltages can be generated in many different ways on an industrial power system. Surges can originate from lightning strokes on or near overhead power lines serving the plant, and internally from forced-current zero switching, the blowing of current-limiting fuses, or restriking interruption of circuit switching devices. A lightning-induced surge will have the form of a steep-front wave that will travel away from the stricken point in both directions along the power lines. As the surge travels along the power conductors, it will gradually decay. Properly rated arresters at the plant terminal of the incoming line can generally reduce the overvoltage to a level within the withstand rating of most station apparatus. Surges due to switching are generally less severe. However, certain types of apparatus are more susceptible to voltage surges than are other types. Hence it is advisable to investigate the damaging voltage surges of such apparatus.

The appearance of abnormal applied voltage stresses, either transient, short time, or sustained steady state, contributes to permature insulation failure. The insulation failure results not only from impressed overvoltages, but also from the sum of total duration of such overvoltages. Lightning is a major source of transient overvoltages, which may be introduced into the industrial distribution system via open wire overhead lines. Steep-wave-front transient overvoltages are also generated in plant wiring by switching actions that change the circuit operation from one steady-state condition to another. Switches that tend to chop the normal ac wave, such as thyristors, vacuum switches, current-limiting fuses, and high-speed circuit breakers, force the current to zero, which accelerates collapse of the magnetic field around the conductor, generating a transient overvoltage. Figure 7.14 shows the initial overvoltage spike resulting from the interruption action of a current-limiting fuse. There are other means of switching which will also generate transient overvoltages. The transient overvoltages are propagated along the electric power conductors to create insulation distress far away from the origin of the voltage surge.

7.5.2 Concepts of Protection Against Surge Voltage

An acceptable system of insulation protection will be influenced by a number of factors. The most important factor is knowledge of the insulation system withstand capability and endurance qualities. These properties are indicated by insulation designations and specified high-potential and surge-voltage test withstand capabilities. Another facet of the problem relates to the identification of probable sources of overvoltage exposure and the character, magnitude, duration, and repetition rates that are likely to be impressed upon the apparatus and circuits. The appropriate application of surge protective devices will lessen the magnitude and duration and is considered to be the most effective tool to achieve the desired insulation security. One of the confusing aspects of an insulation system capability and its protection is the progressive accumulation of deterioration within the dielectric that results from the complete history of voltage stress exposure.

Design of the electric distribution system, including the use of surge suppression devices (to assume adequate insulation security), should correctly interpret the effect of the inverse relationship between the imposed voltage magnitude and the allowable duration. A 30% increase in the applied ac voltage for most equipment will result in a tenfold reduction in insulation life. System design engineers must set the margin of safety based on their knowledge of the probable character and repetition rate of troublesome surge voltage transients.

7.5.3 Arrester Characteristics and Classes

Historically, various types of surge arresters have been used for power system protection. The use of pellet and expulsion-type

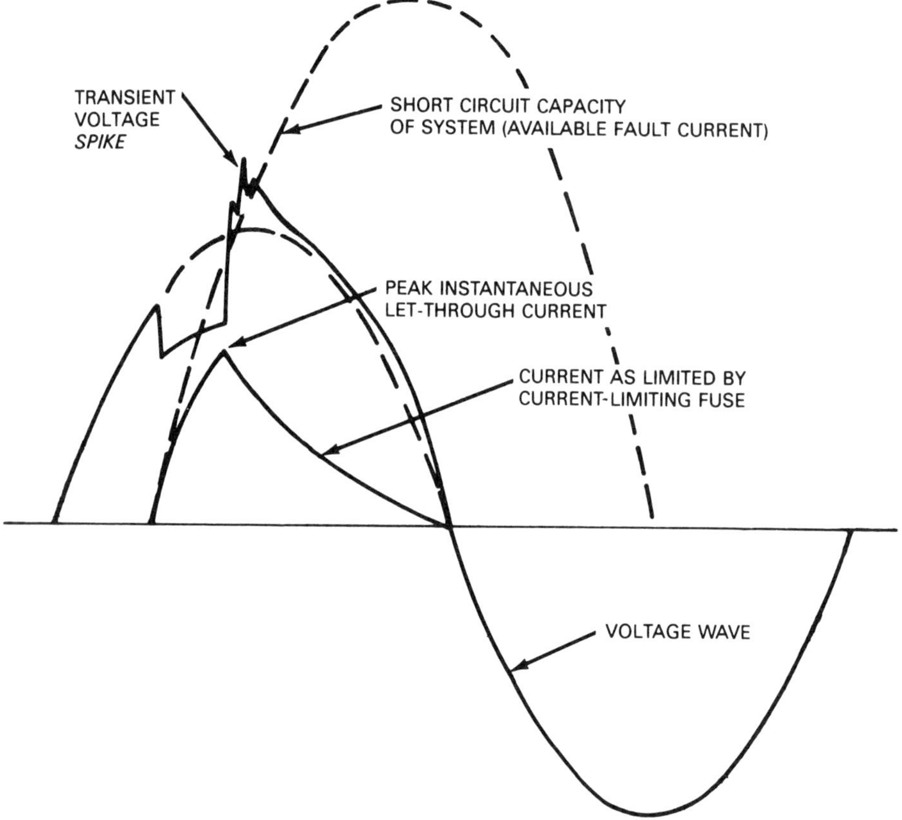

Figure 7.14 Typical transient overvoltage produced by short-circuit current-limiting action of fuse.

arresters diminished when valve-type arresters were introduced. In valve-type arresters, the basic elements are the gap unit (used in some design) and the so-called valve element. The valve element consists of a nonlinear resistance that exhibits a relatively high resistance at low voltage and current, and a much lower resistance at high (surge) voltage and current. This nonlinear property greatly enhances overall arrester performance and assists the gap unit in resealing after surge discharge to prevent continued flow of power follow current.

Use of a conventional valve-type arrester, which permits the most significant savings in the insulation levels of transmission and subtransmission equipment, has not made it possible to achieve desired surge protective margins in some very important industrial and commercial applications, specifically certain motors and certain dry-type transformers. The valve element material used, silicon carbide, does not possess sufficient nonlinearity to be self-protecting in the presence of continuous line-to-ground voltage, and series gaps are required to provide isolation. The newly developed metal oxide valve element material (specifically, a zinc oxide base) is now available in an arrester design that has sufficient nonlinearity that a series gap will not be required. An arrester employing a gap in series with a valve element is referred to as a valve-type arrester or a silicon carbide arrester. The newer arrester may be referred to as a metal oxide, zinc oxide, or gapless arrester. The characteristics and classes of arresters are discussed in more detail in the following sections.

Volt-Ampere Characteristics of Arresters

The nonlinear volt-ampere characteristic of silicon carbide is primarily a temperature-activated phenomenon throughout much of its operating range. A current increase of five orders of magnitude (from 10^{-1} to 10^2 A/cm^2) has an associated voltage increase of only slightly greater than one-half. By comparison, for the silicon carbide material an increase of only one order of magnitude in current is associated with a voltage increase by a factor ranging from approximately 1-3/4 to 3. At moderately high current densities (greater than 10 A/cm^2), the volt-ampere characteristic develops a significant turn-up that is attributable to the resistivity of the zinc oxide gains. A similar turn-up occurs in silicon carbide valve elements, which is somewhat greater in degree at the highest current densities encountered in arresters. As a consequence of the differences in the volt-ampere characteristics of the two materials, metal oxide arresters will exhibit higher discharge voltages at low discharge currents and lower discharge voltages at high discharge currents than will silicon carbide arresters.

Arrester Classes

Three classes of valve-type arresters are recognized by industry standards (ANSI/IEEE C62.1-1984). In order of decreasing cost and overall protective quality and durability, they are: (1) station class, (2) intermediate class, and (3) distribution class. As a general guide to arrester class usage versus equipment size, the following appears to prevail as typical practice:

1. *Station class:* component protection of 7.5 MVA and above, large or essential rotating machines
2. *Intermediate class:* component protection of 1 to 20 MVA substation and rotating machines
3. *Distribution class:* distribution class apparatus, small rotating machines, and dry-type transformers

There is considerable overlap of these categories, tending toward the use of higher-class arresters at higher voltages. Switching surges are of particular concern in systems experiencing frequent switching. Particularly when capacitors are connected, switching transients may impose a more severe time-current duty on arresters. Such applications should utilize station class arresters. Metal oxide arresters, which can provide superior protective levels with far less energy absorption, are ideal for frequently switched applications such as switched capacitor banks and arc furnace transformers.

7.5.4 Surge Protection Applications

Protection of Distribution Transformers

Dry transformers present relatively difficult lightning protective problems due to their usual low basic impulse level (BIL) values compared to those of liquid-filled transformers. When surge exposure is by directly connected overhead lines, arresters are required in direct shunt with the dry-type transformer. When an arrester is required at the transformer, a special low sparkover distribution class valve-type arrester will suffice. For the most used wye-delta and delta-wye-connected supply transformers, arresters are generally not required at the dry-type transformer. However, for those dry-type transformer applications where the transformer can be subjected to internally generated surges due to current-limiting fuse operation or chopping or switching effects of circuit breakers, surge protection should consist of arresters and possibly surge capacitors that reduce the effective load surge impedance and transient voltage rate of rise in those applications where an inadequate length of cable is present to achieve this.

Rotating Machine Protection

The basic winding design patterns of motors and generators involve large capacitance coupling between the conductor of the winding of each coil and the grounded core iron that surrounds it. A fast-rising surge voltage at the motor terminal raises the potential of the terminal turns, but with no immediate response for the deeper turns. This tends to result in severe voltage stress on the turn-to-turn insulation of the terminal coil. The turn insulation is fairly thin. It is the protection of the turn insulation that becomes critical in avoiding winding failure. Much documentation exists on the subject of surge protection of rotating machines. An ideally protected installation requires:

1. An effectively shielded environment
2. Arresters at terminals of machine
3. Surge capacitors at terminals of machine
4. Strict adherence of good grounding practices

Overhead-Line Protection (4 to 69 kV)

Historically, relatively little consideration has been given to the protection of open wire overhead-line insulation. This leads to line insulator flashover, which in turn results in a momentary or extended circuit interruption. Therefore, insulator flashover should be minimized with appropriate protections. A recent comprehensive study indicates that arresters protecting each phase at economically spaced intervals along the line will often provide improved protection and reliability of service over that of the overhead-wire-shield method. There is a growing practice among electric utility companies to use the new approach in protecting overhead circuits in the range 4 to 69 kV. This is an important step toward improving protection on overhead circuits that serve sensitive industrial complexes.

7.6 COORDINATION

The coordination study provides data useful for the selection of instrument transformer ratios, protective relay characteristics and settings, fuse ratings, low-voltage circuit breaker ratings, characteristics, and settings. It also provides other information pertinent to the provision of optimum protection and selectivity or coordination of these devices.

7.6.1 Primary Consideration

Short-Circuit Currents

To obtain complete coordination of the protective equipment applied, it may be necessary to obtain some or all of the following information:

1. Maximum and minimum zero- to three-cycle (momentary) total rms short-circuit current
2. Maximum and minimum three-cycle to 1 s (interrupting duty) total rms short-circuit current
3. Maximum and minimum ground-fault currents

The maximum and minimum zero- to three-cycle (momentary) currents are used to determine the maximum and minimum currents to which instantaneous and direct-acting trip devices respond, and to verify the capability of the apparatus applied, such as circuit breakers, fuses, switches, and reactor and bus bracings. The maximum three-cycle to 1 s (interrupting) current at minimum generation is needed to determine whether the circuit-protection sensitivity of the circuits is adequate.

Coordination Time Intervals

When plotting coordination curves, certain time intervals must be maintained between the curves of various protective devices to ensure corrective sequential operation of the devices. These intervals are required because relays have overtravel, fuses have damage characteristics, and circuit breakers have certain speeds of operation. These intervals are sometimes called margins. Some intervals are as follows:

Circuit breaker opening time (five cycles): 0.08 s
Overtravel: 0.10 s
Safety factor: 0.12–0.22 s

This margin may be decreased if field tests of relays and circuit breakers indicate that the system still coordinates with the decreased margins. When solid-state relays are used, overtravel is eliminated and the time may be reduced by the amount included for overtravel.

7.6.2 Data Required for a Coordination Study

1. *One-line diagram*
   a. Apparent power and voltage ratings as well as the impedance and connections of all transformers
   b. Normal and emergency switching conditions

   c. Nameplate ratings and subtransient reactance of all major motors and generators as well as transient reactances of synchronous motors and generators, plus synchronous reactances of generators
   d. Conductor sizes, type, and configurations
   e. Current transformer ratios
   f. Relay, direct-acting trip, and fuse ratings, characteristics, and ranges of adjustment
2. *Short-circuit study.* Short-circuit calculations are discussed in Chapter 5. This study should include short-circuit current values for both first cycle and interrupting duties of the circuit switching devices.
3. *Time-current characteristics.* These must be designated for all protective devices under consideration.
4. *Maximum loading.* Maximum loading on any circuit should be considered carefully with respect to the utility settings on relays, so that coordination for the entire distribution system can be achieved.

7.7 EXAMPLES OF COORDINATION STUDY OF PROTECTIVE DEVICES

Refer to Figure 7.15 for a typical medium-voltage industrial distribution system. In considering a relatively large system with more than one voltage transformation, the characteristic curve of the smallest device is plotted as far to the left of the paper as possible. The maximum short-circuit level on the system is the limit of the curves to the right. A minimum number of trip characteristics should be plotted on one sheet of paper (log-log paper). Indexing the various curves to a common scale needs some explanation.

Let us consider a system in which a 750 kVA transformer with 4160 V delta primary and a 480 V wye secondary is the largest device. Assume that the transformer is equipped with a primary circuit breaker and a main secondary circuit breaker supplying some feeder circuit breakers. In the system, the full-load current of the transformer at 480 V is 902 A. The corresponding primary current is 104 A. Plotting current on the time-current plot, 902 A at 480 V is the same as plotting 104 A at 4160 V. This type of manipulation permits the study of devices on several different system voltage levels on one coordination curve if the proper current scales are chosen. In plotting, the following points are usually considered:

1. *Scale selection.* Select a scale that will minimize multiplications and manipulations on devices where a range of settings is available. Since the load-end device is fixed, settings will be selected for two devices at 480 V and two at 4160 V in addition to determining

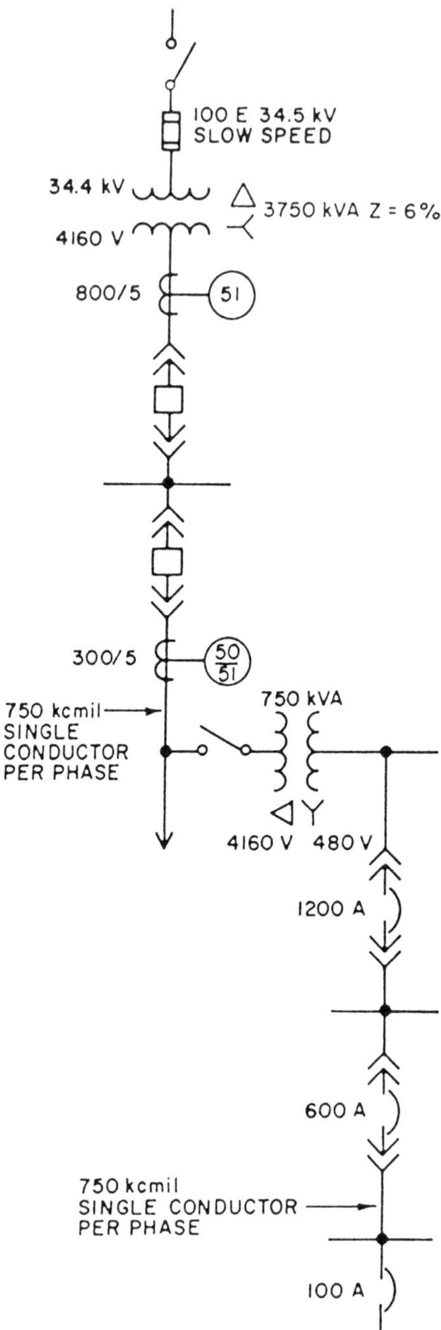

Figure 7.15 One line diagram for a typical medium-voltage industrial distribution system.

cable sizes. Use a multiple of 10 for 4160 V currents, a multiple of 87 for 480 V currents, and a multiple of 1.21 for 34,400 V current.

2. *Fixed points.* Plot the following on log-log paper (refer to Figure 7.15):

a. ANSI inrush and six times full-load points for transformers
b. Short-circuit currents
c. 100 A low-voltage circuit breaker (load device)

3. *High-voltage fuse.* According to published tables, a standard-speed 100E fuse will protect the 3750-kVA transformer. However, an examination of the curves plotted show that the coordination will be close for fitting all the devices necessary between this rating fuse and the largest load device. Therefore, a slow-speed characteristic is selected.

4. *Low-voltage circuit breakers.* By examining the ANSI and inrush points, the limits of the curve for relay protection of the 750 kVA transformer can be determined. A low pickup for this relay is better for cable protection. A 750 MCM cable has an ampacity of about 500 A; hance a trip device set at 500 A adequately protects the cable. A short-time-delay trip device is selected to be selective with the downstream molded-case circuit breaker. Select a 600 A medium-time trip element set at 80% (480 A) and a short-time trip element set at four times (2400 A) with a minimum time characteristic. The next circuit breaker to be selected is the 750-kVA transformer secondary circuit breaker. For 902-A full load, a 1200-A trip is selected with the maximum time characteristic on both long- and short-time-delay elements. Set the short-time setting at three times (3600 A).

5. *Medium-voltage feeder relays.* Allowing a 16% current margin between the short-time setting of the main circuit breaker (3600 A at 480 V), select a pickup for the medium-voltage feeder overcurrent relay. This should be less than 624 A at 4160 V and more than 3600(480/4160)(1.16) = 480 A. With a 300/5 current transformer, 480 A is 480(5/300) = 8 A, a standard tap on induction relay. The 8-A tap allows for the addition of future load. Selecting a characteristic of this relay must be very carefully done since there may be conflict with the main 4160-V circuit breaker relays. Picking a very inverse characteristic instead of an inverse characteristic is sometimes recommended. The instantaneous element is set above the available asymmetrical short-circuit current on the 480 V bus so that it does not trip for 480 V faults. This value is 12,800(480/4160)(5/300)(1.6) = 39.4 A.

6. *Medium-voltage main relays.* select a pickup for the 3750 kVA transformer secondary circuit breaker relays no lower than 125% of full load (520 × 1.25 = 650 A) and no higher than 300% of full load (520 × 3 = 1560 A). A good choice is 800 A with an 800/5 current

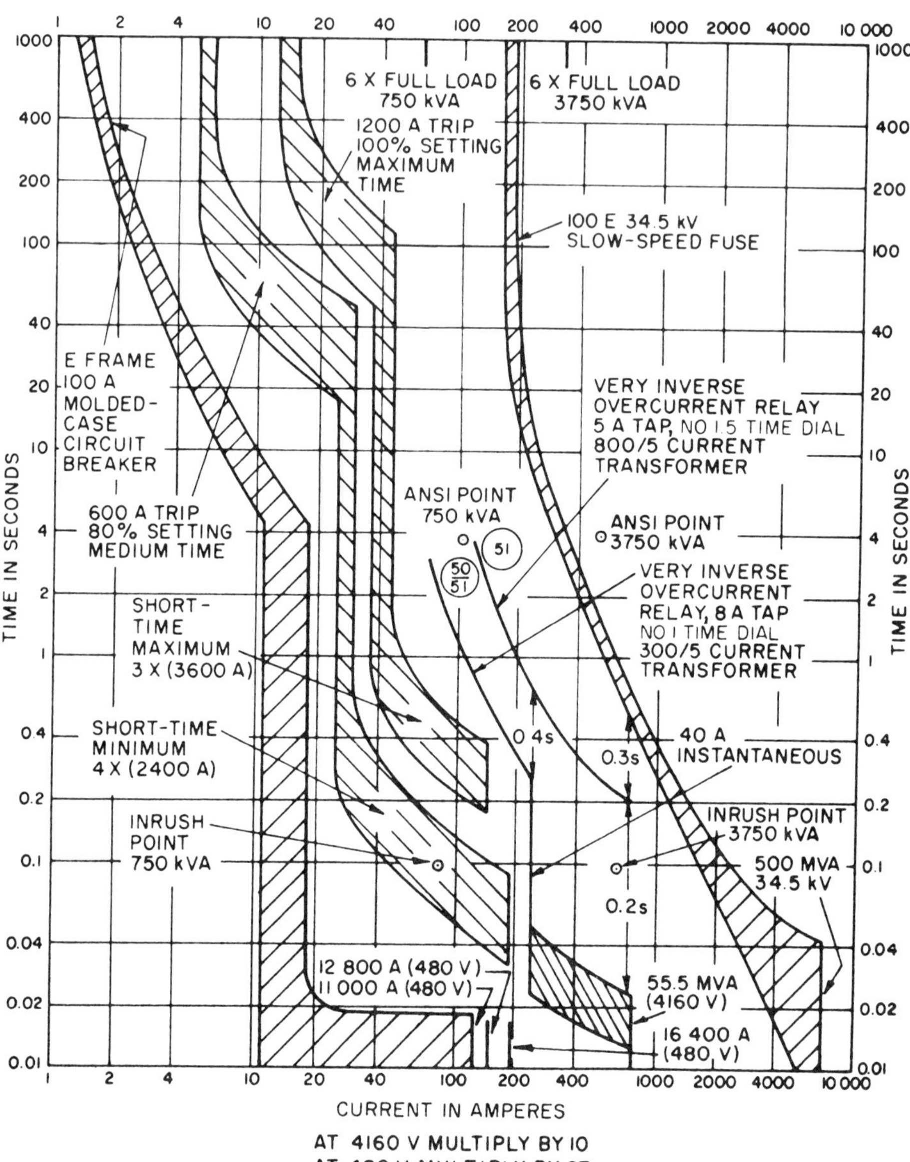

Figure 7.16 (a) Coordinated protection curves for a typical medium-voltage distribution system including fixed points.

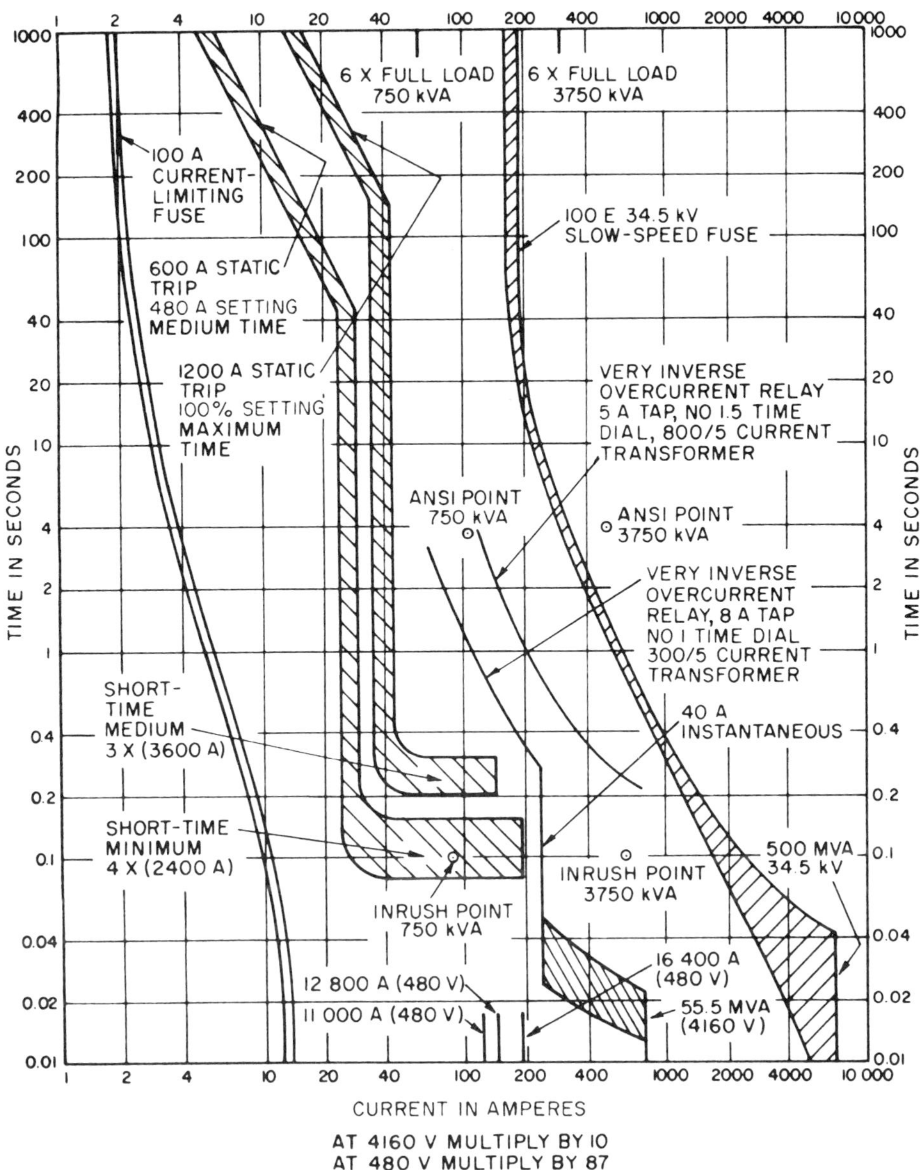

Figure 7.16 (b) Replotted using solid-state trip devices.

transformer. Do not use an instantaneous attachment on this relay since it cannot be made selective with the feeder instantaneous element. A time dial setting such as 0.3 to 0.4 s is obtained between this relay and the feeder relay at the instantaneous setting of the feeder relay (40 × 300/5 = 2400 A). The setting shown in Figure 7.16 allows only 0.2 s at the theoretical 100% fault current point between the main secondary breaker and the feeder circuit breaker. This compromise is usually satisfactory in order to maintain the margin between the transformer primary fuse and main secondary circuit breaker setting. A margin of 0.2 to 0.4 s should be allowed between the primary fuse minimum melting time curve and transformer main secondary breaker relay characteristic at the maximum 4160-V value of short-circuit current. This value is 55.5 mVA or 7600 A.

BIBLIOGRAPHY

AC Motor Protection Guide, Industrial and Commercial Power System Application Series, PRSC-2A, Westinghouse Relay-Instrument Division, Coral Springs, Florida, Jan. 1981.

ANSI C2-1987, American National Standard National Safety Code.

ANSI/IEEE C37.2-1979, IEEE Standard Electrical Power System Device Function Numbers.

ANSI/IEEE C37.91-1985, IEEE Guide for Protective Relay Applications to Power Transformers.

ANSI/IEEE C37.96-1988, IEEE Guide for AC Motor Protection.

ANSI/IEEE C57.12.01-1986, IEEE Standard General Requirements for Dry Type Distribution and Power Transformers.

ANSI/IEEE C62.1-1984, IEEE Standard for Surge Arresters for AC Power Circuits.

ANSI/IEEE Standard 141-1986, IEEE Recommended Practice for Electric Power Distribution for Industrial Plants.

ANSI/IEEE Standard 242-1986, IEEE Recommended Practice for Protection and Coordination of Industrial and Commercial Power Systems.

ANSI/NFPA 70-1987, National Electrical Code.

Brightman, F. P., More About Setting Industrial Relays, *AIEE Transactions on Power Apparatus and Systems*, vol. PAS-73, pt. III-A, 1954, pp. 397–406.

Brightman, F. P., Selecting AC Overcurrent Protective Device Settings for Industrial Plants, *AIEE Transactions on Industry Applications*, vol. IA-71, pt. II, Sept. 1952, pp. 203–211.

Electric Utility Engineering Reference Book, Vol. 3: Distribution Systems, Westinghouse Electric Corp., Trafford, Pa., 1965.

Lathrop, C. M., and Schleckser, C. E., Protective Relaying on Industrial Power Systems, *AIEE Transactions*, vol. 70, pt. II, 1951, pp. 1341–1345.

Transformer Protection Guide, Industrial and Commercial Power System Application Series, PRSC-3B, Westinghouse Relay-Instrument Division, Coral Springs, Florida, Jan. 1980.

Smith, R. L., Shortcuts to Selecting and Coordinating Electrical Trip Devices, *Plant Engineering*, July 27, 1972.

Waldron, J. E., Innovations in Solid-State Protective Relays, *IEEE Transactions on Industry Applications*, vol. IA-14, no. 1, Jan./Feb. 1978, pp. 39–47.

8

Power and Switching Equipment

8.1 INTRODUCTION

In this chapter we present information on the requirements for and application of major apparatus used in an industrial power distribution system. Design engineers should make basic decisions in their choice of equipment for a specific electric system. They should consider all facets of the project, such as continuity of service, reliability, safety, protection, coordination, installation costs, costs for operating and maintenance, security, and delivery time to meet schedules. Energy cost and conservation should also be considered in the initial planning.

8.2 SWITCHING EQUIPMENT FOR POWER CIRCUITS

Switching apparatus is defined as devices used to open and close or to change the connections of a circuit. The general classification of switching equipment used here includes switches, fuses, circuit breakers, and contactors.

8.2.1 Switches

The types of switches normally used in industrial power distribution systems are the following: (1) disconnecting switches, (2) safety switches for 600 V and below, including bolted pressure switches, (3) load interrupters, and (4) transfer switches. Detailed information for each of the foregoing categories follows.

Disconnecting Switches

Disconnecting switches are used to isolate a circuit or equipment from the source of power. Because it has no interrupting rating, it is intended to be operated only after the circuit has been opened by other means. Interlocking is generally required to prevent accidental opening of the switch under the load.

Safety Switches

Safety switches are used for services of 600 V and below. They are generally enclosed with or without fuses, and operated by means of a handle from outside the enclosure. Interlock is usually provided so that the enclosure cannot be opened unless the switch is open or the interlock defeater is operated. Safety switches for motors are rated in horsepower and voltage. The stalled-rotor current of the same horsepower is used as the switch rating at the rated voltage. The National Electrical Code (NEC) recognizes six times full-load motor current as the stalled-rotor current. The NEC limits the application of fused switches to a constant-current rating of at least 115% of the full-load current rating of the motor.

Safety switches with current-limiting fuses can be applied to circuits with up to 200,000 A symmetrical rms fault current if the switch-fuse combination has been tested properly. A bolted pressure switch consists of movable blades and stationary contacts with arcing contacts and a simple toggle mechanism for applying bolted pressure to both the hinge and jaw contacts in a manner similar to a bolted bus joint. The operating mechanism consists of a spring that is compressed by the operating handle and released at the end of the operating stroke to provide quick-make and quick-break switching action.

The electrical-trip bolted-pressure switch has a stored-energy latch mechanism and a solenoid trip release; otherwise, it is basically the same as the manually operated switch. These switches are available in ratings of 800, 1200, 1600, 2000, 2500, 3000, 4000, 5000, and 6000 A, 480 V ac, and are suitable for use on circuits having available fault currents of 200,000 A symmetrical rms when used in combination with current-limiting fuses.

Load Interrupter Switches

For service above 600 V, an interrupter or load-break switch, generally associated with a unit substation, is a switch combining the functions of a disconnecting switch and a load interrupter for interrupting currents at rated voltage not exceeding the continuous-current rating of the switch. Load-break switches are of the air or fluid-immersed type. The interrupter is usually operated manually, has a quick-make, quick-break mechanism, and usually has a close and latch rating to provide maximum safety in the event of closing in an a faulted circuit.

The load interrupter if combined with a fuse can provide fast fault clearing and circuit isolation. The combination may be more economical than a circuit breaker. However, it will be desirable from a safety standpoint to interlock the operation of an interrupter with the secondary circuit breaker to minimize the chance of operating the switch over its interrupting rating. Many interrupter switches have interrupting ratings greater than the transformer full-load current; in such cases, interlocking is not required.

Transfer Switches

The automatic transfer switch is actually a transfer system comprising two major elements, an electrically operated double-throw transfer switch and a control panel (Figure 8.1). The control panel performs the required voltage sensing and time-delay functions and provides the signals needed for operation of the transfer switch. These switches do not normally incorporate overcurrent protection. They are designed and applied in accordance with the NEC articles, and they are commonly available in ratings from 30 to 4000 A. For maximum reliability most transfer switches rated above 100 A are mechanically held and electrically operated from the power source to which the load is to be transferred.

Because of its nature of application, the transfer switch must be designed with unique and more rigorous properties than are required for some other types of switching devices. These special characteristics are:

1. Ability to close against high inrush currents
2. Ability to carry full-rated current continuously from either of two sources
3. Ability to withstand fault currents and to interrupt at least six times the full-load current
4. Capability for withstanding the possible stress produced by two out-of-phase power sources connected to it

In planning a transfer scheme to protect against failure of the utility source, consideration must be given to the following:

1. Open circuit on the inplant system on the load side of the incoming utility service
2. Overload or fault condition
3. Electrical or mechanical failure on the inplant power distribution system

It is therefore recommended to locate transfer switches close to the load and to have operation of the transfer switches independent of overcurrent protection. It is also often desirable to use multiple, smaller transfer switches near individual loads rather than one large transfer switch at the point of incoming service.

Both standby and emergency power systems require a means of transfer between the normal and alternate source of power. Standby power systems are increasingly being installed in industrial plants to safeguard human life; prevent panic, accidents, and theft; and preserve goodwill and revenue. On medium-voltage systems, the interface between the normal and standby source is usually accomplished with power switchgear. The transfer scheme can call for nonautomatic switching of tie breakers, or it can be accomplished automati-

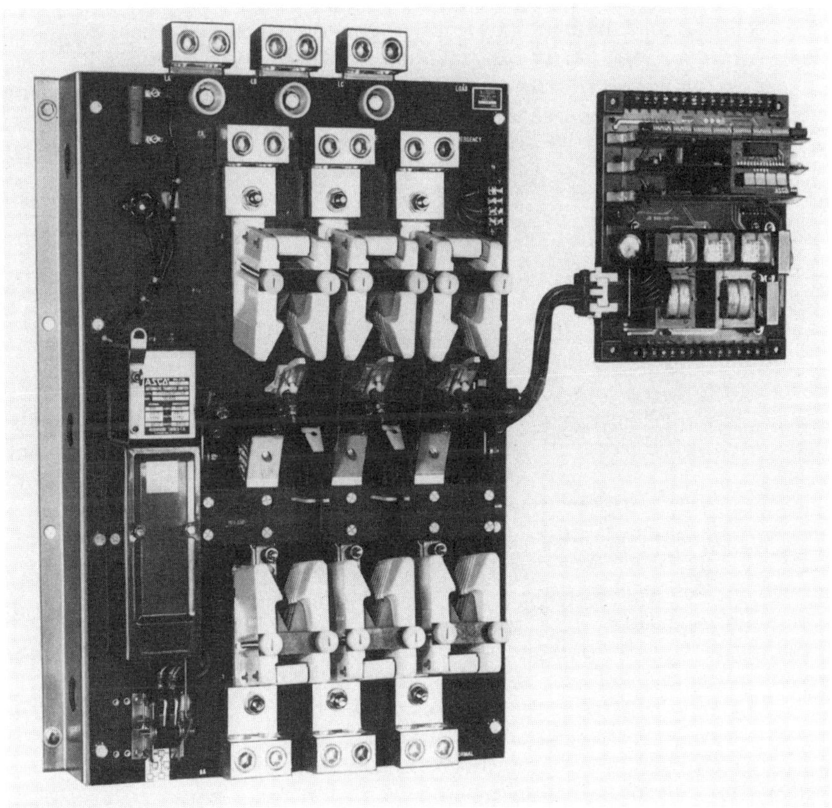

Figure 8.1 An automatic transfer switch (courtesy of Automatic Switch Company).

cally with appropriate interlock safeguards. On systems rated 600 V and below, transfer is normally accomplished with an automatic transfer switch.

8.2.2 Fuses

A fuse is defined as an overcurrent protective device with a circuit-opening fusible part that is heated and severed by the passage of overcurrent through it. The time-current characteristic depends on the rating and type of fuse. Non-time-delay fuses have no intentional built-in time delay. They are generally employed in other than motor circuits or in combination with circuit breakers. A circuit breaker provides protection in the overload current range, and a fuse provides protection in the short-circuit range.

Time-delay fuses have intentional built-in time delay in the overload range. This characteristic often permits the selection of fuse ratings closer to full-load currents. The dual-element time-delay fuse is widely used, as it has adequate time delay to permit its use as a motor overcurrent running protection. These fuses can provide protection for both motors and circuits. Different types of fuses are discussed below. These fuses are covered by the standards such as ANSI C97.1-1972 and ANSI/UL 198.

Low-Voltage Fuses

Fuse Ratings. Low-voltage fuses have current, voltage, and interrupting ratings that should not be exceeded in practical applications. In addition, some fuses are also rated according to their current-limiting capability as established by NEMA or UL standards and designated by class marking on the fuse label (class L, K, J, etc.). Current-limiting capabilities are established according to the maximum peak currect let-through and the maximum I^2t let-through of the fuse upon clearing a fault.

1. *Current rating:* maximum direct current or alternating current, in amperes, at rated frequency which it will carry without exceeding specified limits of temperature rise. Ranges from milliamperes up to 6000 A.

2. *Voltage rating:* the alternating- or direct-current voltage at which the fuse is designated to operate. Low-voltage fuses are given a voltage rating of 600, 300, 250, or 125 V ac or dc or both.

3. *Interrupting rating:* the assigned maximum short-circuit current at rated voltage that the fuse will safely interrupt. Low-voltage fuses may have an interrupting rating of 10,000, 50,000, 100,000, or 200,000 A symmetrical rms. The fuse is given a symmetrical interrupting rating based on the maximum asymmetrical value of the current for the first half-cycle after fault occurrence. For low-voltage systems, this is computed as 1.4 times the symmetrical fault current.

4. *Current limiting:* allows less than available current to flow into a fault for a relatively low ratio of available to rated current of the fuse. The design of the fuse is such that in the current-limiting range, a high arc voltage is developed as a fusible element melts and the current is prevented from reaching the magnitude it otherwise would. The current-limiting action limits the total energy flowing into a fault and thus minimizes mechanical and electrical stresses in the element of the faulted circuit.

Fuse Classes:

1. *NEC categories.* The NEC recognizes two principal categories of fuses, plug fuses and cartridge fuses. In addition, NEC mentions the following fuses: time-delay, time-lag, current-limiting, non-current-limiting, fuses over 600 V, and primary fuses. Plug fuses are rated 125 V and are available with current ratings up to 30 A. The NEC requires type S fuses in all new installations of plug fuses because they are tamper resistant. Cartridge fuses are classified by NEC in ratings 0 to 600 A and 601 to 6000 A. All fuses recognized by the NEC that have an interrupting rating exceeding 10,000 A must be marked on the fuse label with the designated interrupting rating.

2. *UL listings.* In conjunction with NEMA, UL has established standards for the classification of fuses by letter rather than by type. The class letter may designate interrupting rating, physical dimensions, degree of current limitation (maximum peak let-through current), and maximum clearing energy (I^2t) under specific test conditions or a combination of these characteristics. Descriptions of these classes follow.

a. *Class G fuses (0 to 60 A).* These are miniature fuses rated at 300 V, developed primarily for use on 480/277-V systems for connection phase to ground. These fuses are available in ratings up to 60 A, and carry an interrupting rating of 100,000 A symmetrical rms. Case sizes for 15, 20, 30, and 60 A are each of a different length. Class G fuses are time-delay fuses, having a minimum time delay of 12 s at 200% of their current rating.

b. *Class H fuses (0 to 600 A).* These are rated at 250 or 600 V. Class H fuses still have their place in residential and small commercial use, but for the most part they are inadequate for the industrial use. Their interrupting rating is only 10,000 A. Class H fuses are either nonrenewable or renewable. Renewable fuses can be taken apart after interrupting a circuit and the fusible zinc link replaced.

c, *Class J fuses (0 to 600 A).* The main advantage of the class J fuse is that it provides current limitation and high interrupting capacity in a much smaller physical package than class H and class K. This prevents class H fuses to be installed in fuseholders designed for class J fuses. Class J fuses have an interrupting rating of 200,000 A. They are available only in 600-V ratings, but can be applied on 208/120-V and 480/277-V systems.

d. *Class K fuses (0 to 600 A).* Class K fuses were developed to permit class H fuse installation to be upgraded by replacing the existing fuses with current-limiting fuses. Class K fuses are avilable in voltage ratings of 250 and 600 V and in single- and dual-element versions. Interrupting rating choices are 50,000, 100,000, and 200,000 A symmetrical rms. Class K fuses are offered in three degrees of current limitation, K1, K5, and K9, with K1 having the greatest current-limiting ability and K9 the least. Class K fuses are required by UL to have a minimum time delay of 10 s at 500% of rated current.

e. *Class L fuses (601 to 6000 A).* Class L fuses have specific physical dimensions and bolt-type terminals. They are rated 600 V and carry an interrupting rating of 200,000 A symmetrical rms. Class L fuses are current-limiting and UL has specified maximum values of peak let-through current and I^2t for each rating. Standards for time-delay characteristics have not been established for class L fuses. Most available fuses have a minimum time delay of 4.5 s at 500% of rated current.

3. *Cable limiters.* Cable limiters are used in multiple-cable circuits to provide short-circuit protection for cables. They are rated up to 600 V, with interrupting ratings as high as 200,000 A summetrical rms. They are rated in accordance with cable sizes (i.e., 4/0, 500MCM, etc.) and have numerous types of terminations. These limiters are designed to provide short-circuit protection for cables. They are used primarily in low-voltage networks or in service entrance circuits where more than two cables per phase are brought into a switchboard. The limiter does not have the characteristics associated with fuses, but will limit the extent of fault while preserving service to the balance of the system.

Medium- and High-Voltage Fuses

Medium- and high-voltage fuses are used in the industrial and commercial power distribution systems. Applicable standards are ANSI C37.46-1981 and NEMA SG2-1981. Power fuses are identified by the following characteristics:

1. Dielectric withstand (BIL) strengths at power levels
2. Application primarily in stations and substations.
3. Mechanical construction basically adapted to station and substation mountings

A power fuse consists of a fuse support plus a fuse unit, or alternately a fuse holder that accepts a refill unit or fuse link. The two basic types of power fuses are expulsion type and current-limiting type.

Expulsion Power Fuses. This type of power fuse interrupts overcurrents through the deionizing action of the gases liberated from the lining of the interrupting chamber of the fuse by the heat of the arc established when the fusible element melts. It possesses operating characteristics similar to those of a distribution cutout. This type of fuse has been restricted to outdoor usage, and generally to substations remotely located from human habitation.

The limited interrupting capacity of the early expulsion power fuses and their inability to be used within buildings led to the development in the United States of a new boric-acid or solid-material fuse. This new type of fuse permitted a great expansion in the use of power fuses in utility, industrial, and commercial power distribution systems. They are available in two styles: the fuse unit style and the fuseholder and refill unit style. The former is used primarily outdoors at subtransmission voltages: 100, 200, and 300 A maximum at 34.5, 46, and 69 kV, and 100, 250 A at 115 and 138 kV. The latter is used either indoors or outdoors in continuous-current ratings of 200, 400, and 720 A maximum for voltages up to 14.4 kV, and of 200 and 300 A maximum for voltages at 23 and 34.5 kV.

Current-Limiting Power Fuses. These fuses have three features that have led to their extensive usage on high-capacity medium- and high-voltage power distribution circuits:

1. Interruption of overcurrent is accomplished quickly without the expulsion of arc products. This enables this type of fuse to be used indoors or in enclosures of small size.
2. Current-limiting action may reduce the stresses and possible damage to the circuit up to the fault or to the faulted equipment itself. In the case of current-limiting fuses used with a high-voltage motor starter, the contactor is only required to have momentary-current and making-current capabilities equal to the maximum let-through current of the largest current rating of fuse that is to be used in the starter.
3. Very high interrupting ratings are achieved by virtue of its current-limiting action so that they can be applied on medium- or high-voltage distribution circuits of very high short-circuit capacity.

Current-limiting power fuses are suitable for the protection of potential transformers, auxiliary power transformers, small power

Table 8.1 Preferred Ratings for Indoor Oil-less Circuit Breakers

	Rated Values										Related Required Capabilities		
	Voltage		Insulation Level		Current		Transient Recovery Voltage				Current Values		
			Rated Withstand Test Voltage								K Times Rated Short-Circuit Current		
Line No.	Rated Max Voltage (1) kV, rms	Rated Voltage Range Factor, K (2)	Low Frequency kV, rms	Impulse (3) kV, Crest	Rated Continuous Current at 60 Hz (4) Amperes, rms	Rated Short-Circuit Current (at Rated Max kV) (5) (6) kA, rms	Rated Time to Point P T_2 (11) μs	Rated Interrupting Time (7) Cycles	Rated Permissible Tripping Delay Y Seconds	Rated Max Voltage Divided by K kV, rms	Max Symmetrical Interrupting Capability (8) kA, rms	3-Second Short-Time Current Carrying Capability (9) kA, rms	Closing and Latching Capability 1.6 K Times Rated Short-Circuit Current (9) (10) kA, rms
	Col 1	Col 2	Col 3	Col 4	Col 5	Col 6	Col 7	Col 8	Col 9	Col 10	Col 11	Col 12	Col 13
1	4.76	1.36	19	60	1200	8.8		5	2	3.5	12	12	19
2	4.76	1.24	19	60	1200	29		5	2	3.85	36	36	58
3	4.76	1.24	19	60	2000	29		5	2	3.85	36	36	58
4	4.76	1.19	19	60	1200	41		5	2	4.0	49	49	78
5	4.76	1.19	19	60	2000	41		5	2	4.0	49	49	78
6	4.76	1.19	19	60	3000	41		5	2	4.0	49	49	78
7	8.25	1.25	36	95	1200	33		5	2	6.6	41	41	66
8	8.25	1.25	36	95	2000	33		5	2	6.6	41	41	66
9	15.0	1.30	36	95	1200	18		5	2	11.5	23	23	37
10	15.0	1.30	36	95	2000	18		5	2	11.5	23	23	37
11	15.0	1.30	36	95	1200	28		5	2	11.5	36	36	58
12	15.0	1.30	36	95	2000	28		5	2	11.5	36	36	58
13	15.0	1.30	36	95	1200	37		5	2	11.5	48	48	77
14	15.0	1.30	36	95	2000	37		5	2	11.5	48	48	77
15	15.0	1.30	36	95	3000	37		5	2	11.5	48	48	77
16	38.0	1.65	80	150	1200	21		5	2	23.0	35	35	56
17	38.0	1.65	80	150	2000	21		5	2	23.0	35	35	56
18	38.0	1.65	80	150	3000	21		5	2	23.0	35	35	56
19	38.0	1.0	80	150	1200	40		5	2	38.0	40	40	64
20	38.0	1.0	80	150	3000	40		5	2	38.0	40	40	64

Notes to Table 8.1: These ratings were prepared by EEI-AEIC-NEMA Joint Committee on Power Circuit Breakers. The interrupting ratings are for 60-Hz systems. Current values have been rounded off to the nearest kiloampere (kA) except below 10 kA, where two significant figures are used.

1. The voltage rating is based on American National Standard Voltage Ratings for Electric Power Systems and Equipment (60 Hz), ANSI C84.1-1982, where applicable, and is the maximum voltage for which the breaker is designed and the upper limit for operation.
2. The rated voltage range factor, K, is the ratio of rated maximum voltage to the lower limit of the range of operating voltage in which the required symmetrical and asymmetrical current interrupting capabilities vary in inverse proportion to the operating voltage.
3. $1.2 \times 50\ \mu s$ positive and negative wave. All impulse values are phase-to-phase and phase-to-ground and across the open contacts.
4. The 25-Hz continuous-current ratings in amperes are given herewith following the respective 60-Hz rating: 600—700; 1200—1400; 2000—2250; 3000—3500.
5. To obtain the required symmetrical current interrupting capability of a circuit breaker at an operating voltage between 1/K times rated maximum voltage and rated maximum voltage, the following formula is used:

$$\text{required symmetrical current interrupting capability} = \text{rated short-circuit current} \times \frac{\text{rated maximum voltage}}{\text{operating voltage}}$$

   For operating voltages below 1/K times rated maximum voltage, the required symmetrical current interrupting capability of the circuit breaker is equal to K times rated short-circuit current.
6. With the limitation stated in Section 5.10 of ANSI/IEEE C37.04-1979, all values apply for polyphase and line-to-line faults. For single phase-to-ground faults, the specific conditions stated in Section 5.10.2.3 of ANSI/IEEE C37.04-1979 apply.
7. The ratings in this column are on a 60-Hz basis and are the maximum time interval to be expected during a breaker opening operation between the instant of energizing the trip circuit and interruption of the main circuit on the primary arcing contacts under certain specified conditions. The values may be exceeded under certain conditions as specified in Section 5.7 of ANSI/IEEE C37.04-1979.
8. Current values in this column are not to be exceeded even for operating below 1/K times rated maximum voltage. For voltages between rated maximum voltage and 1/K times rated maximum voltage, follow item 5.
9. Current values in this column are independent of operating voltage up to and including rated maximum voltage.
10. If currents are to be expressed in peak amperes, multiply values in this column by a factor of 1.69, which is a ratio of 2.7/1.6.
11. The rated values for T_2 are not standardized for indoor oil-less circuit breakers; however, E_2 = 1.88 times rated maximum voltage.

transformers, and capacitor banks for systems up to 34.5 kV. Current-limiting power fuses are available in various frequency, voltage, and continuous-current carrying capacities, and with interrupts ratings that conform to the standards, ANSI/IEEE C37.40-1981, ANSI/IEEE C37.41-1981, ANSI/IEEE C37.46-1981, and ANSI/IEEE C37.47-1981.

8.2.3 Circuit Breakers

A circuit breaker is a device designed to open and close a circuit by nonautomatic means, and to open the circuit automatically on a predetermined overload of current when properly applied within its rating without damage to itself. The interrupting waves and momentary ratings of a circuit breaker must be equal to or greater than the available system short-circuit currents. Power circuit brakers are used extensively on utility and on industrial power distribution systems (over 600 V) and on industrial and commercial systems predominantly 600 V and below to provide essential switching flexibility and circuit protection. Circuit breakers are available for the entire voltage range and may be provided single-, double-, or triple-pole, and arranged for indoor or outdoor use. Circuit breakers above 34.5 kV services are generally available only for outdoor application. Various types of circuit breakers are discussed in detail below.

Circuit Breakers over 600 V

In applying circuit breakers over 600 V, the rated close and latch and interrupting current capabilities are very important factors and should be considered carefully. The close and latch capability is a measure of the equipment's ability to withstand the mechanical stresses produced by the asymmetrical short-circuit current during the first cycle without mechanical damage, and is normally expressed as total rms current. An asymmetrical current is a dc component superimposed on an ac component. The dc component decays with time depending on the X/R ratio of the circuit. The initial value of the dc component of the short-circuit current depends on the point of the normal voltage wave at which the fault occurs. Application data are given in ANSI/IEEE C37.010-1979.

For the rating of power circuit breakers in the over-600 V class, refer to ANSI C37.06-1979. Circuit breakers currently being manufactured are rated on the symmetrical basis. In specifying these circuit breakers, considerations should be given to the related values and required capabilities listed as headings in Table 8.1. This table lists preferred ratings for indoor oil-less circuit breakers. These ratings are applicable for services at altitudes up to 3300 ft. For services above that altitude, derating factors must be applied in accordance with ANSI/IEEE C37.04-1979. Table 8.2 lists several altitude correction factors (ACFs). Table 8.3 lists preferred ratings for outdoor oil-less circuit breakers. Different types of power circuit breakers are discussed in further detail below.

Air-Magnetic Circuit Breakers. Power circuit breakers used for applications through 15 kV are predominantly of the air-magnetic type, with increasing use of vacuum interrupters and some SF types. For voltages above 15 kV, the available types of circuit breakers include oil, compressed air or gas, and vacuum interrupters.

Vacuum-Interrupter Circuit Breakers. In general, vacuum power circuit breakers are applied in accordance with the specific continuous and short-circuit current requirements in the same manner as air-magnetic circuit breakers. However, vacuum interrupters do have characteristics that are different from air-magnetic circuit breakers. Vacuum interrupters will force a premature current zero by opening the circuit in an unusually short time. When this occurs, a higher-then-normal transient recovery voltage can impose excessive dielectric stress on the equipment connected, which could fail as a result.

Experimental vacuum interrupters were pioneered in the 1920s. Modern vacuum interrupters were placed in service in the early 1960s. Since then, thousands have been operating successfully. These units are rated from 4.16 kV/250 mVA through 13.8 kV/1000 mVA nominal at 1200, 2000, and 3000 A. These vacuum interrupters are compact, reducing the size and weight of the circuit breaker and the switchgear. Interruption is fast and silent, and no arc products are released. The high-vacuum seal protects the contacts from exposure to the pollutants to prevent interruptions from being affected by the environment. Figure 8.2 shows a typical vacuum-interrupted circuit breaker with explicit view of its inside construction. A vacuum interrupter requires less maintenance because it is sealed in a high-vacuum environment free from contamination. Use of a vacuum interrupter avoids the special equipment needed to purify the interrupting medium used in SF_6 and oil switchgear design.

SF_6 Interrupter Circuit Breakers. The first 550 kV breakers installed in the United States in 1964 used type SF interrupter modules filled with sulfur hexafluoride (SF_6) gas as a dielectric and interrupting medium. The interrupting module, the basic breaker component has two sets of serially connected contacts, as shown in Figure 8.3. The moving contacts, mounted on opposite ends of a rocker arm, engage and disengage stationary contacts as the rocker arm is rotated about its axis. The rocker arm assembly is linked to the breaker mechanism by an operating rod. When the breaker is in the closed position, the operating rod is held in tension by a charged torsion bar in the module and a latch in the mechanism. The force of a permanent magnet keeps the mechanism latched. When

Table 8.2 Altitude Correction Factors[a]

Altitude [ft (m)]	ACF for voltages[b]	ACF for continuous current
3300 (1000)	1.00	1.00
5000 (1500)	0.95	0.99
10000 (3000)	0.80	0.96

[a]Interpolated values are used in determining correction factors for intermediate altitudes.
[b]For some types of circuit breakers (e.g., those with sealed interrupters), it may not be necessary to apply the ACF for voltage to rated maximum voltage. The manufacturer should be consulted.

the high-speed trip coil is energized, the resulting flux decreases the holding force of the permanent magnet, thereby unlatching the mechanism and allowing the charged torsion bar to rotate the rocker arm assembly. This action simultaneously opens a gas blast valve mounted in the hub of the rocker arm. High-pressure SF_6 gas is released from the auxiliary reservoir, through the hollow rocker arm, through the tubular moving and stationary contacts, and into the module tank. As an arc is produced when the contacts of each break separate, the gas flow transfers the arc from the fingers to an arcing horn to minimize arc erosion of the members that carry continuous current. The electronegative SF_6 gas absorbs electron from the arc discharge and extinguished the arc. The blast valve is reset for the next opening operation. The entire sequence is carried out in less than 2 cycles.

The development of a removable breaker element for 23 to 34.5 kV service makes it possible to use metal-clad switchgear. The breaker element is a single-pressure dead-tank SF_6 circuit breaker that uses the fault current to drive a puffer element magnetically, forcing SF_6 gas through the arc. The breaker has an interrupting capacity of 1500 mVA and an interrupting time of three cycles. Continuous current-carrying capacities are 1200 and 3000 A. A puffer interrupter consists essentially of a pair of contacts, a piston, and a cylinder, all mounted in a reservoir containing a suitable interrupting gas. When the contacts are separated, the piston moves in the cylinder to drive gas through the arc and interrupt it.

Use of SF_6 gas as both the insulating and interrupting medium in the circuit breakers affords several side advantages.

Table 8.3 Preferred Ratings for Outdoor Oil-less Circuit Breakers

	Rated Values									Related Required Capabilities			
	Voltage		Insulation Level		Current		Transient Recovery Voltage				Current Values		
			Rated Withstand Test Voltage								K Times Rated Short-Circuit Current		
Line No.	Rated Max Voltage (1)* kV, rms	Rated Voltage Range Factor, K (2)	Low Frequency kV, rms	Impulse (3) kV, Crest	Rated Continuous Current at 60 Hz (4) Amperes, rms	Rated Short-Circuit Current (at Rated Max kV) (5) (6) kA, rms	Rated Time to Point P T_2 μs	Rated Interrupting Time (7) Cycles	Rated Permissible Tripping Delay Y Seconds	Rated Max Voltage Divided by K kV, rms	Max Symmetrical Interrupting Capability (8) kA, rms	3-Second Short-Time Current Carrying Capability (9) kA, rms	Closing and Latching Capability 1.6 K Times Rated Short-Circuit Current (9) (10) kA, rms
	Col. 1	Col. 2	Col. 3	Col. 4	Col. 5	Col. 6	Col. 7	Col. 8	Col. 9	Col. 10	Col. 11	Col. 12	Col. 13
1	38	1.65	See Table 6		1200	22	63	5	2	23	36	36	58
2	72.5	1.10			2000	37	97	3	2	66	41	41	65

*See notes p. 159.

Figure 8.2 A vacuum-interrupter circuit breaker (courtesy of Westinghouse Electric Corporation).

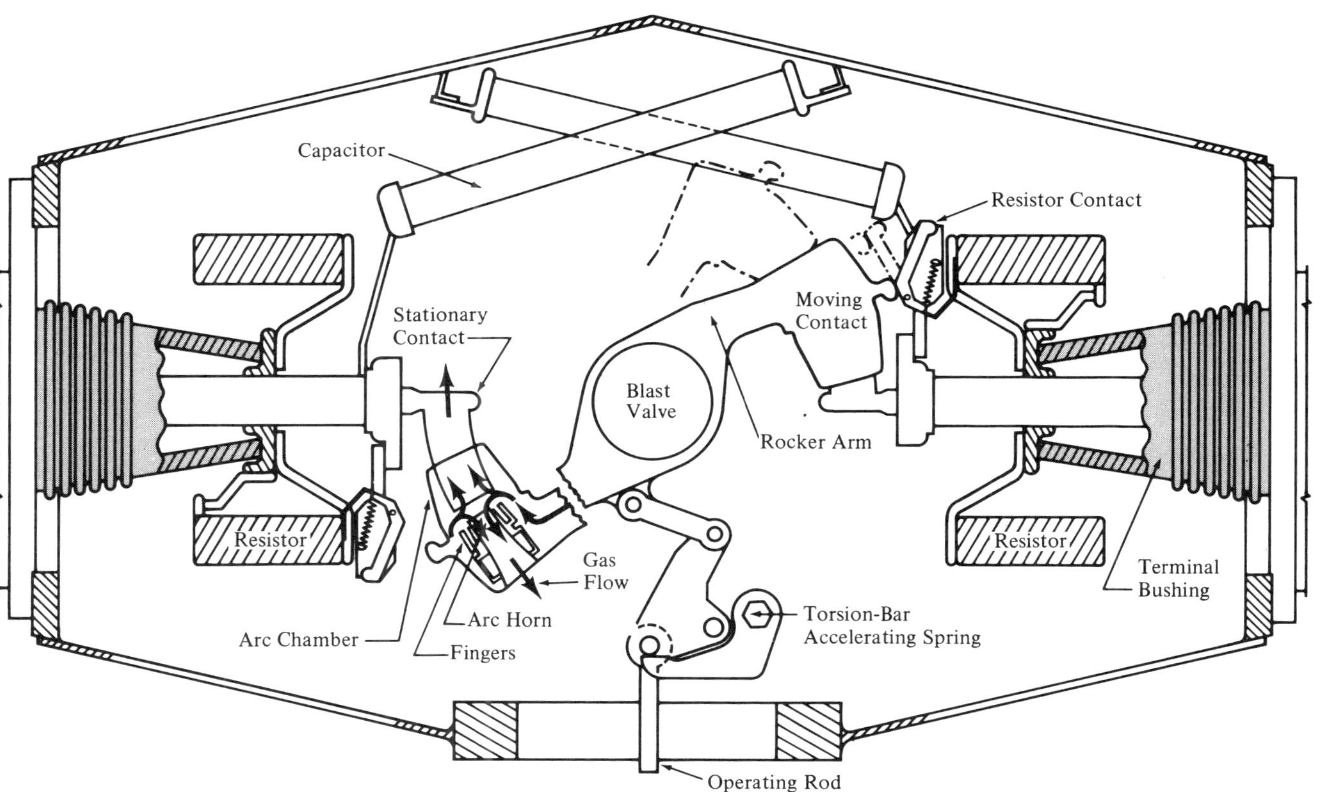

Figure 8.3 Diagram of the SF_6 interrupter module.

The entire gas system is closed to the atmosphere, eliminating continual maintenance of costly high-pressure air supply and dryer systems that air-blast breakers require to remove condensate. Air-blast breakers also require special sumps in the earth and external piping to handle moisture and oil residue from the compressor blow-down.

Oil Circuit Breakers. The "de-ion" grid interrupters used on these breakers consist of a stack of horn-fiber plates arranged to provide orifice plates, oil pockets, and exhaust vents in such a manner that efficient high-speed arc interruption is effected. When the contacts part on a faulted operation, the heat of the arc causes the oil in the pockets to disintegrate and form nonionized gas particles. The pressure built up inside the grid provides a turbulent action that causes these nonionized particles to mix with the ionized particles in the arc stream and at the same time forces this mixture through the orifice plates and out the exhaust vents to cause rapid deionization of the arc stream. When sufficient dielectric strength is established between the separating contacts to withstand the system recovery voltage, the arc will be extinguished at the next current zero.

Figure 8.4 shows a 14.4 kV 600 A continuous-current 250 mVA interrupting capacity de-ion grid oil circuit breaker installed in the switchyard of an industrial plant. A control cabinet is mounted on the front of the frame, which supports the oil circuit breaker. The cabinet houses all control components for the circuit breaker. The shunt trip magnet is used to trip the breaker electrically. It may be equipped with a coil for dc, ac, or capacitor tripping. When the shunt trip magnet is energized, the core is drawn up into the magnet yoke. A plunger attached to the core strikes the trip level directly above it, which disengages the primary latch to trip the breaker. Figure 8.5 shows a typical ac trip control diagram for this type of oil circuit breaker. Table 8.4 lists preferred ratings for outdoor oil circuit breakers.

Circuit Breakers 600 V and Below

Circuit breakers rated 600 V and below have traditionally been divided into two types: power circuit breakers, sometimes known as metal frame breakers, and molded-case circuit breakers.

Power Circuit Breakers. Power circuit breakers 600 V and below are open-construction assemblies on metal frames with all parts designed for accessible maintenance, repair, and ease of replacement. Tripping units are field adjustable over a wide range and interchangeable within their frame sizes. The tripping units used have been the electromegnetic overcurrent direct-acting type; however, solid-state tripping units are also available from most manufacturers and are widely used today. With both electromagnetic and solid-state trips, they can provide the following types of protection: long-time overcurrent, short-time overcurrent, and short circuit. Trip units can be specified to include any combination of these three forms of protection. With solid-state units, ground-fault protection can also be provided without recourse to external relays.

Power circuit breakers 600 V and below can be used with integral current-limiting fuses in drawout construction to meet interrupting current requirements up to 200,000 A symmetrical rms. In current design, 600 V and below air circuit breaker contacts often begin to part during the first cycle of short-circuit current, but have a

Figure 8.4 14.4 kV, 600 A, "de-ion" grid oil circuit breaker.

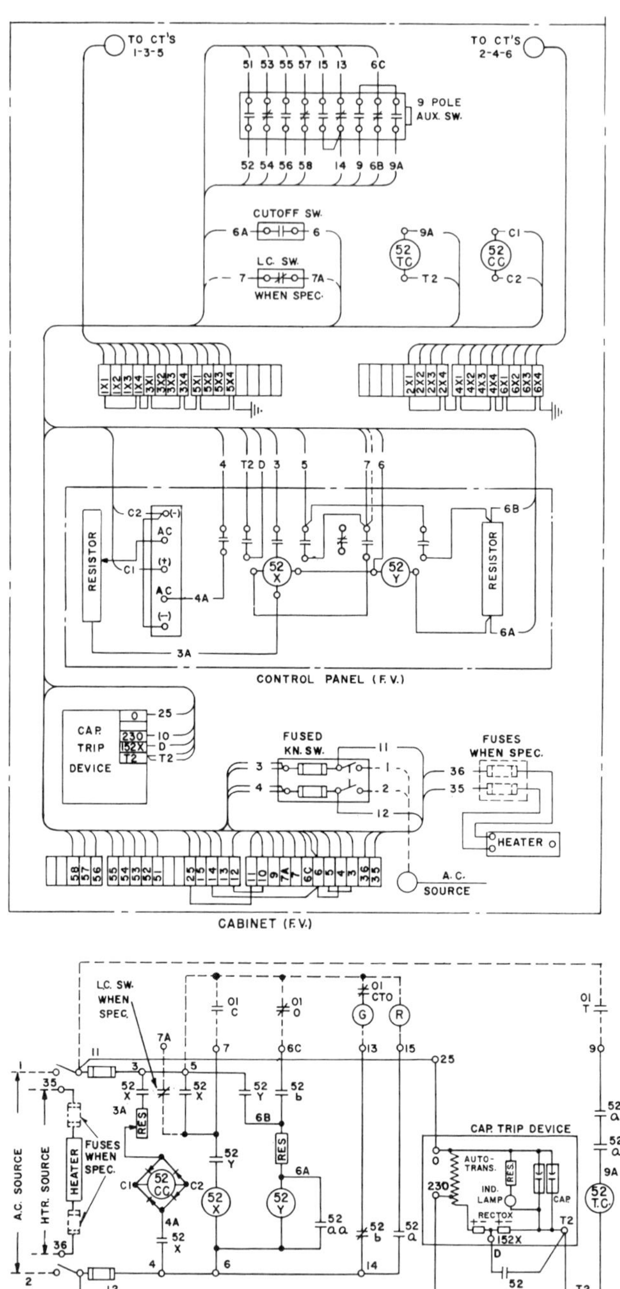

Figure 8.5 A typical AC trip control diagram for the "de-ion" grid O. C. B.

Table 8.4 Preferred Ratings for Outdoor Oil Circuit Breakers

	Rated Values									Related Required Capabilities			
	Voltage		Insulation Level		Current		Transient Recovery Voltage				Current Values		
			Rated Withstand Test Voltage								Max Symmetrical Interrupting Capability (8)	3-Second Short-Time Current Carrying Capability (9)	
											K Times Rated Short-Circuit Current		
Line No.	Rated Max Voltage (1)* kV, rms	Rated Voltage Range Factor, K (2)	Low Frequency kV, rms	Impulse (3) kV, Crest	Rated Continuous Current at 60 Hz (4) Amperes, rms	Rated Short-Circuit Current (at Rated Max kV) (5) (6) kA, rms	Rated Time to Point P T_2 μs	Rated Interrupting Time (7) Cycles	Rated Permissible Tripping Delay Y Seconds	Rated Max Voltage Divided by K kV, rms	kA, rms	kA, rms	Closing and Latching Capability 1.6 K Times Rated Short-Circuit Current (9) (10) kA, rms
	Col. 1	Col. 2	Col. 3	Col. 4	Col. 5	Col. 6	Col. 7	Col. 8	Col. 9	Col. 10	Col. 11	Col. 12	Col. 13
1	15.5	2.67			600	8.9	36	5	2	5.8	24	24	38
2	15.5	1.29			1200	18	33	5	2	12	23	23	37
3	25.8	2.15			1200	11	52	5	2	12	24	24	38
4	38	1.65			1200	22	63	5	2	23	36	36	58
5	48.3	1.21			1200	17	80	5	2	40	21	21	33
6	72.5	1.21			1200	19	106	5	2	60	23	23	37

*See notes p. 159.

multicycle total clearing time. Therefore, these breakers should be designed to interrupt the maximum available quarter-cycle asymmetrical current. However, since these breakers are rated on a symmetrical current basis, the need for applying dc offset multipliers to determine their interrupting rating is eliminated. Caution should be taken when the air circuit breakers are supplied with short-time delay trips because an increase in short-circuit stress on the breaker could result in extensive damage if withstand ratings were exceeded. The manufacturer's literature should be consulted.

Power circuit breakers 600 V and below can be applied on a symmetrical basis if the system X/R ratio does not exceed 6.6. If the system X/R is higher, the asymmetrical capability in the pole of the circuit breaker having maximum offset surrent should be checked against the maximum phase asymmetrical current available at the circuit breaker location.

All circuit breakers must be capable of closing, carrying, and interrupting the highest fault current within its rating at that location. It is important for the system design engineers to select a circuit breaker whose interrupting rating at the circuit voltage is equal to or greater than the available short-circuit current at the point of installation. The significance of short-circuit selection of circuit breakers rated 600 V and below is covered in great detail in Chapter 5.

Manufacturers' publications give specific information on mechanical and electrical features of cirucit breakers rated 600 V and below. Tables 8.5 and 8.6 show standard ratings for power circuit breakers rated 600 V and below and molded-case circuit breakers. For service at altitudes above 6600 ft above sea level, derating factors must be applied in accordance with ANSI/IEEE C37.13-1981.

Molded-Case Circuit Breakers. A molded-case circuit breaker based on NEMA AB1-1975 is a switching device and an automatic protective device assembled in an integral housing of insulating material. These breakers are generally capable of clearing a fault more rapidly than do power circuit breakers and are available in the following general types:

1. *Thermal-magnetic.* This type uses thermal tripping for overloads and instantaneous magnetic tripping for short circuits. They are the most widely applicable molded-case circuit breakers.
2. *Magnetic.* This type uses only instantaneous magnetic tripping where only short-circuit interruption is required.
3. *Integrally fused.* This type combines regular thermal-magnetic protection against overloads and lower value short-circuit faults with current-limiting fuses responding to higher short-circuit currents. Interlocks are usually provided to ensure safe and proper operation.

Table 8.5 Preferred Ratings for Low-Voltage AC Power Circuit Breakers with Instantaneous Direct-Acting Phase Trip Elements (See ANSI/IEEE C37.13-1981)

Line No	System Nominal Voltage (volts)	Rated Maximum Voltage (volts)	Insulation (Dielectric) Withstand (volts)	Three-Phase Short-Circuit Current Rating or Short-Time Current Rating (symmetrical amperes)*†	Frame Size (amperes)	Range of Trip-Device Current Ratings (amperes) — Setting of Short-Time-Delay Trip Element: Minimum Time Band	Intermediate Time Band	Maximum Time Band
	Col 1	Col 2	Col 3	Col 4	Col 5	Col 6	Col 7	Col 8
1	600	635	2200	14 000	225	100–225	125–225	150–225
2	600	635	2200	22 000	600	175–600	200–600	250–600
3	600	635	2200	22 000	800	175–800	200–800	250–800
4	600	635	2200	42 000	1600	350–1600	400–1600	500–1600
5	600	635	2200	42 000	2000	350–2000	400–2000	500–2000
6	600	635	2200	65 000	3000	2000–3000	2000–3000	2000–3000
7	600	635	2200	85 000	4000	4000	4000	4000
8	480	508	2200	14 000	225	100–225	125–225	150–225
9	480	508	2200	22 000	600	175–600	200–600	250–600
10	480	508	2200	22 000	800	175–800	200–800	250–800
11	480	508	2200	42 000	1600	350–1600	400–1600	500–1600
12	480	508	2200	50 000	2000	350–2000	400–2000	500–2000
13	480	508	2200	65 000	3000	2000–3000	2000–3000	2000–3000
14	480	508	2200	85 000	4000	4000	4000	4000
15	240	254	2200	14 000	225	100–225	125–225	150–225
16	240	254	2200	22 000	600	175–600	200–600	250–600
17	240	254	2200	22 000	800	175–800	200–800	250–800
18	240	254	2200	42 000	1600	350–1600	400–1600	500–1600
19	240	254	2200	50 000	2000	350–2000	400–2000	500–2000
20	240	254	2200	65 000	3000	2000–3000	2000–3000	2000–3000
21	240	254	2200	85 000	4000	4000	4000	4000

*Short-circuit current ratings for circuit breakers without direct-acting trip devices, opened by a remote relay, are the same as those listed here.
†Single-phase short-circuit current ratings are 87 percent of these values.

Table 8.6 Preferred Ratings for Low-Voltage AC Power Circuit Breakers Without Instantaneous Direct-Acting Phase Trip Elements (Short- Time-Delay Element or Remote Relay) (See ANSI/IEEE C37.13-1981)

Line No.	System Nominal Voltage (volts)	Rated Maximum Voltage (volts)	Insulation (Dielectric) Withstand (volts)	Three-Phase Short-Circuit Current Rating, Symmetrical (amperes)*	Frame Size (amperes)	Range of Trip-Device Current Ratings (amperes)
	Col 1	Col 2	Col 3	Col 4	Col 5	Col 6
1	600	635	2200	14 000	225	40–225
2	600	635	2200	22 000	600	40–600
3	600	635	2200	42 000	1600	200–1600
4	600	635	2200	42 000	2000	200–2000
5	600	635	2200	65 000	3000	2000–3000
6	600	635	2200	85 000	4000	4000
7	480	508	2200	22 000	225	40–225
8	480	508	2200	30 000	600	100–600
9	480	508	2200	50 000	1600	400–1600
10	480	508	2200	50 000	2000	400–2000
11	480	508	2200	65 000	3000	2000–3000
12	480	508	2200	85 000	4000	4000
13	240	254	2200	25 000	225	40–225
14	240	254	2200	42 000	600	150–600
15	240	254	2200	65 000	1600	600–1600
16	240	254	2200	65 000	2000	600–2000
17	240	254	2200	85 000	3000	2000–3000
18	240	254	2200	130 000	4000	4000

***Single-phase short-circuit current ratings are 87% of these values.**

This material is reproduced with permission from American National Standard Preferred Ratings, Related Requirements, and Application Recommendations for Low-Voltage Power Circuit Breakers and AC Power Circuit Protectors, ANSI C37.16-1980, copyright 1980 by the American National Standards Institute. Copies of this standard may be purchased from ANSI, 1430 Broadway, New York, NY 10018.

4. *High interrupting capacity.* This fuseless type provides interrupting capacities for higher short-circuit currents than does standard construction of contacts and mechanism, plus a special high-impact molded casing.
5. *Current-limiting.* This type provides high interrupting capacity protection, and it limits let-through current and energy to a value significantly lower than the corresponding value for a conventional molded-case circuit breaker. Restoration of services is possible by resetting without replacement of any fusible elements or other parts.

Molded-case circuit breakers are not designed to be maintained in the field as are power circuit breakers. They are mostly sealed to prevent tampering. Manufacturers recommend total replacement if a defect appears. The larger sizes are not suitable for repetitive switching. The newer-style molded-case circuit breaker incorporates a new type of solid-state trip circuitry mounted with the associated current monitors inside the breaker case. They are physically and electrically interchangeable with conventional molded-case circuit breakers of the same frame sizes. Different continuous-current ratings are easily obtained for those breakers merely by changing the rating plug.

Figure 8.6 shows such a breaker with replaceable plugs within a certain frame size. Conventional thermal-magnetic breakers have two ranges of operation. The thermal trip range provides a nonadjustable inverse time-current trip characteristic with delays of seconds to minutes, depending on the magnitude of the overload. The magnetic trip range provides instantaneous trip at a preset current level, typically 5 to 10 times the trip units continuous-current rating.

The new-style circuit breakers have three ranges of operation: the long-delay, magnetic (or short-delay), and instantaneous trip ranges. The magnetic trip range provides a short delay, on the order of a few cycles. This short delay ensures service continuity for an electrical system by permitting coordination with downstream devices for selective tripping; that is, the breakers can be set so that the protective device nearest the fault clears the circuit, thus maximizing the uptime for the system as a whole.

Service Protectors. A service protector consists of a current-limiting fuse and nonautomatic circuit breaker switching device in a single enclosure. It utilizes basic circuit breaker principles and permits frequent repetitive operation under normal and abnormal current conditions up to 12 times the device's continuous-current rating. In combination with current-limiting fuses, it is capable of closing and latching against fault currents up to 200,000 A symmetrical rms. During fault interruption, the service protector will withstand the stresses created by the let-through current of the fuses.

a

Figure 8.6 Molded-case circuit breakers with replaceable plugs (courtesy of Westinghouse Electric Corp.).

b

Figure 8.6 (Continued).

They are available at continuous-current ratings of 800, 1200, 1600, 2000, 3000, 4000, 5000, and 6000 A for use on 240 and 480 V ac systems, in two-pole or three-pole construction. An open-fuse trip device to prevent the occurrence of single phasing is included in the design of this type of service protector.

8.3 SWITCHGEAR

Switchgear is a general term covering switching and interrupting devices alone, or their combination with associated control, metering, protective, and regulating equipment. A power switchgear assembly consists of a complete assembly of one or more of the above-mentioned devices and main bus bars, interconnecting wiring, accessories, supporting structures, and enclosure. Power switchgear is usually installed throughout an industrial plant, but is used principally for incoming line service and to control and protect load centers, panelboards, and other secondary distribution equipment. Outdoor switchgear can be of the non-walk-in (without enclosed maintenance aisle) or walk-in (with an enclosed maintenance aisle) type.

8.3.1 Types of Metal-Enclosed Power Switchgear

Specific types of metal-enclosed power switchgear used in the industrial plants are classified as (1) metal-clad switchgear, (2) interrupter switchgear, (3) low-voltage power circuit breaker switchgear, and (4) low-voltage metal-enclosed distribution switchboards. Detailed discussions of each type are given below.

Metal-Clad Switchgear (5 to 34.5 kV)

Metal-clad switchgear is available with voltage ratings of 4.16 to 34.5 kV and with circuit breakers having interrupting ratings from 8.8 kA at 4.16 kV to 40 kA at 34.5 kV. Continuous ratings are 1200, 2000, and 3000 A. Industrial standards for power switchgear are ANSI C37.20-1969, Section 6.2, and NEMA SG5-1981, Part 9.03. Metal-clad switchgear can be characterized by the following essential features:

1. The main circuit switching and interrupting device is of the removable type arranged with a mechanism for moving it physically between connected and disconnected positions and equipped with self-aligning and self-coupling primary and secondary disconnecting devices.
2. Major parts of the primary circuit, such as the circuit switching or interrupting devices, buses, and potential transformers, are enclosed by grounded metal barriers.
3. All live parts are enclosed within grounded metal compartments. Automatic shutters prevent exposure of primary circuit elements when the removable element is in the test, disconnected, or fully withdrawn position.
4. Primary bus conductors and connections are covered with insulating material throughout.
5. Mechanical interlocks are provided to ensure a proper and safe operating sequence.
6. Instrument, meters, relays, secondary control devices, and their wiring are isolated by grounded metal barrier from all primary circuit elements.
7. The door through which the circuit interrupting device is inserted into the housing may serve as an instrument or relay panel and provide access to a secondary or control compartment within the housing.

The term "metal-clad switchgear" can be properly used only if metal-enclosed switchgear conforms to the foregoing characteristics. All metal-clad switchgear is metal-enclosed, but not all metal-enclosed switchgear can be designated as metal-clad. Figure 8.7 shows a 15 kV walk-in outdoor power circuit breaker switchgear which was installed on the secondary side of two 5000/6250 kVA OA/FA substation transformers. It consists of two main circuit breakers and several plant feeder breakers, control power, and protective relays associated with the system. All circuit breakers are of the draw-out type, which have a continuous-current rating of 1200 A, short-time momentary rating of 40,000 A, and three-phase interrupting capacities of 500 mVA. These breakers are closed by a 48-V ac stored energy mechanism and tripped by a 48-V dc battery-activated spring magnet release.

Interrupter Switchgear (5 to 34.5 kV)

Metal-enclosed interrupter switchgear is also metal-enclosed power switchgear, which is widely used to protect both switching and short circuits through the use of interrupter switches and power fuses. Standard voltage ratings are 4.16 and 13.8 kV; 23 and 34.5 kV

Figure 8.7 (a) 15 kV walk-in power circuit breaker switchgear outdoor housing.

switchgear is also available from a few manufacturers. The main bus is rated 600, 1200, or 2000 A. The power fuses have interrupting ratings up to 270 mVA at 4.16 kV and 1000 mVA at 34.5 kV. Metal-enclosed interrupter switchgear should include the following components:

1. Interrupter switches
2. Power fuses
3. Bare bus and connections
4. Instrument and control power transformers
5. Control wiring and accessory devices

The interrupter switches and power fuses may be of the stationary or removable type. For the removable type, mechanical interlocks are provided to ensure a proper safe operating sequence. The low cost of metal-enclosed interrupter switchgear enhances the use of a larger number of radial circuits to serve individual loads or to provide extensive sectionalizing of a single radial circuit serving multiple loads, thus elevating radial circuits to a level of circuit continuity, which may help to eliminate the need of secondary selective or secondary-network systems. (See Chapter 4 for definitions of various types of

Figure 8.7 (b) 15 kV walk-in power circuit breaker switchgear (inside the housing).

power distribution systems for industrial plants.) Available standards are ANSI/IEEE C37.20-1969, Section 6.4, and NEMA SG5-1981, Part 9.07.

Figure 8.8 shows 5-kV outdoor power interrupter switchgear which consists of three fused load-interrupter switches and associated control and protective equipment. The main switch cubicles receive a 4.16-kV three-phase aerial power cable from the utility pole. The other two switches are used to protect the plant primary feeders.

Figure 8.8 5 kV outdoor power interrupter siwtchgear.

Low-Voltage Power Circuit Breaker Switchgear

The low-voltage power circuit breakers are covered in Section 8.2.3. Power circuit breakers 600 V and below are intended for service in switchgear compartments or other enclosures of dead front construction. These power cirucit breakers are either drawout type or molded-case type. The power circuit breakers can be operated electrically or manually and equipped with added devices, such as shunt trip, undervoltage, and auxiliary switches. They are available either with conventional electromagnetic overcurrent direct-acting tripping devices or static tripping devices.

Drawout circuit breakers and compartments have separate main and secondary disconnect contacts to achieve connected, test, disconnect, and fully withdrawn positions. Separate compartments are provided for required meters, relays, instruments, and so on. Potential and control power transformers are normally mounted in these compartments to be front accessible. Current transformers may be mounted around the stationary power primary leads within the circuit breaker compartment or in the rear bus area. The rear section is isolated from the front section and accommodates the main bus, feeder terminals, small wiring, and terminal blocks. Bus is fabricated from high-grade aluminum, designed for an allowable temperature rise of 65°C above an average 40°C ambient. Copper bus is usually obtainable at a higher cost. Control wiring is extended to terminal blocks mounted in the rear section.

The assembled switchgear can be used alone, but is most widely used as an integral part of a unit substation, which usually consists of a primary disconnect switch, a distribution transformer, and low-voltage circuit breaker cubicles. Figure 8.9 shows a typical 750-kVA 12.47 kV/480 V unit substation with some of the panels removed during installation. The primary disconnect switch is housed in the left-side cubicle, the transformer in the middle cubicle, and three drawout circuit breakers, meters, and control switches in the right-side cubicle.

Low-Voltage Metal-Enclosed Distribution Switchboards

Low-voltage metal-enclosed distribution switchboards are frequently used in commercial buildings at 600 V and below for service entrance, industrial plants power and lighting distribution, and as the secondary section of unit substations. A wide range of protective devices and single- or multisection assemblies are available for large service from 400 to 4000 A. Equipment ground-fault protection is recommended where the switchboard is applied on grounded wye systems. These switchboards can be built with either front accessible or rear accessible, and for wall mounting or for floor standing construction, which is usually applied as the secondary section in a unit substation.

Figure 8.9 750 kVA and 12.47 kV/480 V unit substation.

Figure 8.10 Secondary section of a unit substation.

Figure 8.10 shows the secondary section of a unit substation in the form of a switchboard with molded-case breakers clearly visible.

8.3.2 Switchgear Ratings

The ratings of switchgear assemblies and metal-enclosed buses are designations of the operational limits of the particular equipment under specific conditions of ambient temperature, altitude, frequency, duty cycle, and so on. Table 8.7 lists the rated voltages and insulation levels for ac switchgear assemblies discussed in this section. Table 8.8 lists similar ratings for metal-enclosed buses. The definitions of the ratings listed in Tables 8.7 and 8.8 and others discussed subsequently can be found in ANSI/IEEE C37.100-1981. Standard self-cooled continuous-current ratings of the main bus in metal-clad power switchgear are listed in Table 8.9. Metal-enclosed but standard self-cooled continuous-current ratings are shown in Table 8.10. The

Table 8.7 Rated Voltage and Insulation Levels for AC Switchgear Assemblies

Rated Voltage (rms)		Insulation Levels (kV)		
Rated Nominal Voltage	Rated Maximum Voltage	Power Frequency Withstand (rms)	DC Withstand*	Impulse Withstand
Metal-Enclosed Low-Voltage Power Circuit Breaker Switchgear				
Volts	Volts			
240	250	2.2	3.1	—
480	500	2.2	3.1	—
600	630	2.2	3.1	—
Metal-Clad Switchgear				
kV	kV			
4.16	4.76	19	27	60
7.2	8.25	36	50	95
13.8	15.0	36	50	95
34.5	38.0	80	†	150
Metal-Enclosed Interrupter Switchgear				
kV	kV			
4.16	4.76	19	27	60
7.2	8.25	26	37	75
13.8	15.0	36	50	95
14.4	15.5	50	70	110
23.0	25.8	60	†	125
34.5	38.0	80	†	150
Station-Type Cubicle Switchgear				
kV	kV			
14.4	15.5	50	†	110
34.5	38.0	80	†	150
69.0	72.5	160	†	350

*The column headed "DC Withstand" is given as a reference only for those using direct-current tests and represents values believed to be appropriate and approximately equivalent to the corresponding power frequency withstand test values specified for each voltage class of switchgear. The presence of this column in no way implies any requirement for a direct-current withstand test on alternating-current equipment. When making direct-current tests, the voltage should be rasied to the test value in discrete steps and held for a period of 1 min.

†Because of the variable voltage distribution encountered when making direct-current withstand tests, the manufacturer should be contacted for recommendations before applying direct-current withstand tests to the switchgear. Potential transformers above 34.5 kV should be disconnected when testing with direct current.

Table 8.8 Voltage Ratings for Metal-Enclosed Bus

Rated AC Voltage (kV rms)		Insulation Level (kV)			
		Power Frequency Withstand (rms)			
Nominal	Rated Maximum	(Dry 1 Minute)	(Dew 10 Seconds)*	DC Withstand (Dry)†	Impulse Withstand
0.6	0.63	2.2	—	3.1	—
4.16	4.76	19.0	15	27.0	60
13.8	15.00	36.0	24 (36)	50.0	95
14.4	15.50	50.0	30 (50)	70.0	110
23.0	25.80	60.0	40 (60)	85.0	150
34.5	38.00	80.0	70 (80)	‡	200
69.0	72.50	160.0	140 (160)	‡	350

For applications of isolated phase bus to generators, the following voltage ratings apply:§

Rated kV of Generator (rms)	Power Frequency Withstand (rms) (Dry 1 Minute)	Power Frequency Withstand (rms) (Dew 10 Seconds)	DC Withstand (Dry)	Impulse Withstand
14.4 to 24	50	50	70	110

*Applied to porcelain insulation only. Values in parentheses apply to "high creepage" designs.

†The column headed "DC Withstand" is given as a reference only for those using direct-current tests and represents equivalent to the corresponding power frequency withstand test values specified for each voltage class of bus. The presence of this column in no way implies any requirement for a direct-current withstand test on alternating-current equipment. When making direct-current tests the voltage should be raised to the test value in discrete steps and held for a period of 1 min.

‡Because of the variable voltage and distribution encountered when making direct-current withstand tests, the manufacturer should be contacted for recommendations before applying direct-current withstand tests to these voltage ratings. Potential transformers above 34 5 kV should be disconnected when testing with direct current.

§These ratings are applicable to generators rated 14.4 to 24 kV which are directly connected to transformers without intermediate circuit breakers and where adequate surge protection is provided. These bus withstand ratings are compatible with or in excess of required withstand values of the generators.

Table 8.9 Continuous-Current Ratings of Buses for Metal-Enclosed Power Switchgear

Type of Assembly	Rated Continuous Current of Buses (amperes)
Metal-clad switchgear	1200, 2000, 3000
Metal-enclosed interrupter switchgear	600, 1200, 2000
Metal-enclosed bus	
Station-type cubicle switchgear	2000, 3000, 4000, 5000

NOTE: The numerous combinations of circuit breaker sizes and ratings in similar housings make it impractical to provide current ratings of buses for metal-enclosed low-voltage power circuit breaker switchgear.

Table 8.10 Current Ratings for Metal-Enclosed Bus (Amperes)

0.6 AC and All DC	Voltage Ratings (kV)					
	2.4, 4.16	13.8	14.4	23.0	34.5	69.0
600	—	—	—	—	—	—
1200	1200	1200	1 200	1200	1200	1200
1600	—	—	—	—	—	—
2000	2000	2000	2 000	2000	2000	2000
—	—	—	2 500	2500	2500	2500
3000	3000	3000	3 000	3000	3000	3000
—	—	—	3 500	3500	—	—
4000	—	—	4 000	4000	—	—
—	—	—	4 500	4500	—	—
5000	—	—	5 000	5000	—	—
—	—	—	5 500	5500	—	—
6000	—	—	6 000	6000	—	—
—	—	—	6 500	—	—	—
—	—	—	7 000	—	—	—
—	—	—	7 500	—	—	—
—	—	—	8 000	—	—	—
—	—	—	9 000	—	—	—
—	—	—	10 000	—	—	—
—	—	—	11 000	—	—	—
—	—	—	12 000	—	—	—

momentary and short-circuit ratings of power switchgear assemblies must correspond to the equivalent ratings of the switching or interrupting devices used.

8.3.3 Control Power

Successful operation of switchgear depends much on a reliable source of control power that will maintain voltage at the terminals of such devices within their rated operating voltage range. Table 8.11 lists the preferred control voltages in both dc and ac. Two primary uses of the control power in switchgear are tripping power and closing power. The source of tripping power must always be available. For circuit breakers rated 1000 V and below, manual closing up to 1600 A frame is also practical. In general, four practical sources of tripping power are:

1. Direct current from a storage battery
2. Direct current from a charged capacitor

Table 8.11 Preferred Control Voltage and Their Ranges for Power Circuit Breakers Rated 600 V and Below

		Power Supply (volts)		Tripping
Rated Voltage (volts)	Control Voltage (volts)	Solenoid or Motor Operator	Stored Energy Operator†	Voltage Range (volts)
Direct Current*				28
24‡	—	—	—	14-30‡
48	38-56	—	—	28-56
125	100-140	90-130§	100-140	70-140
250	200-280	180-260§	200-280	140-280
Alternating Current				
120	104-127**	—	104-127	104-127
240	208-254**	190-250§	208-254	208-254
480	416-508**	380-500§	416-508	416-508

NOTE: It is recommended that trip, closing, relay coils, etc, normally connected continuously to one direct-current potential should be connected to the negative wire of the control circuit to minimize electrolytic deterioration.

*Control from exciter circuits is not recommended.
†For driving motor for air compressors and compressed spring mechanisms.
‡Unless the circuit breaker is located close to the battery and relay and adequate electric conductors are provided between the battery and trip coil, 24 V tripping is not recomended.
**Includes heater circuits.
§Some operating mechanisms will not meet all the closing requirements over the full control voltage range. In such cases it will be necessary to provide for two ranges of closing voltage. Where applicable, the preferred method of obtaining the double range of closing voltage is by the use of tapped coils.

3. Alternating current from the secondaries of current transformers in the protected power circuit
4. Direct or alternating current in the primary circuit passing through direct-acting trip devices

Where a storage battery has been chosen as a source of tripping power, it can also supply closing power. Batteries offer an extremely reliable source of energy but require regular attention and care. Also, their capacity in ampere-hours is limited. Inspection and testing of individual cells must be made at regular intervals to assure that electrolyte level and correct charge are maintained. Batteries of the nickel-lead-alkaline type or the nickel-cadmium type are more suitable for standby service. Battery charging equipment will be determined by the battery characteristics and the type of load being served.

8.3.4 How to Select Proper Switchgear

Since there is a great variety of switchgear to select from, design engineers must gather all necessary data before they can make an intelligent selection of proper switchgear for their needs. For the primary system, the choice is between circuit breaker and switch-fuse combinations. For the secondary system, the choice is between fused and unfused power circuit breakers. The following steps are normally helpful in selecting switchgear equipment:

1. Prepare a one-line diagram.
2. Determine rating of power switching apparatus.
3. Select main bus rating.
4. Select current and potential transformers.
5. Select metering, relaying, and control power.
6. Determine closing, tripping, and other control power requirements.
7. Consider special applications.

Metal-enclosed switchgear is available for application at voltages up through 34.5 kV. Metal-clad switchgear is available for application at voltages from 2.4 through 34.5 kV; however, for economic reasons, it is seldom used above 15 kV.

Essentially, all recognized basic bus arrangements—transfer bus, sectionalized bus, synchronizing bus, and ring bus—are available in metal-enclosed switchgear to ensure the desired system reliability and flexibility. A choice is made based on an evaluation of initial cost, installation cost, required operating procedures, and total system requirements. Because power system facilities are always increased to serve larger loads, it is advisable to consider future expansion when selecting the bus continuous-current rating and the momentary and interrupting ratings of the power switching apparatus.

8.4 PANELBOARDS

Electrical systems in the industrial buildings usually utilize panelboards (either fuse or circuit breaker) as a convenient means of power distribution.

8.4.1 Types of Panelboards

Panelboards are generally classified into several categories:

1. *Power distribution panelboards.* This category includes all other panelboards not defined as lighting and appliance panelboards. The 42 overcurrent protective device limitation does not apply. However, care should be exercised not to exceed practical physical limitations, such as standard box heights and widths. Ratings are single-phase two- or three-wire; three-phase three- or four-wire; 120/240 through 600 V ac, 250 V dc; 50 to 1200 A, 1200 A maximum branch.

2. *Lighting and applicance panelboards.* This type is defined as one having more than 10% of its overcurrent devices rated 30 A or less, for which neutral connections are provided. The number of overcurrent devices (branch circuit poles) is limited to a maximum of 42 in any one box. Ratings of these panels are single-phase two- or three-wire; three-phase three- or four-wire; 120/240 V ac, 50 to 60 A, 125 A maximum branch.

3. *Multisection panelboards.* Both lighting and appliance panelboards and power distribution panelboards requiring more than one box are called multisection panelboards. Unless a main overcurrent device is provided in each section, each section must be furnished with main bus and terminals of the same rating for connecting to the single feeder. Three methods commonly used for interconnecting multisection panelboards are as follows:

   a. *Gutter tapping.* Increased gutter width may be required. Tap devices are not furnished with the panelboards.
   b. *Subfeeding.* A second set of main lugs (subfeed) are provided directly beside the main lugs of each panelboard section, except the last in the lineup.
   c. *Through feeding.* A second set of main lugs is provided on the main bus at the opposite end from the main lugs of each section, except the last in the lineup.

8.4.2 Panelboard Data

To assist design engineers in planning an installation, manufacturers' catalogs usually provide a wide choice of panelboards for specific applications. Several important rules governing the application of panelboards are outlined in the National Electrical Code:

1. *42-circuit rule.* The NEC section 384-14 defines a lighting and appliance branch-circuit panelboard as one having more than 10% of its overcurrent devices rated 30 A or less, for which neutral connectors are provided. Section 384-15 states that not more than 42 overcurrent devices of a lighting and appliance branch-circuit panelboard may be installed in any one cabinet.

2. *6-circuit rule.* The NEC section 230-71 provides that a device may be suitable for service equipment when not more than six main disconnecting means are provided. In addition, a disconnecting means must be provided for the ground conductor.

3. *Gutter tap rule.* The NEC section 240-21 states that overcurrent devices must be located at the point where the conductor to be protected receives its supply. Exception 5 to this paragraph permits omission of the main overcurrent device if (a) the smaller conductor has a current-carrying capacity of not less than the sum of the allowable current-carrying capacities of the one or more circuits or loads supplied, and (b) the tap is not over 10 ft long and does not extend beyond the panelboard it supplies. Gutter taps are permitted under this ruling.

8.5 TRANSFORMERS

Transformers are employed to change a voltage level from utility transmission or distribution voltage to a voltage that is usable within the industrial buildings. They are also used to reduce the building distribution voltage to a level that can be utilized by specific equipment, such as lighting services.

8.5.1 Transformer Types

The following types of transformers are normally used in the industrial buildings: (1) substation, (2) primary unit substation, (3) secondary unit substation (power center), (4) network, (5) pad mounted, and (6) indoor distribution. Other types of transformers are also manufactured for special applications, but discussions of these special transformers are beyond the scope of this chapter. Detailed discussions of each of the foregoing types are presented below.

Substation Transformers

Substation transformers are used in outdoor switchyards and are rated 750 to 5000 kVA for single-phase units and 750 to 10,000 kVA for three-phase units. The primary voltage range is 2400 V and up. Taps are usually manually operated, but automatic load tap changing may also be obtained. The secondary voltage range is 480 to 34,500

Figure 8.11 69/6.9 kV, 5000/6250 kVA, 3 phase substation transformer.

V. Primaries are usually delta-connected and secondaries wye-connected. The cooling medium is usually oil. The voltage connections are on cover-mounted bushings. Low-voltage connections may be on cover bushings or an air terminal chamber. Figure 8.11 shows a 69/6.9 kV 5000/6250 kVA three-phase substation transformer with three surge arresters on its low side.

Primary Unit Substation Transformers

Primary unit substation transformers are three-phase units, rated 750 to 10,000 kVA. The primary voltage range is 8320 V and up. Taps are usually manually operated, but automatic load tap changing is also available. The secondary voltage range is 2400 to 34,500 V. Primaries are usually delta-connected, and secondaries, wye-connected. The cooling medium may be oil, askarel,* air, or gas. The high-voltage connections may be cover bushings, an air terminal chamber, throat, or flange. The low-voltage connection is a throat or flange.

---

*Askarel, which is a generic name for a fluid consisting of about 70% polychlorinated biphenyl (PCB), has been marketed under several trade names. During the past decade it was determined that askarel was harmful to the environment and human beings. Therefore, the EPA (Environmental Protective Agency) banned its use in the new transformers and capacitors and set very strict rules on the use of existing equipment containing askarel.

Secondary Unit Substation Transformers

Secondary unit substation transformers are three-phase units, rated 112.5 to 2500 kVA. The primary voltage range is 2400 to 15,000 V. The taps are manually operated. The secondary voltage range is 208 to 600 V. Primaries are usually delta-connected, and secondaries, wye-connected. The cooling medium may be oil, askarel, air, or gas. The high-voltage connections may be cover bushings, an air terminal chamber, a throat, or a flange. The low-voltage connection is a throat or flange.

Network Transformers

Network transformers are used in secondary network systems. They are rated 300 to 2500 kVA. The primary voltage range is 4160 to 34,500 V. Taps are manually operated. The secondary voltages are 208Y/120 and 480Y/277 V. The cooling medium may be oil, askarel, or air. The primary is delta-connected, the secondary, wye-connected. The high-voltage connection is generally a network switch (on-off-ground). The secondary connection is generally a network protector.

Pad-Mounted Transformers

Pad-mounted transformers are usually three-phase units to be used outside buildings. These units are rated 75 to 2500 kVA. The primary voltage range is 2400 to 34,500 V, and the secondary voltage range is 208 to 600 V. Primaries may be delta-connected, and secondaries are usually wye-connected. The cooling medium may be oil, askarel, or air. The high-voltage connection is in an air terminal chamber, which may contain just pressure-type connectors or may have a disconnecting device either fused or unfused. The low-voltage connection is usually by cable at the bottom.

Indoor Distribution Transformers

Indoor distribution transformers are rated 1 to 333 kVA for single-phase units, and 3 to 500 kVA for three-phase units. Both primaries and secondaries are 600 V and below (the most common ratio is 480 to 208Y/120 V). The cooling medium is air. High- and low-voltage connections are pressure-type connections for cables. Impedances of these types of transformers are usually lower than those of substation or secondary unit substation transformers.

8.5.2 Transformer Specifications

In specifying a transformer for a particular application, the following items should be considered and included:

1. Rating in kVA or mVA
2. Single-phase or three-phase
3. Frequency
4. Voltage ratings and taps
5. Winding connections, delta or wye
6. Impedance (base rating)
7. Basic impulse level (BIL)
8. Temperature rise

In addition, the following details pertaining to construction should be specified:

1. Insulation medium, dry or liquid type
2. Indoor or outdoor
3. Type and location of terminal facilities
4. Permissible sound level
5. Manual or automatic load tap changing
6. Grounding requirement
7. Provisions for future cooling if required

Consideration should also be given to energy conservation features in the transformer specification. After an overall life-cycle cost analysis is made, the following information about the transformer should be given to the prospective supplier:

1. Cost in dollars/kW at which load and no-load loses are valued
2. The percentage of the transformer rating at which losses will be evaluated during the bid comparison process

Some of the more important items in the list above will be discussed more fully as follows:

Power and Voltage Ratings

Ratings in kVA or mVA will include the self-cooled rating at a specified temperature rise, as well as the forced-cooling rating if required. See Tables 8.12 and 8.13 for data. The standard average winding temperature rise (by resistance test) for the modern liquid-filled power transformer is 65°C, based on average ambient of 30°C for any 24-hour period. Liquid-filled transformers may be specified with a 55°C/65°C rise to permit 100% loading with a 55°C rise, and 112% loading at a 65°C rise. Even for a 115°C rise high-fire-point liquid-insulated transformers are available from some manufacturers.

In NEMA TR1-1980 and ANSI/IEEE C57.12.00-1980 and ANSI/IEEE C57.12.01-1979, the average winding temperature rise (by resistance) for the modern dry transformer is 150°C, based on an average ambient of 30°C (40°C maximum) for any 24-h period. Low-loss high-efficiency

Table 8.12 Transformer Standard Base kVA Ratings

Single-Phase				Three-Phase			
3	75	1250	10 000	15	300	3750	25 000
5	100	1667	12 500	30	500	5000	30 000
10	167	2500	16 667	45	750	7500	37 500
15	250	3333	20 000	75	1000	10 000	50 000
25	333	5000	25 000	112½	1500	12 000	60 000
37½	500	6667	33 333	150	2000	15 000	75 000
50	833	8333		225	2500	20 000	100 000

dry-type transformers can be specified with a 115°C or an 80°C rise. The lower-temperature-rise units have a longer life expectancy and greater overload capacities than do the regular 150°C rise design. A 115°C rise dry-type transformer has approximately 10 times the life expectancy of a 150°C rise unit and has a 15% overload capability. A 80°C rise unit has a 30% overload capability.

Both liquid-insulated and dry-type transformers are available with low core and coil watt loss designs at higher initial prices, but with significantly lower overall operating costs due to the higher energy efficiency. Many transformer specifiers believe that a lowered operating temperature is synonymous with improved efficiency, or conversely, a piece of equipment that runs hot is inefficient. Based on a recent study report, this is not necessarily correct, especially if the transformer design is optimized. If the efficiencies of the 115 and 150°C nominal rise units are held to that of the 80°C rise unit, and the units are redesigned to minimum cost, both units at equal total efficiency have a lower material cost than that of the lower-temperature-rise unit. Thus a high-temperature-rise, lower-cost unit is not synonymous with inefficiency.

Table 8.13 Classes of Transformer Cooling Systems

Type Letters	Method of Cooling
OA	Oil-immersed, self-cooled
OW	Oil-immersed, water-cooled
OW/A	Oil-immersed, water-cooled/self-cooled
OA/FA	Oil-immersed, self-cooled/forced-air-cooled
OA/FA/FA	Oil-immersed, self-cooled/forced-air-cooled/forced-air-cooled
OA/FA/FOA	Oil-immersed, self-cooled/forced-air-cooled/forced air—forced-oil-cooled
OA/FOA/FOA	Oil-immersed, self-cooled/forced-air-cooled/forced air—forced-oil-cooled
FOA	Oil-immersed, forced-oil-cooled with forced-air cooler
FOW	Oil-immersed, forced-oil-cooled with forced-water cooler
AA	Dry-type, self-cooled
AFA	Dry-type, forced-air-cooled
AA/FA	Dry-type, self-cooled/forced-air-cooled

Table 8.14 Temperature Versus Cost for Transformers

Temperature rise	80°C	115°C	150°C
Efficiency (%)	98.49	98.52	98.49
Total losses (watts)	7666	7505	7642
Iron losses (watts)	1571	1565	1454
Material cost	$1273	$1265	$1228
Cost savings vs 80°C	—	$8	$45

In Table 8.14 note that the material savings are less when the efficiencies are equalized, but the higher-rise units still have a lower material cost than does the minimum-cost 80°C-rise unit. A 150°C rise transformer (220°C hottest spot) can be designed at equal efficiency to, and manufactured at lower cost than, a minimum-cost 80°C rise unit. Higher-rise transformers have lower-iron, core-excitation losses, and during periods of low loading, are less expensive to operate than low-rise units, even though both types have equal total efficiency at their rated loads. The transformer voltage ratings include the primary and secondary continuous duty levels at the specified frequency, as well as the basic impulse level for each winding. Standard values of basic impulse level established for each nominal voltage class are listed in Table 8.15.

Voltage Taps and Connections

Voltage taps are usually necessary to compensate for small changes in the primary supply to the transformer, or to vary the secondary voltage level with changes in the load requirements. The most commonly used tap arrangement is the manually adjustable no-load type, consisting of four 2½% steps or variations from the nominal primary voltage rating. The taps are usually numbered 1 through 5, with number 1 position providing the lowest output voltage on a specific incoming voltage. In addition to the no-load taps, automatic tap changing under load is also available. This provides an additional ±10% voltage adjustment automatically in incremental steps, with continuous monitoring of the secondary terminal voltage.

Connections for standard two-winding power transformers are preferably delta-primary and wye-secondary. The wye-secondary, specified with external neutral bushing, provides a convenient neutral point for establishing a system ground. The delta-connected primary isolates the two systems with respect to the flow of zero-sequence currents resulting from third-harmonic exciting current.

Table 8.15 Transformer Standard Basic Impulse Insulation Levels

Nominal System Line-to-Line Voltage (volts)	Insulation Class (kilovolts)	Basic Impulse Insulation Level (kilovolts): Liquid Insulated		Dry Type	
		Power	Distribution	Ventilated	TENV Cast Coil Gas-Filled Sealed
120-1 200	1.2	45	30†	10	10
2 400	2.5	60	45†	20	20
4 160	5.0	75	60†	30	30
4 800	5.0	75	60†	30	30
6 900	8.7	95	75†	45	45
7 200	8.7	95	75†	45	45
12 470	15.0	110	95†	60	60
13 200	15.0	110	95†	60	60
13 800	15.0	110	95†	60	60
14 400	15.0	110	95†	60	60
22 900	25.0	150	150	110	110
23 000	25.0	150	150	110	110
26 400	34.5	200	200	125	125
34 500	34.5	200	200	150	150
43 800	46.0	250	250	—	—
46 000	46.0	250	250	—	—
67 000	69.0	350	350	—	—
69 000	69.0	350	350	—	—
92 000	92.0	450	—	—	—
115 000	115.0	550	—	—	—
		350‡	—	—	—
		450‡	—	—	—
138 000	138.0	650	—	—	—
		550‡	—	—	—
		450‡	—	—	—
161 000	161.0	750	—	—	—
		650‡	—	—	—
		550‡	—	—	—

†Ratings are also applicable to primary and secondary unit substation transformers.

‡Optional reduced levels applicable if equivalent reduced rating arresters are properly applied on the system.

Ventilated dry-type transformers, totally-enclosed nonventilated dry-type transformers, cast coil dry-type transformers, gas-filled sealed dry-type transformers, and liquid-insulated-type transformers are available with basic impulse levels higher than indicated at increased prices. For dry-type transformers 15 kV units can be specified to have 95 kV BIL, 25 kV units can be specified to have 150 kV BIL, and 34.5 kV units can be specified to have 200 kV BIL.

In some cases a ground primary wye-wye transformer connection is used to minimize the problem of ferroresonance.

Impedance

Impedance voltage is normally expressed as a percent value of the rated voltage of the winding in which the voltage is measured on the transformer self-cooling rating in kVA. The percent impedance voltage levels considered as standard for two-winding transformers rated up through 10,000 kVA are listed in Table 8.16. A value specified above or below those listed will usually result in higher costs. For transformer ratings above 10,000 kVA or 67 kV, a percent impedance voltage may be considered standard if it lies within a published minimum and maximum range. When design engineers specify a low-impedance voltage level, they should make references to ANSI/IEEE C57.12.00-1980 for short-circuit requirements to ensure that the transformer is capable of withstanding the stresses imposed by the external faults.

Insulation Medium

For outdoor installations, the mineral oil-insulated transformer is widely accepted due to its lowest cost and inherent weatherproof construction. For indoor installations, discontinuance of the use of an askarel (PCB) liquid-filled transformer has increased the use of high-fire-point liquids, such as poly(α-olefins), silicones, and high-molecular-weight hydrocarbons. In general, these liquids increase the cost of the transformer compared to mineral oil. Transformers insulated with a nonflammable dielectric fluid must be permitted to be installed indoors per NEC. When installed in combustible buildings, high-fire-point liquid-insulated transformers require automatic fire extinguishing systems or vaults.

Table 8.16 Standard Impedance Values for Three-Phase Transformers

High-Voltage Rating (volts)	kVA Rating	Percent Impedance Voltage	
Secondary Unit Substation Transformers			
2400—13 800	112.5—225	Not less than 2.0	
2400—13 800	300—500	Not less than 4.5	
2400—13 800	750—2500	5.75	
22 900	All	5.75	
34 400	All	6.25	
Liquid-Immersed Transformers, 501—30 000 kVA			
		Low Voltage, 480 V	Low Voltage, 2400 V and Above
2400—22 900		5.75	5.5
26 400, 34 400		6.25	6.0
43 800		6.75	6.5
67 000			7.0
115 000			7.5
138 000			8.0

NOTES: (1) Ratings separated by hyphens indicate that all intervening standard ratings are included. Ratings separated by a comma indicate that only those listed are included.

(2) Percent impedance voltages are at self-cooled rating and as measured on rated voltage connection.

The ventilated dry-type transformer has application in industrial plants for indoor installation. Since BIL for the ventilated dry-type transformer winding is usually lower than that of the liquid- or gas-filled dry-type, surge arresters should be included for the primary winding.

8.6 UNIT SUBSTATIONS

A unit substation consists of the following sections:

1. *Primary section.* This provides for the connection of one or more incoming high-voltage circuits, each of which may or may not be provided with a switching device or a switching and interrupting device.
2. *Transformer section.* This includes one or more transformers with or without automatic load-tap-changing equipment. Automatic load–tap–changing is not commonly used in unit substations.
3. *Secondary section.* This provides for the connection of one or more secondary feeders, each of which is provided with a switching and interrupting device.

8.6.1 Types of Unit Substations

Sections of unit substations are normally subassemblies for connection in the field. Unit substations are usually designed in the following types. (Application of these to industrial distribution systems is described in Chapter 4.)

1. *Radial:* one primary feeder to a single step-down transformer with a secondary section for connection of one or more outgoing radial feeders (see Figure 4.1).

2. *Secondary selective:* two step-down transformers, each connected to a separate primary source. The secondary of each transformer is connected to a separate bus through a suitable switching and protective device. The two sections of bus are connected by a normally open switching and protective device. Each bus has provisions for one or more secondary radial feeders (see Figure 4.5).

3. *Primary selective and primary loop:* each step-down transformer connected to two separate primary sources through switching equipment to provide a normal and alternate source. Upon failure of the normal source, the transformer is switched to the alternate source (see Figures 4.6 to 4.8).

4. *Secondary spot network:* two step-down transformers, each connected to a separate primary source. The secondary side of each

transformer is connected to a common bus through a network protector that is equipped with relays to trip the protector on reverse power flow to the transformer and reclose it upon restoration of the correct voltage, phase angle, and phase sequence at the transformer secondary. The bus has provisions for one or more secondary radial feeders (see Figure 4.11).

5. *Distributed network:* a single step-down transformer having its secondary side connected to a bus through a network protector, which is equipped with relays to trip it on reverse power flow and reclose upon restoration of the correct voltage, phase angle, and phase sequence at the transformer secondary. The bus has provisions for one or more secondary radial feeders and one or more tie

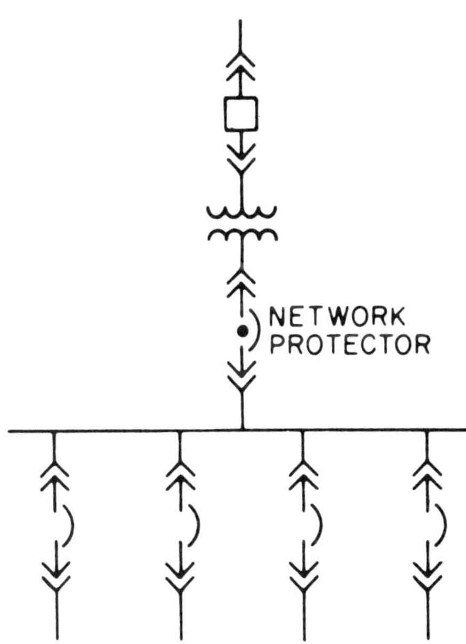

Figure 8.12 Distributed network diagram.

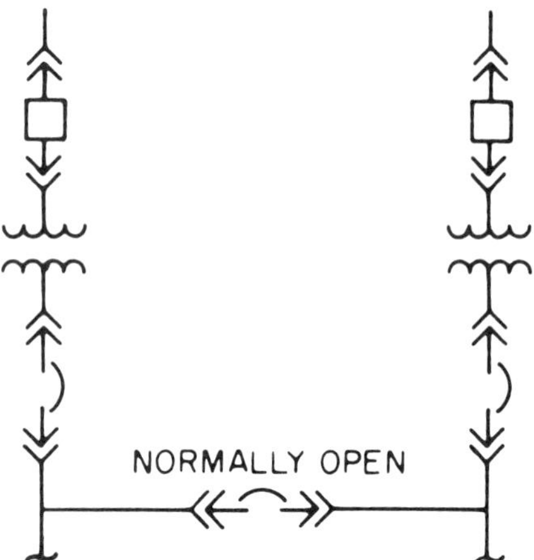

Figure 8.13 Duplex unit substation diagram.

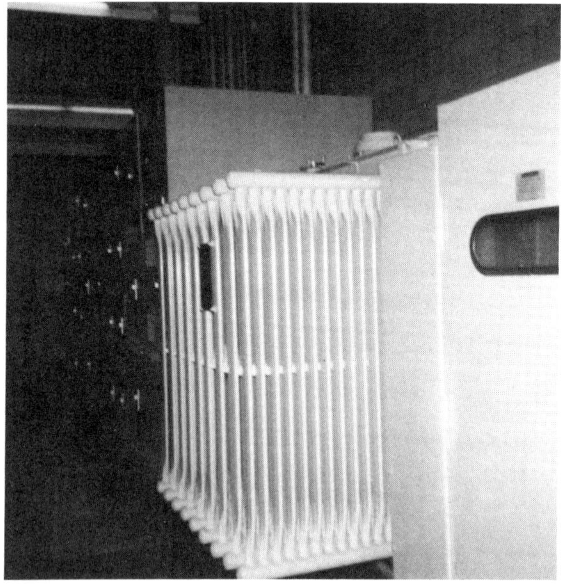

Figure 8.14 A Duplex unit substation with two 2000 kVA oil transformers.

connections to a similar unit substation (see Figure 8.10).

6. *Duplex:* two step-down transformers, each connected to a separate primary source. The secondary side of each transformer is connected to a radial feeder. These feeders are joined on the feeder side of the power circuit breaker by a normally open tie circuit breaker. This type is used primarily on electric utility primary distribution systems (see Figure 8.11). Figure 8.12 shows a duplex unit substation with two 2000-kVA oil transformers.

8.6.2 Advantages of Unit Substations

Unit substation has been widely accepted for industrial power distribution mainly because the engineering of the components is coordinated by the manufacturer, the costs of field labor and installation time are greatly reduced, and it is safer to operate. The operating costs are reduced due to the reduced power losses from shorter secondary feeders. Besides, a unit substation is very flexible and easy to expand.

Unit substations are available for either indoor or outdoor locations. Primary unit substations may be located outdoors, particularly when the primary supply is above 34.5 kV. There is a trend toward metal-enclosed equipment above 34.5 kV in a unit substation arrangement. Most secondary unit substations are located indoors to reduce costs and improve voltage regulation by placing the transformer as close as possible to the center of load of the area. (See Figure 8.9 which shows a typical 12.47 kV/480 V, 750-kVA unit substation.) Figure 8.10 shows a similar unit substation except that the secondary section consists of molded-case circuit breakers arranged in the form of a power distribution panelboard.

8.7 CAPACITORS AND POWER FACTOR

8.7.1 Definition of Power Factor

One definition expresses power factor as the cosine of the phase displacement angle between the circuit voltage and current. The other definition is that power factor is the ratio of active power to apparent power in a circuit. It is generally given in percent. The following formula is most familiar to the engineers:

$$\begin{aligned} \text{kW} &= (\text{kVA})\ (\text{power factor}) \\ &= (\text{kVA})\ (\text{cosine } \phi) \end{aligned}$$

8.7.2 Power-Factor Fundamentals

Most utilization devices require two components of current, active and reactive. The power-producing current (active current) is the current that is converted by the equipment into work, usually in the form of heat, light, or mechanical power. The unit of active power is the watt. The magnetizing current (reactive current) is the current required to produce the flux necessary to the operation of electromagnetic devices. The unit of reactive power is the var. The normal phasor relationship of these two components of current to the total current is shown in Figure 8.15.

$$\text{Total current } I_t = \sqrt{(\text{active current})^2 + (\text{reactive current})^2}$$
$$= \sqrt{(I \cos \phi)^2 + (I \sin \phi)^2}$$

The equation above is based on fundamental frequency and zero harmonic current.

8.7.3 Capacitor Standards and Operating Characteristics

1. *Capacitor ratings.* Early industry standards list ratings for shunt capacitor units from 240 to 26,600 V. Units rated at 1 to 15 kvar are common for applications with motors. Large units are available in ratings up to 600 kvar, 13,200 V, three phase.
2. *Operating characteristics.* The following relationships apply when capacitors are operated at other than their design-rated operating conditions:

    a. The reactive power varies approximately as the square of the applied voltage.
    b. The reactive power varies approximately as the frequency.

8.7.4 Utilization Equipment Applications

The following items are considered to be the major equipment in an industrial plant:

1. *Motors.* The power factor of a lightly loaded induction motor is poor. By using a capacitor of proper rating, attractive improvement over the entire load range can be achieved. The T-frame motor, available since 1964, generally has a lower power factor than that of the U-frame motor. Figure 8.16 shows the power factor comparison for these two different frame motors. Recently introduced high-efficiency motors may have much higher power factor characteristics. Hermetic and wound-rotor motors have a lower operating

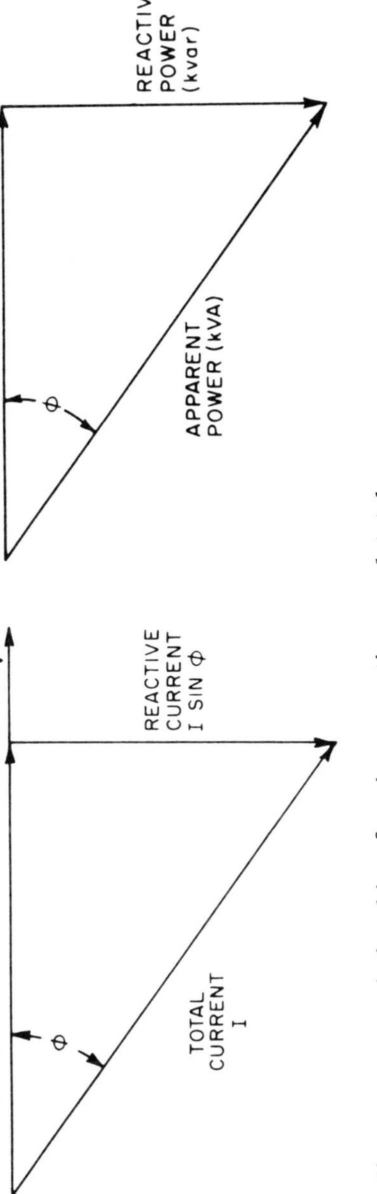

Figure 8.15 Relationship of active, reactive, and total power.

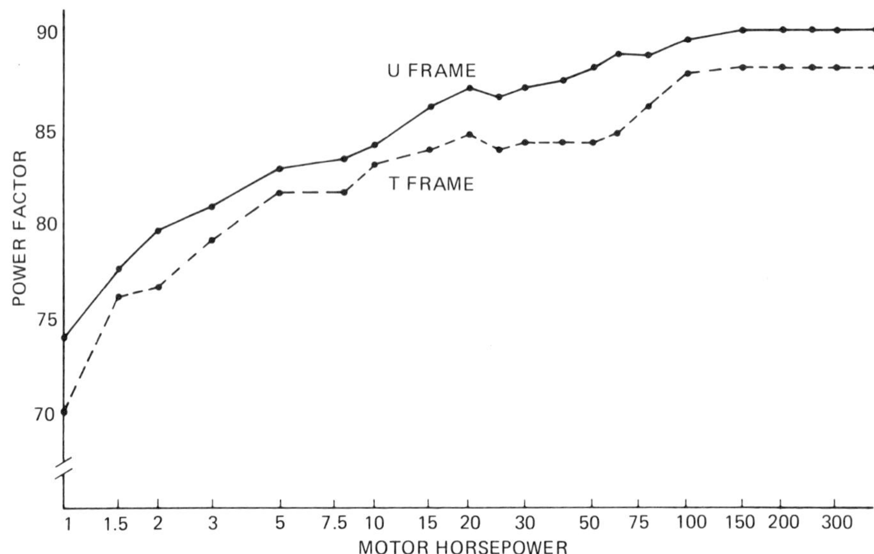

Figure 8.16 Power factor versus motor horsepower rating for U-frame and T-frame designs.

power factor than do other induction motors of the same power and speed ratings.

2. *Electric furnaces.* Arc furnaces have a poor power factor, typically 75 to 90%. Industion furnaces have a power factor of 30 to 70%; switched capacitors are normally used to maintain near-unity power factor.

3. *Transformers.* These are not ordinarily considered as loads, but they do contribute to lowering the system power factor. The transformer exciting current is usually 1 to 2% of the transformer rating in kVA and is independent of load. Reactive power is also required by the transformer leakage reactance. Such reactive power varies as the square of load current. At rated current the leakage reactance requires reactive power equal in magnitude to the transformer rating in kVA times the nameplate impedance in per unit.

Selection of Capacitors with Induction Motors

Economics may not always favor the individual motor-capacitor method because of the higher unit cost of capacitors in small ratings. However, this method is gaining in popularity because of the operational advantages. It puts the right amount of capacitance at the correct location as production equipment is added, taken away, or moved about the plant. It assures that capacitors are on the line when the motor is energized.

The power factor of a squirrel-cage motor at full load is usually 80 to 90%, depending on the motor speed and type. At light loads, the power factor drops rapidly. Even the power factor of an induction motor varies considerably from no load to full load; the motor reactive power does not change very much. This characteristic makes the induction motor an attractive application for capacitors. With a properly sized capacitor, the operating power factor can be excellent over the entire load range of the motor.

Capacitors have been applied to induction motors and switched with the motor as a unit with satisfactory results, except in a few applications. Sometimes difficulties are encountered because too large a capacitor rating has been used or the capacitors were misapplied on reversing applications. A general rule to follow in selecting capacitor ratings is that the total kvar rating of capacitors that are connected on the load side of a motor controller should not exceed the value required to raise the no-load power factor of the motor to unity.

Figure 8.17 shows various locations of capacitors when used with induction motors for power factor improvement. In this figure (a) and (b) represent the fact that the capacitor and motor are switched as a unit. The preferred location from an overall standpoint for

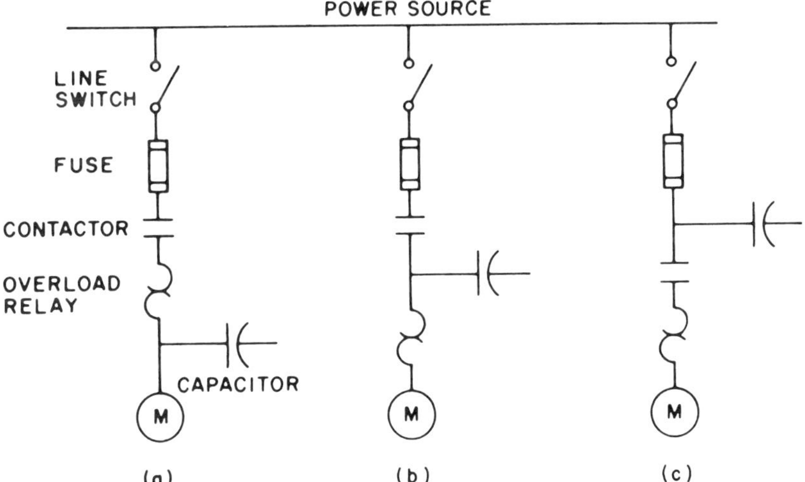

Figure 8.17 Electrical location of capacitors.

application not involving repetitive switching is that of (a) and (b). In either case the capacitor and motor are switched as a unit by the motor starter, so the capacitor is always in service when the motor is in operation. Connection in (b) may be preferred for existing installations because no change in the overload relay is required.

Following are the conditions under which motor-capacitor applications should be dealt with special caution:

1. Motors that are subject to reversing or plugging
2. Motors that are restarted while still running
3. Capacitors that are used with crane or elevator motors where the load may drive the motor
4. Open-transition reduced-voltage starters that are used with wye-delta connections and capacitors (capacitors should be connected on the line side of contactors)

In general, for low-voltage systems the induction motor method is more economical up to about 200 hp with full-voltage starters, and to about 350 hp with reduced-voltage starters. For medium-voltage systems, a synchronous motor is more economical over the entire hp and speed range if a power circuit breaker or a contactor is used to switch the excess capacitors.

The National Electrical Code has now properly omitted tables of capacitor values for motors because of the difference in motor designs among manufacturers. There will also be large differences in the recommended capacitor ratings for motors of different design generations, such as

Pre-U-frame, prior to 1955
U-frame, since 1955
T-frame, since 1964
U- and T-frame (high efficiency), since 1978

8.7.5 Benefits of Power Factor Improvements

Most benefits from power-factor improvements stem from the reduction of reactive power in the system. This results in: (1) lower utility costs if an incentive for power-factor clause is provided in the contract, (2) release of system capacity, (3) voltage improvement, and (4) lower system losses. The benefits derived by installing capacitors, synchronous machines, static power-factor controllers, or any other means for power factor improvement result in the reduction of reactive power circulating in the system. Capacitors and synchronous machines should be installed as close as possible to the load for which the power factor is being improved. A static power-factor controller would be located electrically adjacent to the motor for which power factor improvement is being made. Large plants with extensive distribution systems often install capacitors at the primary voltage bus when utility billing encourages the user to improve power factor. The benefits of power factor improvement are discussed in more detail below.

Release of System Capacity

Release of capacity means that as the power factor is improved, the current in the system will be reduced, thus permitting additional load to be served by the same system. Improvement in the power factor can release both active power and apparent power capacity. This method of releasing system capacity can sometimes be used as an expedient way of obtaining additional power for expansion of product lines. For example, an industrial plant with only 600 kVA spare capacity in its power supply transformer suddenly required 850 kVA of capacity for starting up a crash program in 8 months. The utility company would take a minimum of 15 months to install another substation transformer. A study of power factor at the incoming substation indicated that the plant power factor had been only 92%, and that by improving its power factor to 98%, 400 kVA of additional capacity would be released for service. Further study pointed out that the power factor improvement can be achieved by installing 1350 kvar of capacitors anywhere downstream of the unility substation transformer. All equipment needed for the capacitor installation were delivered from the manufacturer within 5 months. The capacitor bank consists of nine single-phase 7.2 kV 150 kvar units, connected three per phase in a wye configuration. This installation was completed in time for the project to use the released capacity and to start on schedule. In addition, payback for the installation was less than 1 year. Thus a problem was converted to a cost-savings opportunity. Figure 8.18 shows the 1350 kvar capacitor bank installed on a concrete pad in the outdoor switchyard. All protective and control equipment were housed neatly in an outdoor metal enclosure with the capacitors.

Voltage Improvement

It is well known that capacitors raise a circuit's voltage. However, it is rarely economical to apply them in industrial plants for that reason alone. The voltage improvement is usually regarded as a side benefit. The following approximate equation shows the importance of reducing the reactive power component of current in order to reduce the voltage drop:

$$\Delta V \cong RI \cos \phi \pm XI \sin \phi \tag{8.1}$$

Figure 8.18 1350 kvar capacitor bank in outdoor metal-enclosure.

where ΔV may be a drop or rise in voltage and ϕ is the power-factor angle. The plus sign is used when the power factor is lagging and minus when it is leading. ΔV is positive (voltage drop) for a circuit having a lagging power factor and usually negative (voltage rise) for the typical industrial circuit having a leading power factor. $R \cos \phi$ reflects the active power contribution to voltage drop per total current, and $X \sin \phi$ reflects the reactive power contribution to voltage drop. Since the power factor acts directly to reduce reactive power flow, it becomes most effective in reducing voltage drop.

From equation (8.1) it may be rewritten in a simple form to determine the voltage change due to capacitors at a transformer secondary bus:

$$\% \Delta V = \frac{\text{capacitor kvar} \times \% \text{ transformer impedance}}{\text{transformer kVA}}$$

The voltage increases when a capacitor is switched on and decreases when it is switched off. A capacitor permanently connected to the bus will provide a constant rise in voltage. If excessive voltage becomes a problem, the transformer tap should be changed. However, the voltage rise due to capacitors in most industrial plants with modern power distribution systems and a single transformation is rarely more than a few percent.

Power System Losses

System conductor losses are proportional to current squared, and since current is reduced in direct proportion to power factor improvement, the losses are inversely proportional to the square of the power factor:

$$\%\ \text{power loss} \propto 100\left(\frac{\text{original PF}}{\text{improved PF}}\right)^2 \tag{8.2}$$

$$\%\ \text{loss reduction} = 100\left[1 - \left(\frac{\text{original PF}}{\text{improved PF}}\right)^2\right] \tag{8.3}$$

8.7.6 Resonances, Harmonics, and Preventive Remedies

Resonance is a special circuit condition in which the inductive reactance is equal to the capacitive reactance. The frequency at which the circuit is in resonance is called the natural frequency of the circuit. When there is no intentional capacitance added to the circuit, the natural frequency of most power circuits is in the kilohertz range.

The addition of capacitors to the power system can either reduce or increase the harmonic and transient voltage. The nonsinusoidal voltages and currents associated with transformers operating in saturation are a familiar example of the generation of harmonics by a nonlinear impedance. A rectifier is another example of nonlinear impedance. Many other circuit components have varying degrees of nonlinearity inherent in their impedance. Arc furnaces impose a high harmonic duty on the system. Although capacitors in themselves do not generate harmonics, the effects of a capacitor on the circuit impedance may cause the harmonic voltages to either decrease or increase.

If excessive harmonic currents or voltages are suspected, remedies may be used as suggested in the following:

1. Detuning involves changing the capacitance or inductance of the circuit so that the circuit natural frequency will not fall near an expected or integral multiple of the fundamental frequency. This takes the form of removing or adding capacitor units, or addition of tuning reactors in series with the capacitor bank.
2. Wye-connected capacitor banks should be ungrounded, eliminating a path for the zero-sequence harmonics that should flow through a grounded neutral. Furthermore, a grounded bank could interfere with the performance of the plant ground-relaying system.

3. Harmonic input to the system can be reduced by operating at a lower level on the saturation curve of transformers and motors.
4. Increasing the number of phases of a rectifier or converter can reduce the harmonic input.

8.8 MOTORS AND MOTOR CONTROLLERS

8.8.1 Industrial Electric Motors

There are possibly hundreds of types of electric motors in existence today. Dozens of types will be found in a typical industrial plant in various applications. However, only a few types of motors are of direct and significant interest to the industrial plant engineers. These basic motors can be classified in three major classes: (1) three-phase induction, (2) direct current, and (3) three-phase synchronous.

Three-Phase Induction Motors

Induction motors make up the overwhelming bulk of motors in industrial plants. All operate on the principle of the transformer. All of them have a star winding (the primary) connected to the power source. The magnetic field produced by the primary rotates about the stator by virtue of the progression of the ac current through the winding. The voltage induced in the secondary (rotor) is, in turn, accompanied by a magnetic field. Motion is produced by the interaction of the magnetic fluxes. Various types of induction motors are discussed below.

Squirrel-Cage Induction Motors. Unquestionably, these are the workhorses of industry. Simple, rugged, and reliable, they offer the most horsepower per dollar of any type of motor. A fundamental characteristic of all induction motors is that speed remains almost constant from no load to full load. Development of the adjustable-frequency drive several years ago has permitted the squirrel-cage motor to invade a region that was once almost exclusively the domain of dc motors—adjustable-speed applications. In general, any squirrel-cage motor can be converted to an adjustable-speed motor with an adjustable-frequency drive.

Wound-Rotor Induction Motors. The rotor circuit is different from that of a squirrel-cage motor. Instead of solid-bar conductors, the rotor circuit uses wound coils, which are connected to slip rings on the rotor shaft. The rotor circuit is completed through carbon brushes riding on the slip rings, and a variable-resistance bank external to the machine. Slip and torque can be varied by varying the resistance in the control resistance bank. Wound-rotor motors are used primarily on crane and hoist drives; conveyor drives call for limited speed control. The controllable-torque advantage of this type of motor is being challenged by solid-state electronic variable-voltage controllers.

Multispeed Motors. These are squirrel-cage induction motors wound in such a manner and with appropriate winding leads brought out into the motor conduit box that the motor pole configuration can be changed by changing the winding connections at the motor controller. Two-speed motors are most common. Other multispeeds are also available.

DC Motors

The primary advantage of the dc motor is its excellent speed control and performance over the entire range from zero to full speed to top speed. Dc machines are less tolerant than ac machines of severe operating environments. However, they are more tolerant of operating abuse. There are four basic types of industrial dc motors: shunt wound, series wound, compound wound, and permanent magnet. Dc motor speed is controlled by varying the strength of the field or by varying the armature voltage; either method is easily applied by separately exciting the shunt field through a packaged static drive. As the field strength decreases with respect to armature current, the speed increases, and vice versa.

Synchronous Motors

Synchronous motors require both ac and dc power input. They run at exactly synchronous speed over the full range from no load to full load, without any need for speed adjustment. The machine's operating power factor can be varied. It can therefore be used to improve the power factor while driving its assigned load. They are usually practical only in multi-hundred-horsepower sizes, but they are an excellent choice, especially in slow-speed continuous operation. The efficiency of synchronous motors is also higher than that of induction motors. Usually, a point or two gain in efficiency can mean tremendous savings in the power bill.

High-Efficiency Motors

Since the energy crunch, high-efficiency motors are made available by various manufacturers. However, there is price premium for these motors. To justify their selection, a cost-effectiveness study must be made. The amount of money that can be saved annually using more energy-efficient electric motors can be determined by the following formula:

$$S = 0.746 \times H \times C \times t \times \left(\frac{100}{E_1} - \frac{100}{E_2} \right)$$

where

H = motor output, hp
C = energy cost, \$/kWh
t = annual operating time, hours
E_1 = efficiency of less-efficient motor, %
E_2 = efficiency of high-efficiency motor, %

In general, premium-efficiency, premium-priced motors can return the price premium in a very short time, especially if the motor is heavily loaded and energy rate is high. On the other hand, motors for frequent operations would not justify their premium cost.

8.8.2 Selection of Motor Control Equipment

The majority of motors used in industrial plants are integral horsepower of induction squirrel-cage design, powered from three-phase ac distribution systems rated at 600 V and below. To make a proper choice of motor controllers depends on a number of factors:

1. *Power system.* Dc or ac, single-phase or three-phase? What voltage and frequency? Will the system permit large inrush currents during full-voltage starting without excessive voltage drop?
2. *Motor.* Dc, squirrel-cage induction, wound-rotor induction, or synchronous? What is the horsepower rating? Will the motor be plugged or reversed frequently? What is the acceleration time from start to full speed?
3. *Load.* Is the load geared, belt driven, or direct coupled? Loaded or unloaded start?
4. *Operation.* Manual or automatic?
5. *Protection.* Are fuses or circuit breakers to be used? The full-load current of the motor and the ambient temperature at the motor should be known.
6. *Environment.* Is there excessive vibration, dirt, oil, or water? Will the motor be used in a hazardous or corrosive area?
7. *Cable connection space.* Will there be the required space for cable entrance, bending radius, and terminations?

Design engineers should consult with utility, manufacturing, and process engineers to determine the answers to the foregoing questions before specifying the correct type for any application.

8.8.3 Motor Starters

The primary function of a motor starter is to start and stop the motor to which it is connected. The following represent the major types of motor starters available for various applications.

Starters over 600 V

Starters for motors from 2300 to 13,200 V are designed as integrated complete units based on maximum horsepower ratings for use with squirrel-cage, wound-rotor, synchronous, and multispeed motors for full- or reduced-voltage starting. Ac magnetic, fused-type starters, NEMA class E2, employ current-limiting power fuses and magentic air-break contactors. Each starter will be completely self-contained, prewired, and with all components in space. Air-break contactors will be current rated based on motor horsepower. Combination starters will provide an interrupting fault capacity of 260 mVA symmetrical on a 2300 V system, and 520 mVA symmetrical on a 4160 or 4800 V system. The starter will conform to ANSI/NEMA ICS2-1983, class E2 controllers, and other applicable IEEE and ANSI standards. Figure 8.19 shows a typical three-phase 2300-V metal-enclosed motor starter cubicle.

Starters 600 V and Below

In ac motor starters for applications rated at 600 V and below, contactors are generally used for controlling the circuit to the motor. Starters should properly be applied on circuits and in combination with associated short-circuit protective devices (circuit breakers, fusible disconnects) that are able to limit the available fault current and the let-through energy to a level the starter can withstand. These withstand ratings should be in accordance with ANSI/UL 508-1983, ANSI/NEMA ICS1-1983, and ANSI/NEMA ICS2-1983. The following are the most commonly used motor starters rated at 600 V and below in the industry.

Across-the-Line Starters

1. *Manual.* This type of starter provides overload protection, but no undervoltage protection. One- or two-pole single phase for motor ratings to 3 hp; single- or polyphase motor ratings up to 5 hp at 230 V, single phase; $7\frac{1}{2}$ hp at 230 V, three phase; and 10 hp at 460 V, three phase. Operating control is available in toggle, rocker, or pushbutton design.

2. *Magnetic, nonreversing.* This type is for full-voltage frequent across-the-line starting of ac motors. It can be used for remote control with pushbutton station, control switch, or automatic pilot devices. Available in single-phase up to 15 hp at 230 V, and three-phase ratings up to 1600 hp at 460 V. The combination across-

Figure 8.19 A typical 3 phase, 2300 V metal-enclosed motor starter cubicles (courtesy of Westinghouse Electric Corporation).

the-line starter provides branch-circuit protection in addition to motor overcurrent protection. System design engineers must check the fault current available before deciding which fuses or circuit breakers should be used. It also provides undervoltage protection and is suitable for remote control. They are readily available up to NEMA size 5 from most manufacturers and size 9 from several.

3. *Magnetic, reversing.* This type is for full-voltage across-the-line starting of single-phase and polyphase motors where application requires frequent starting and reversing or plugging operation. It contains two contactors wired to provide phase reversal, mechanically and electrically interlocked to prevent both contactors from being closed at the same time. The combination across-the-line reversing starter is the same as the above, except that it is equipped with non-

fusible disconnect, fusible disconnect, or circuit breaker for branch-circuit protection. Figure 8.20 shows a typical combination magnetic reversing starter components arrangement in a metal enclosure.

Reduced-Voltage Starters

1. *Autotransformer, manual.* For limiting starting current and torque on polyphase induction motors to comply with power supply regulations or to avoid shock to the machine being driven. Overload and undervoltage protection are usually provided. Taps are provided on the autotransformer for adjusting starting torque and current.

2. *Autotransformer, magnetic.* Same as manual, but suitable for remote control. It has a timing relay for adjustment of the time at which full voltage is applied. To overcome the objection of the open-circuit transition associated with this type of starter, a circuit known as the Korndorfer connection is in common use. This type of starter requires a two-pole and a three-pole start contactor. The two-pole contactor opens first on the transition from start to run, opening the connections to the neutral of the autotransformer. The windings of the transformer are then momentarily used as series reactors during the transfer, allowing a closed-circuit transition. This is useful on high-inertia centrifugal compressors to obtain the advantages of low line current surges and closed-cricuit transition.

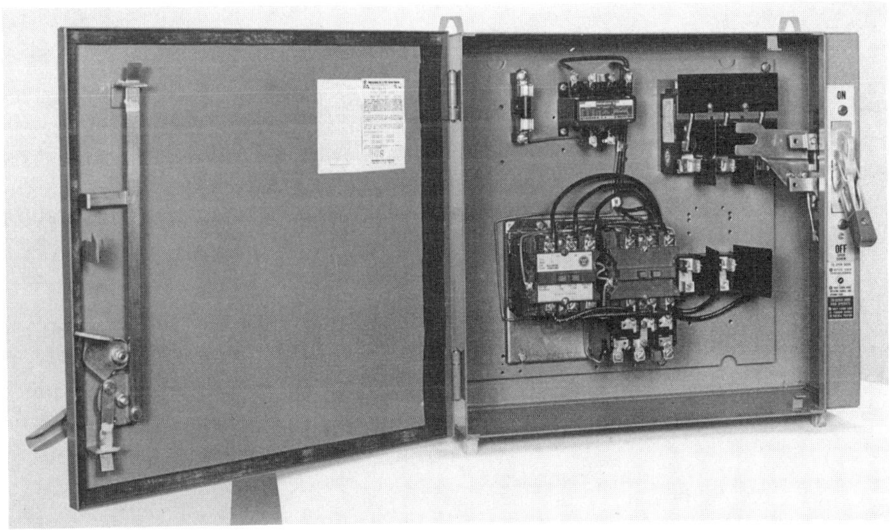

Figure 8.20 A typical combination magnetic reversing starter (courtesy of Westinghouse Electric Corporation).

Figure 8.21 shows an magnetic autotransformer reduced-voltage starter. An sutotransformer is mounted at the bottom. Figure 8.22 shows a typical schematic connections for a magnetic autotransformer starter.

3. *Primary resistor or reactor type.* This type provides the smoothest acceleration of all available reduced-voltage starting methods. During the starting period, resistance or reactance is inserted in series with motor windings. When maximum current is being drawn during the starting period, a maximum voltage drop will appear across the resistance or reactance. As the motor accelerates, less voltage drop will appear across the starting impedance. When the motor reaches full speed, the starting impedance is shorted out. They are inherently of the closed transition type. The primary reactor type is more suitable for high-voltage and/or high-current applications.

4. *Part-winding type.* They are to be used only on special motors with windings divided into two or more equal parts, with terminals of each winding available for external connection. There are two megnetic starters, each selected for one of the two motor windings, and a time-delay relay controlling the time at which the second winding is energized.

5. *Wye-delta type.* This type of starter is most applicable to starting motors that drive high-inertia loads with resulting long acceleration times. When the motor has accelerated on the wye connection, it is automatically reconnected by contactors for normal delta operation. Wye-delta starters more closely approach high starting torque per ampere than any other type of reduced-voltage starter. Both open and closed transition versions are available.

6. *Solid-state type.* Solid-state motor starters can control the starting cycle and provide reduced voltage starting for standard ac squirrel-cage motors. They are available in standard models for motors rated from 10 to 600 hp. One type of reduced-voltage starter uses six silicon-controlled rectifiers (SCRs) in a full-wave configuration to vary the input voltage from zero to full on, so that the motor accelerates smoothly from zero to full running speed. The SCRs are activated by an electronic control section that has an initial step voltage adjustment. This adjustment, when combined with a ramped voltage and current limit override, provides constant current (torque) to the motor until it reaches full speed. Figure 8.23(a) shows a typical schematic diagram of a solid-state motor starter, and a reduced-voltage solid-state motor starter components arrangement. Some variations in the design of the starting circuit are as follows:

a. Three power diodes replace the three return conducting SCRs. Each SCR is protected against reverse voltage by its associated diode. This half-wave configuration could produce harmonics that produce added heat in the motor windings. Thermal protective devices should be properly

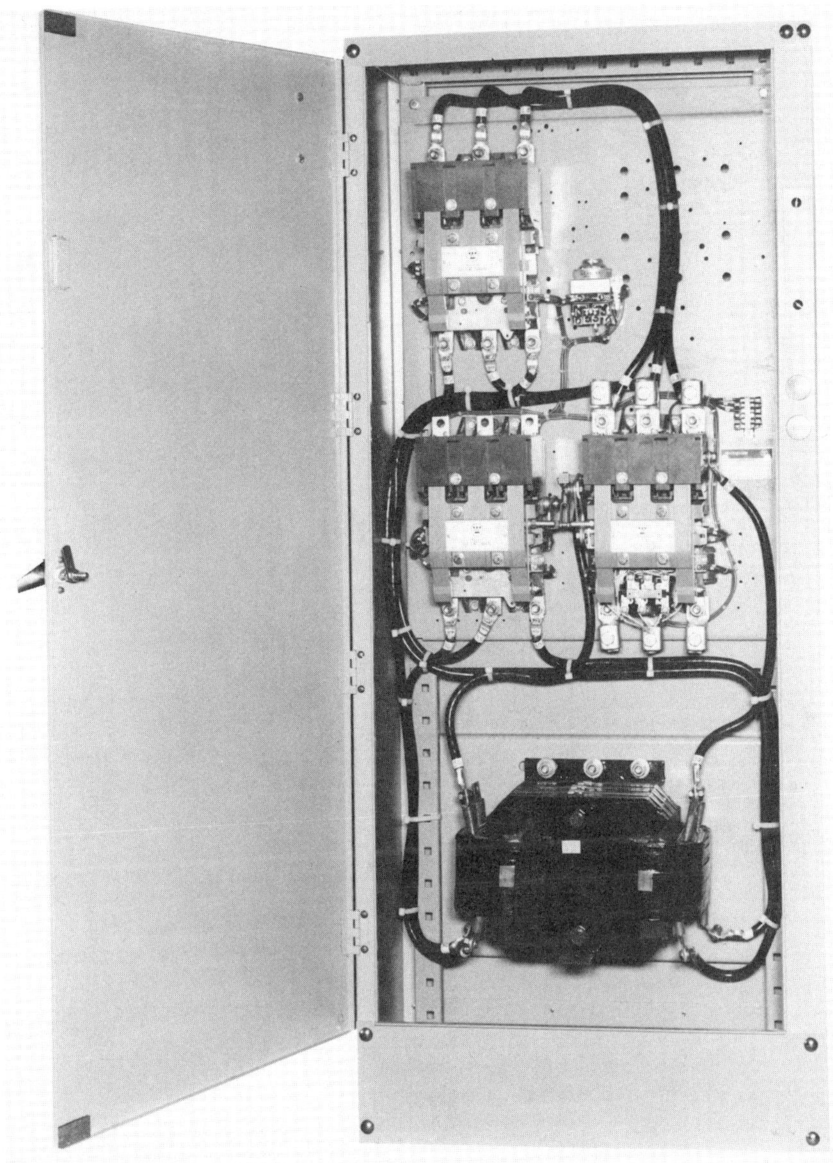

Figure 8.21 A magnetic autotransformer reduced-voltage starter (courtesy of Westinghouse Electric Corporation).

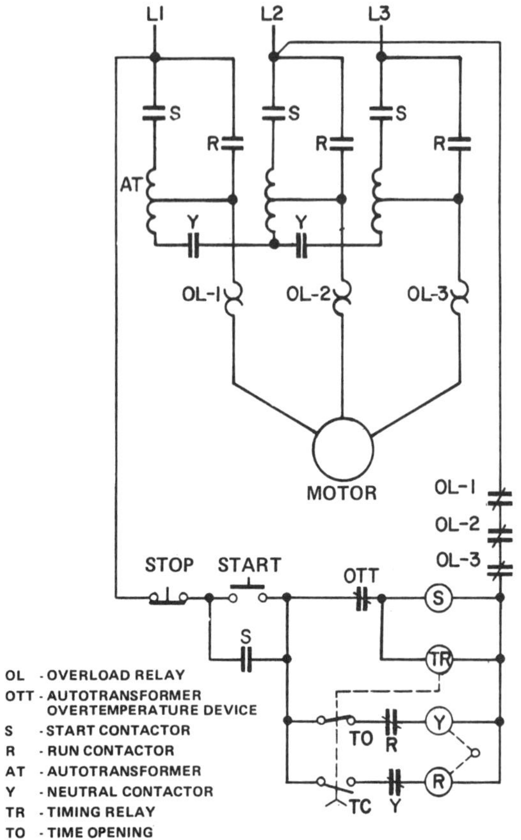

Figure 8.22 A typical schematic connections for a magnetic autotransformer starter.

chosen to prevent this additional heat from damaging the motor.

b. SCRs are used only during the starting phase. At full voltage, a run contactor closes and the circuit operates as a conventional electromechanical starter.

c. A starter with linear times acceleration used a closed-loop feedback system to maintain the motor acceleration at a constant rate. The required feedback signal is provided by a dc tachometer coupled to the motor.

7. *Comparison of various reduced-voltage starters.* Which of the foregoing reduced-voltage starters is best? The answer will

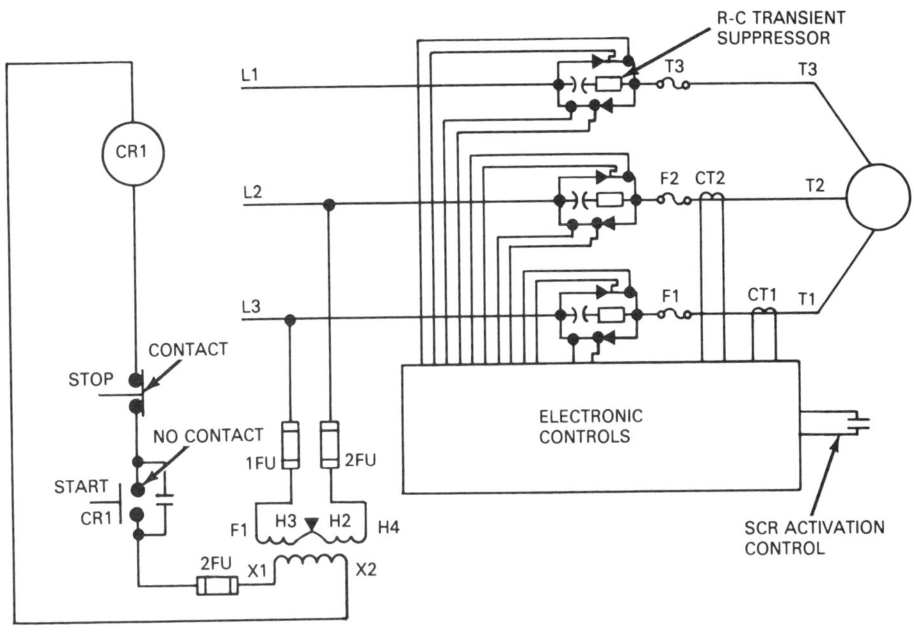

Figure 8.23 (a) A typical schematic diagram of a solid-state motor starter; (b) A reduced voltage solid-state motor starter, showing components arrangement.

vary with the application. Autotransformer type is more flexible, but it is also the most expensive. Table 8.17 is prepared to simplify selection of a most appropriate reduced-voltage starter for one's specific needs. To use this table, the design engineer must first prepare a list of starting requirements: current limitations, torque requirements, smoothness of acceleration, flexibility, allowable acceleration time, manual or magnetic, motor cost, maintainability, and delivery. Having compiled a listing of pertinent considerations, arrange them in order of importance. By consulting the table, proper selection can be quickly narrowed down.

Other Types of Motor Starters

1. *Slip-ring motor starter.* The wound-rotor or slip-ring motor functions in the same manner as the squirrel-cage motor, except that the rotor windings are connected through slip-rings and brushed to external circuits with resistance to vary motor speed. Increasing the resistance in the rotor circuit reduces the motor speed and vice

b

Figure 8.23 (Continued)

versa. There are also variations that use SCRs instead of contactors and resistors.

2. *Multispeed controllers.* These controllers are designed for the automatic control of two-, three-, or four-speed squirrel-cage motors. They are available for constant-horsepower, constant-torque, or variable-torque three-phase motors used on fans, blowers, refrigeration compressors, and similar machinery.

Table 8.17 Comparison of Various Reduced-Voltage Starting Methods

	REDUCED VOLTAGE CONTROLLER CHARACTERISTICS				
FEATURE	AUTOTRANSFORMER CLOSED—TRANSITION	PRIMARY RESISTOR	PRIMARY REACTOR	PART—WINDING	WYE (STAR)—DELTA
Smoothness of acceleration (1—smoothest) (4—least smooth)	2	1	1	4	3
Application Flexibility (1—Most flexible) (5—Least flexible)	1	4	5 (Normally used in high current and/or voltage applications)	3	2
Allowable acceleration time	30 sec. based on NEMA medium duty transformers	5 sec. based on NEMA Class 116 resistors	15 sec. based on NEMA medium duty reactors	2—3 sec. limited by motor design	45—60 sec. limited by motor design
Equipment	3 contactors, timer and starting element.	2 contactors, timer and starting element.	2 contactors, timer and starting element.	2 contactors and timer. Starting element inherent in motor design.	3 contactors and timer on open transition. 4 contactors, timer and resistor on closed transition. Starting element inherent in motor design
Approximate cost comparison of starter (1—Lowest cost) (4—Highest cost)	4	3	3	1	2
Approximate motor cost comparison	Standard squirrel cage induction motor.	Standard squirrel cage induction motor.	Standard squirrel cage induction motor.	Over one-third more than standard squirrel cage induction motor.	Over one-third more than standard squirrel cage induction motor.

8.8.4 Motor Control Centers

The National Electrical Manufacturers Association (NEMA) defines a motor control center as "a floor mounted assembly of one or more enclosed vertical sections having a horizontal common power bus and principally containing combination motor starting units. These units are mounted one above the other in the vertical section. The sections may incorporate vertical buses connected to the common power bus, thus extending the common power supply yo the individual units." ANSI/NEMA ICS2-1983 governs the type of enclosure and wiring; NEMA type 1, 2, 3, and 12 enclosures are generally available. Wiring of motor control centers conforms to two NEMA classes and three types. Class I provides for no wiring by the manufacturer between compartments of the center. Class II requires prewiring by the manufacturer, with interlocking and other control wiring completed between compartments of the center. With type A, no terminal blocks are provided; with type B, all connections within individual compartments are made to terminal blocks; and with type C, all connections are made to a master terminal block located in the horizontal wiring trough at the top or bottom of the center. The wiring specification for minimum field installation time and labor is NEMA class II, type C wiring. However, the wiring specification most frequently used by industrial contractors is NEMA class I, type B wiring. Figure 8.24 shows a typical motor control center which consists of three vertical sections, and several different-sized combination starters in each section.

ANSI/NEMA ICS2-1983 specifies that a control center must carry a short-circuit rating defined as the maximum available symmetrical rms current in amperes permissible at the line terminals. The available short-circuit current at the line terminals of motor control center is computed as the sum of maximum available current of the system at the point of connection and short-circuit current contribution of the motors connected to the control center. Most manufacturers show only the short-circuit rating of the bus work on the nameplate. It is therefore very important to establish the actual rating of the entire unit and, in particular, the plug-in units (circuit breakers, disconnects, starters, etc.).

8.8.5 Control Circuits for Motor Starters

Conventional starters for 600 V and below are factory wired with coils of the same voltage rating as the phase voltage to the motor. Where it is desirable or necessary to use control circuits and devices of lower voltage rating than the motor, control transformers are used to step the voltage down to permit the use of lower-voltage coil circuits. The control transformer is normally incorporated in the controller enclosure and wired in with an operating coil of proper voltage rating. Such transformers can be obtained with fused or otherwise protected secondaries to satisfy code requirements. The line voltage of the supply to the motor determines the required primary rating of the transformer. The control transformer should have sufficient capacity to supply power to control devices, including indicating lamps and solenoids.

Two forms of protection, undervoltage release and undervoltage protection, can be provided in the motor starter. In the former, if the voltage drops below a set minimum, or if the control voltage fails, the contactor will drop out but will reclose as soon as the voltage is restored. In the latter, low voltage will cause the contactor to drop out, but it will not reclose upon restoration of voltage.

8.8.6 Motor Protection

Many factors combine to determine the best type of motor protection for a specific application. However, there are two basic categories of motor protection. In the first category, equipment is intended to protect the motor from damage by sensing impending conditions

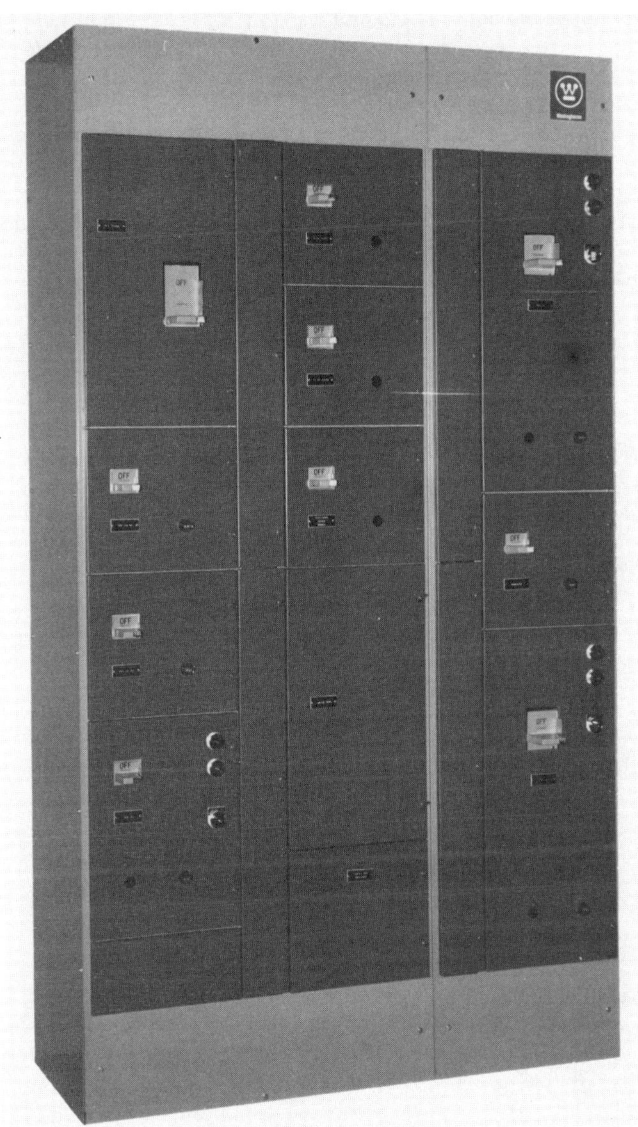

Figure 8.24 A typical motor control center (courtesy of Westinghouse Electric Corporation).

that could result in motor damage if no correction action is taken. Devices in the second category protect by minimizing damage to the motor and power system equipment, such as cables and motor starters, once a fault developes in the motor.

Devices in the first category are undervoltage and overvoltage relays, temperature detectors, and long-time overcurrent devices. Principal devices in the second category are instantaneous overcurrent relays and ground-fault devices. Table 8.18 lists most of the devices that can be applied for motor protection. Quite often, more than one type of protection is incorporated in the same relay. For example, one common type of relay incorporates undervoltage protection, loss-of-phase protection, and phase-reversal protection in the same relay enclosure.

One of the most common causes of motor burnout is single phasing, which will result in extreme voltage unbalance at the motor terminals and consequent motor failure. Undervoltage relays will prevent motor starting if one phase of the system has been opened, but they cannot detect loss of phase if the motor is already running. It is a common practice to depend on motor controller thermal protective devices to guard against single phasing when a motor is operating near full load. At reduced loads, a 120-Hz harmonic current can develop, causing rotor damage. This can be protected against only by applying a negative-sequence voltage relay. Another means of detecting single phasing is to use a current balance relay; this permits detection of a single-phasing condition at values of motor loading down to 12% of full-load current. Good single-phasing protection can be obtained by installing a negative-sequence voltage relay on the incoming power line, and providing a current balance relay on each important motor circuit.

8.9 INSTRUMENTS AND METERS

8.9.1 Definitions and Objectives

An instrument is defined as a device for measuring the value of a quantity under observation. Instruments may be either indicating or recording type. A meter is defined as a device that measures and registers the integral of a quantity with respect to time. The term "meter" is commonly used with other words, such as varmeter, voltmeter, frequency meter, even though these devices should be classified as instruments. Instruments and meters are used in industrial plants for the purpose of operating, monitoring, billing, accounting, planning, conserving energy, and maintaining equipment. They provide information concerning an electrical load, energy consumption, load factor, power factor, voltage, and so on. Care must be exercised to ensure the compatibility of the instruments and meters to

Table 8.18 Applications of Motor Protective Devices

Type of Protection or Relay	Motor HP Rating: 600 VOLTS & BELOW, Fractional to 5 HP	Motor HP Rating: 600 VOLTS & BELOW, To 300 HP	Motor HP Rating: 2300 VOLTS & ABOVE, To 1500 HP	Motor HP Rating: 2300 VOLTS & ABOVE, Over 1500 HP	Additional Protection for Syn. Motors
Inherent Thermal Protection (Linebreak)	●				
Short Circuit (Circuit Breakers or Fuses)	●	●	●	●	
Over Current (Time)	●	●	●	●	
Over Current (Time & Inst.)		●	●	●	
Temperature Sensing (Bimetallic and Thermister)	●	●			
Temperature Sensing (RTD or thermocouple)		●	●	●	
Ground Fault		●	●	●	
Undervoltage	●	●	●	●	
Overvoltage			●	●	
Loss of Phase	●	●	●	●	
Phase Reversal		●	●	●	
Locked Rotor	●	●	●	●	
Differential				●	
Current Unbalance				●	
Under Frequency				●	
Power Factor		●	●	●	
Negative Sequence Voltage			●	●	
Reverse Power				●	
Bearing Temperature		●	●	●	
Lightning & Voltage Surges			●	●	
Vibration			●	●	
Damper Winding					●
Field Current Failure					●
Field Voltage Failure					●
Pullout					●
Incomplete Sequence					●

their application so that the user is not injured or the equipment damaged.

8.9.2 Switchboard and Panel Instruments

Switchboard and panel instruments are permanently mounted and used in the continuing operation of a plant. In general, switchboard instruments are physically larger, more tolerant of transients and vibrations, and more accurate than an equivalent panel instrument. The current coils of most instruments are rated 5 A; their potential coils are rated 120 V. Current and potential transformers are often used to provide the required input of the instruments. Some of the most common instruments are discussed below.

1. *Voltmeters:* used to measure the potential difference between conductors and terminals, and connected directly across the points where a potential difference reading is desired. In general, when the voltage is higher than 120 V, a potential transformer would be required.
2. *Ammeters:* used to measure the current that flows in a circuit. If the current is high, a current transformer is often required. Selector switches are also installed to switch from one phase to another.
3. *Wattmeters:* used to measure the magnitude of electric power being delivered to a load. Proper installation of this instrument requires that the polarity of voltages, phasing of voltages, and current applied to it be taken care of correctly.
4. *Varmeters:* used for measuring reactive power. They have an advantage over a power-factor meter because their scale is linear. Small variations can be detected.
5. *Power-factor meters:* indicate unity power factor at the center scale, leading PF to the left of center, and lagging PF to the right of center. It can monitor the PF of only one phase at a time. The proper choice of a power-factor meter depends on the system to be monitored: three-phase, three-wire or three-phase, four-wire wye; three-phase, four-wire delta; and so on. Figure 8.25 shows a sample metering scheme (three-phase, four-wire, high current and voltage).
6. *Frequency meters:* measure the frequency of an ac power supply. Two common types are the pointer-indicating and the vibrating-reed. They are connected in the same way as the voltmeters.

8.9.3 Portable Instruments

A portable instrument has the same function as a switchboard instrument, but it can be moved easily. The most commonly used portable instruments are:

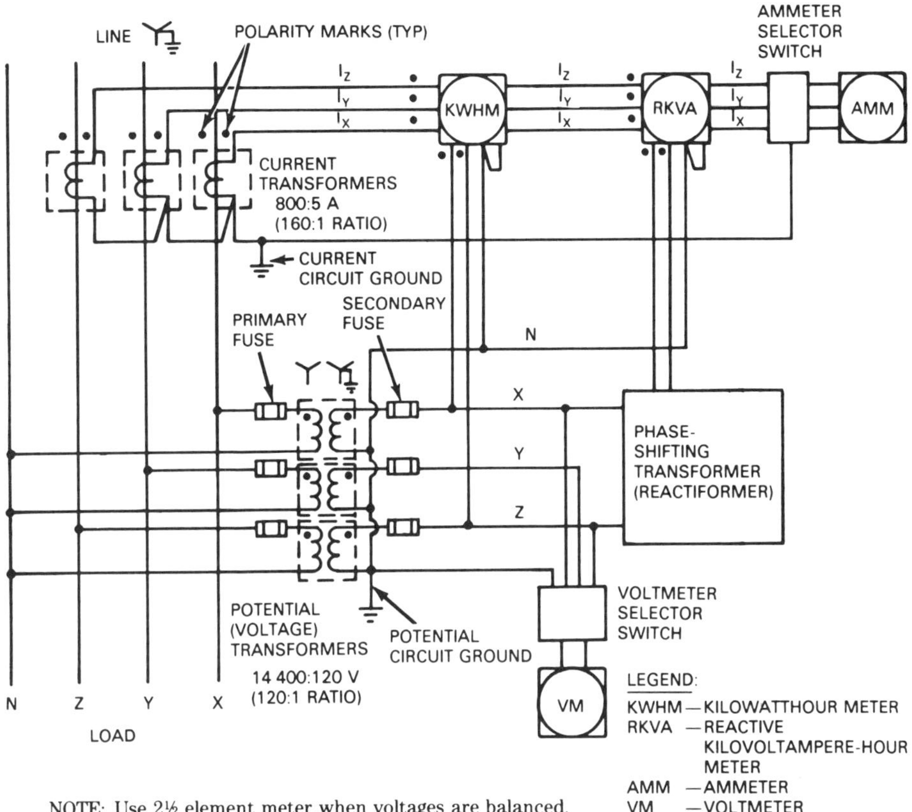

Figure 8.25 A sample metering scheme.

1. *Volt-ohmmeters:* can indicate a wide range of voltage, resistance (in ohms), and current (in milliamperes). They are very useful for investigating circuit problems.
2. *Clamp-on ammeters:* use a split-core current transformer to encircle a conductor and determine the amount of ac flowing. They are usually calibrated with several current ranges.

8.9.4 Miscellaneous Instruments

1. *Megohmmeters:* test the insulation resistance of cables, buses, motors, and other electric equipment. They consist of a hand-cranked or motor-driven dc generator and a resistance indicator. They are calibrated in megohms and available in 500, 1000, or

2500 V dc. A high reading does not always mean that the equipment's insulation can withstand rated potential. A high-potential test will be required to ascertain the equipment's withstand capability.

2. *Oscilloscopes:* electronic instruments used to study very high frequencies (up to millions of hertz) or phenomena of short duration. They can be used to study transients that occur in power circuits. A storage scope will display this waveform and a camera can be used to record the waveform.
3. *Recording instruments:* mostly direct-reading instruments; available as recording or curve-drawing instruments for portable or switchboard use. Charts are either strip or circular. The record may be continuous, or readings can be taken at regular intervals.

8.9.5 Meters

Kilowatthour Meters

Kilowatthour meters measure the amount of energy consumed by a load. Ac kilowatthour meters use an induction-disk type of mechanism; the disk revolves at a speed proportional to the rate at which energy passes through the meter. The metered kilowatthours are indicated on a set of dials driven by the revolving disk through a gear train. Recently, solid-state kilowatthour meters have been developed. Kilowatthour meters come in several classes. Following is a list of the common classes together with the maximum current that each can safely monitor:

Class 10	10 A
Class 20	20 A
Class 100	100 A
Class 200	200 A
Class 320	320 A

High-current services would require a class 10 or class 20 meter employed with current transformers. Table 8.19 lists application data as a general guideline for selection of kilowatthour meters for a variety of system. Other factors used in selection include:

Type of mountings: socket, bottom-connected
Voltage: 120, 240, 480, 240/120, etc.
Register: clock, cyclometer
Type of load current bypass: automatic, manual

Table 8.19 Data for Selecting a Kilowatthour Meter for a System

Service voltage	Stators	CTs	PTs
One-phase, two-wire	1	1	1
One-phase, three-wire	1	2	1
One-phase, three-wire (wye)	2	2	2
One-phase, three-wire (wye)	2	2	2
Three-phase, three-wire (delta)	2	2	2
Three-phase, four-wire (wye)			
Balanced conditions	$2\frac{1}{2}$	3	2
Unbalanced conditions	3	3	3

Kilovarhour Meters

Kilovarhour meters measure the amount of reactive energy drawn by a load. Their internal mechanism is similar to that of a kilowatthour meter, but the potential applied is shifted 90 degrees. Most kilovarhour meters have a ratchet-type assembly to prevent them from running backward. Therefore, they can record only lagging or leading flow, depending on the connection.

Demand Meters

Demans meters register the average use of power during a specified interval. It also indicates the maximum demand that has occurred since the meter was last reset. A printing demand meter records the average power during a specific interval. It records the total number of impulses received during a given interval. The record may be on printed paper tape, a chart, punched tape, magnetic tape, or a computer memory chip. A computer is required to process the information from the magnetic tape or memory chip.

8.9.6 Auxiliary Devices

Potential Transformers

Potential transformers provide a secondary voltage compatible to the rating of the instrument's potential coil. Switches should be provided in the secondary circuit of the potential transformer to disconnect the instrument for testing. For safety, the secondary winding of a potential transformer should be grounded. In most applications, both the primary and secondary circuits are fused.

Current Transformers

Current transformers insulate the instrument circuit from the primary voltage. Care should be exercised to ensure that the current transformer is insulated for the full system voltage. They reduce current to values within the rating of the instrument elements—usually 5 A. A current transformer can generate a dangerously high potential when the secondary circuit is opened. Therefore, a shorting bar, test switch, or jack is used to short-circuit the transformer secondary when the connected instrument is being tested. Current transformers must have a secondary circuit ground to restrict the building of static voltages caused by the high-voltage conductors. The accuracy of a current transformer or potential transformer is usually stated as a percentage at a rated burden.

Shunts

In dc measurements of current or energy, shunts are used to carry the main current to be measured. The dc ammeter actually measures the millivolt drop across its shunt and is calibrated in terms of current rating of its associated shunt.

Transducers

Transducers are devices used to transfer one or more analog inputs into another analog value that will be more suitable for use in instrumentation.

BIBLIOGRAPHY

ANSI C12.1-1982, Code for Electricity Metering.

ANSI C37.46-1981, American National Standard Specifications for Power Fuses and Fuse Disconnecting Switches.

ANSI C97.1-1972, American National Standard Low-Voltage Cartridge Fuses 600 V and Less.

ANSI/IEEE C37.010-1979 (include Supplement ANSI/IEEE C37.010d-1984), Application Guide for AC High-Voltage Circuit Breakers Rated on a Symmetrical Current Basis.

ANSI/IEEE C37.04-1979, IEEE Standard Rating Structure for AC High-Voltage Circuit Breakers Rated on a Symmetrical Current Basis.

ANSI/IEEE C37.06-1979, American National Standard Preferred Ratings and Related Required Capabilities for AC High-Voltage Circuit Breakers Based on a Symmetrical Current Basis.

ANSI/IEEE C37.13-1981, IEEE Standard for Low-Voltage AC Power Circuit Breakers Used in Enclosures.

ANSI/IEEE C37.20-1969 (include Supplements ANSI/IEEE C37.20a-1970, C37.20b-1972, C37.20c-1974, and C37.20d-1978), Standard for Switchgear Assemblies Including Metal-Enclosed Bus.

ANSI/IEEE C37.40-1981, IEEE Standard Service Conditions and Definitions for High Voltage Fuses, Distribution Enclosed Single-Pole Air Switches, Fuse Disconnecting Switches, and Accessories.

ANSI/IEEE C37.41-1981, IEEE Standard Design Tests for High-Voltage Fuses, Distribution Enclosed Single-Pole Air Switches, Fuse Disconnecting Switches, and Accessories.

ANSI/IEEE C37.47-1981, American National Standard Specifications for Distribution Fuse Disconnecting Switches, Fuse Supports, and Current-Limiting Fuses.

ANSI/IEEE C37.100-1981, IEEE Standard Definition for Power Switchgear.

ANSI/IEEE C57.12.00-1980, IEEE Standard General Requirements for Liquid-Immersed Distribution, Power and Regulating Transformers.

ANSI/IEEE C57.12.01-1981, IEEE Standard General Requirements for Dry-Type Distribution and Power Transformers.

ANSI/IEEE C57.13-1978 (R 1986), IEEE Standard Requirements for Instrument Transformers.

ANSI/IEEE Standard 141-1986, Recommended Practice for Electric Power Distribution for Industrial Plants.

ANSI/NEMA ICS1-1983, General Standards for Industrial Control and Systems.

ANSI/NEMA ICS2-1983, Industrial Control Deivces, Controllers, and Assemblies.

ANSI/NFPA 70-1987, National Electrical Code.

ANSI/UL 508-1983, Safety Standards for Industrial Control Equipment

Bowers, George H., and Goethe, Paul K., Importance of Efficiency and Temperature Rise on Transformer Selection, *Electrical Construction and Maintenance,* Feb. 1981, pp. 77–78.

Castenschold, R., Understanding Automatic Transfer Switches, *Plant Engineering,* Mar. 3, 1983, pp. 175–179.

Chen, Kao, Capacitor Installation Saves the Day; Saves Money Too, *Plant Engineering,* Nov. 1977, pp. 188–189.

Chen, Kao, Selection, Installation, Testing, and Calibration of Power Distribution Equipment, *Plant Management and Engineering,* Feb. 1961, pp. 38–41.

Dalasta, F. B., AC Motor Protection, *Plant Engineering,* Mar. 7, 1974, pp. 180–186.

NEMA AB1-1975 (R 1981), Molded-Case Circuit Breakers.

NEMA SG2-1981, High-Voltage Fuses.

NEMA SG3-1981, Low-Voltage Power Circuit Breakers.

NEMA SG4-1975, (R 1980), AC High-Voltage Circuit Breakers.

Prince, Frank J., and Gariepy, Robert E., How to Select Reduced-Voltage Starters, *Plant Engineering,* Aug. 21, 1969, pp. 58–61.

9

Power-Carrying Devices

9.1 CABLE

The principal requisite of an industrial power distribution system is to transfer electric power from its source to utilization loads. Cable can serve such a requisite safely and efficiently provided that the selection of correct wiring method is carried out properly. Cables may be installed in raceway, tray, underground duct, or direct buried, preassembled on a messenger, in a cable bus, or as open runs.

Selection of conductor size requires consideration of load current to be carried and loading cycle, emergency overload requirements, fault clearing time, interrupting capacity of the cable overcurrent protection, and voltage drop for a particular installation. Provisions should be made for proper terminating, splicing, and grounding of cables. The application and sizing of cables rated up to 35 kV is governed by ANSI/NFPA 70-1987 (National Electrical Code). Cable use may also be covered in state and local regulations. The various tables presented in this chapter are intended to help the design engineers to determine cable requirements.

9.1.1 Cable Construction

Conductors

The two conductor materials in common use are copper and aluminum. Annealed copper is the metal most generally used because of its combined excellent electrical and mechanical properties and resistance to corrosion. A aluminum conductor will have a cross-sectional area

1.59 times that of a tinned copper conductor to have equivalent dc resistance. The difference in area is approximately equal to two American Wire Gauge (AWG) sizes. The current-carrying capacity of an insulated cable with an aluminum conductor ranges from approximately 78 to 84% of the ampacity of a copper conductor of corresponding size.

The need for mechanical flexibility usually determines whether a solid or a stranded conductor is used. The NEC requires that conductors of size 8 and larger to be stranded. Three common classes of stranding are concentric-lay, bunched, and rope-lay. Figure 9.1(a) shows concentric layer strands and part (b) shows concentric rope-lay strands.

Insulation

Insulation in common use includes:

1. Thermosetting compounds
2. Thermoplastic compounds
3. Paper-laminated tapes
4. Varnished cloth, laminated tapes
5. Mineral insulation

Extruding and taping are two common processes for applying insulation on an electric conductor. Two tape insulations with long successful service records for bulk distribution of power in industrial plants are varnished cloth and impregnated paper. However, both of those tape insulations are vulnerable to moisture and must be protected with an impervious metallic corrugated sheath, such as lead or a continuous metallic corrugated sheath. The demand for these insulations has greatly declined since the new and superior insulations become available.

Thermosetting and thermoplastic insulations are applied to the conductor by an extrusion process. Thermoplastic methods soften to essentially a liquid state with increasing temperature, and return to their solid state unchanged on cooling. Thermosetting materials tend to retain their dimensional stability with increasing temperature up to their actual decomposition temperature. Table 9.1 shows various insulations which are classified under the thermosetting or thermoplastic categories discussed above, and their electrical and physical properties.

The performance record of the synthetic rubber insulation butyl in industrial plants has been outstanding. This insulation has been used for over 40 years. Excellent stability of physical, thermal, chemical, and electrical characteristics with age and heat makes butyl-insulated cable attractive to the plant engineer. It is suitable for 90°C copper temperature service and up to 35 kV usage.

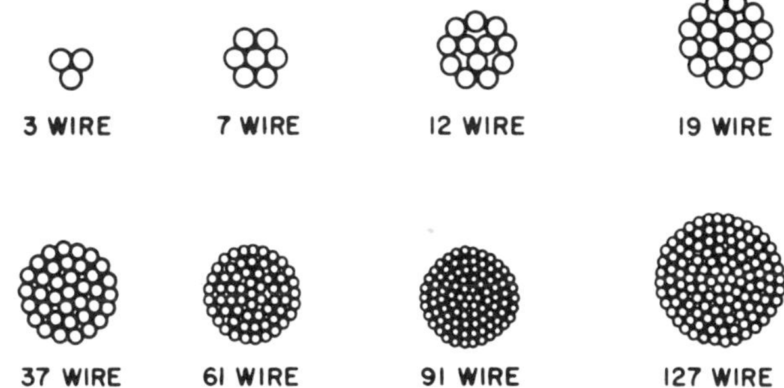

(a) Concentric Layer Strands

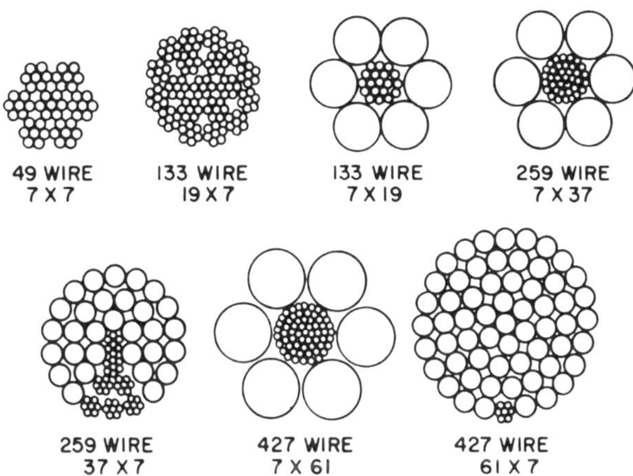

(b) Concentric Rope-Lay Strands

Figure 9.1 Conductor stranding.

Table 9.1 Commonly Used Insulating Materials

Common Name	Chemical Composition	Properties of Insulation Electrical	Physical
Thermosetting			
Crosslinked polyethylene	Polyethylene	Excellent	Excellent
EPR	Ethylene propylene rubber (copolymer and terpolymer)	Excellent	Excellent
Butyl	Isobutylene isoprene	Excellent	Good
SBR	Styrene butadiene rubber	Excellent	Good
Oil base	Complex rubber-like compound	Excellent	Good
Silicone	Methyl chlorosilane	Good	Good
TFE*	Tetrafluoroethylene	Excellent	Good
ETFE†	Ethylene Tetrafluoroethylene	Excellent	Excellent
Neoprene	Chloroprene	Fair	Good
Class CP rubber‡	Chlorosulfonated polyethylene	Good	Good
Thermoplastic			
Polyethylene	Polyethylene	Excellent	Good
Polyvinyl chloride	Polyvinyl chloride	Good	Good
Nylon	Polyamide	Fair	Excellent

*For example, Teflon or Halon.
†For example, Tefzel.
‡For example, Hypalon.

Early in the 1960s, the cross-linking process of polyethylene was introduced. It converts polyethylene from a thermoplastic to a thermosetting material and results in a compound with a unique combination of properties. In general, cross-linked polyethylene retains many of the fine electrical and physical properties of conventional polyethylene. In addition, its permissible operating temperature is increased over that of conventional polyethylene. It also has greater stability than rubber to oxidative degradation. Its recommended maximum operating temperature is 90°C. For applications in the class 600 V and below, cross-linked polyethylene insulation represents both the insulation and the jacket.

In the thermoplastic category, polyethylene insulated cables have had a long and successful service record at 600, 5000, and 15,000 V. Specifications for polyethylene power cables at rates up to 15,000 V are contained in industry standards such as IPECA S-61-402 and NEMA WC-5. The first commercial 69-kV polyethylene cable was installed in a direct buried installation in 1962. The excellent record of this cable has led to further work on cable development for voltages higher than 69 kV. Industry standards (IPECA) for polyethylene insulation specify a wall thickness approximately 72% that of rubber-insulated cables. Because of its improved dielectric strength, this type of cable requires an insulation thickness of only about 59% of that required for rubber-insulated cable.

Outer Coverings

The function of an outer covering is to protect the conductors and insulation during and after installation from mechanical damage and chemical deterioration. Materials adaptable for cable coverings fall in two groups, metallic and nonmetallic. In an industrial power distribution system, the metallic coverings most commonly used are lead sheath, interlocked metallic tape, and impervious corrugated metallic sheath. Different types of metals are available, such as steel, bronze, aluminum, and so on. Supplemental protection with corrosion-resistant material on the outside may be required, such as neoprene tapes or polyethylene jacket over the lead sheath, and PVC or polyethylene over interlocked armor.

The impervious, helically corrugated, lightweight metallic sheath not only protects the insulated conductors from gases and fluids but also from mechanical damage, yet does not result in a stiff heavy cable. Nonmetallic jackets commonly used are: neoprene, polyethylene, polyvinyl chloride, and nylon. When properly compounded and vulvanized, neoprene jackets are flexible, resistant to abrasion and tear, and to other chemicals. Poylethylene- and polyvinyl chloride—jacketed cables offer excellent electrical and physical characteristics and are applicable for service in contaminated areas. Nylon serves very favroably as a thin jacket over PVC to provide physical protection to the insulation. Type THWN (600 V) is an example of a small-diameter nylon-covered wire.

Shielding

Shielding of an electric power cable is the practice of confining its dielectric field to the inside of the cable insulation or insulated conductor assembly by surrounding the insulation or assembly with a grounded conducting medium called a shield. For operating voltages below 2 kV, nonshield constructions are normally used, while above 2 kV, cables are required to be shielded to comply with the NEC and ICEA. The NEC does provide for the use of nonshielded cables up to 8 kV, provided that the conductors are listed by a nationally recognized testing laboratory and are approved for the purpose.

Since shielded cable is usually more expensive than nonshielded cable, and the more complex terminations require a larger space in the terminal boxes, the nonshielded cable has been used extensively at 2400 and 4160 V, and occasionally at 7200 V.

The voltage distribution between a nonshielded cable and a grounded plane is shown in Figure 9.2. Here it is assumed that the air is the same, electrically, as the insulation, so that the cable is in a uniform dielectric above the ground plane to permit a simpler illustration of the voltage distribution and field associated with the cable.

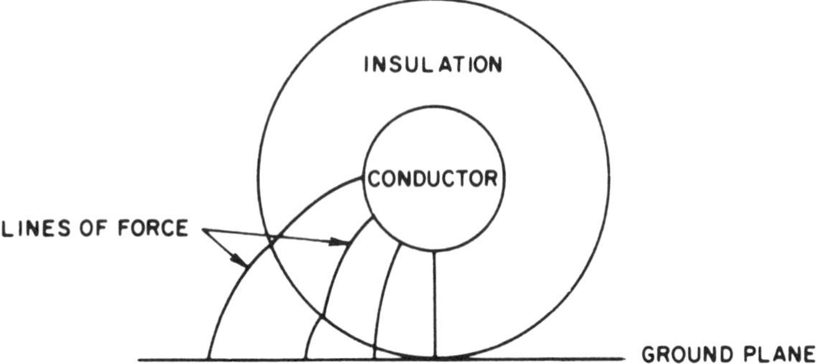

Figure 9.2 Electric field of nonshielded cable on ground plane.

In a shielded cable, the equipotential surfaces are concentric cylinders between conductor and shield (Figure 9.3). The lines of force and stress are uniform and radial, and cross the equipotential surfaces at right angles, eliminating any tangential or longitudinal stresses within the insulation or on its surface. The equipotential surfaces for the nonshielded system are cylindrical but not concentric with the cylinder, and cross the cable surface at many different potentials. Surface tracking, burning, and destructive discharges to ground could occur under these conditions.

There are two distinct types of shields in use today:

1. *Metallic shields:* made of metal tape, metal braid, metallized paper, or serving of wire.
2. *Nonmetallic (semiconducting) shields:* made of rubber or synthetic polymers. These have an impedance such that without fully grounding the surface of the insulation, they provide protection for cable insulation by restricting longitudinal and tangential voltage stress on the cable surface adjacent to miscellaneous ground points to a gradient below that necessary to produce corona.

Shielding is used for the following purposes:

1. To obtain symmetrical radial stress distribution within the insulation
2. To eliminate tangential and longitudinal stress on the surface of the insulation
3. To exclude from the dielectric field materials not intended for insulation, such as braids, tapes, and fillers
4. To increase safety to human life and to remove the fire and explosion risk attending electrical discharge in gaseous locations

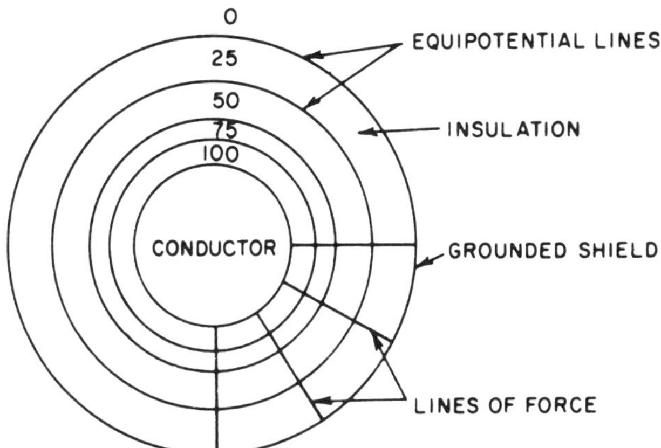

Figure 9.3 Electric field of shielded cable.

Sound grounding practice (see Chapter 6) must be exercised on all shielded cables in accordance with proper system and cable design. Stress cones should be used at all terminations of shielding, such as potheads, according to industry standards.

9.1.2 Types of Cables

Low-Voltage Cables

Low-voltage cables are generally rated at 600 V, regardless of the use voltage, whether 120, 208, 240, 277, 480, or 600 V. The 600-V compounds of XLPE are usually filled to further enhance the relatively good toughness of conventional polyethylene. The combination of cross-linking the polyethylene molecules through vulcanization plus fillers produces superior mechanical properties. Rubber-like insulation, such as EPR and SBR, has been provided with outer jackets for mechanical protection, usually of polyvinyl chloride, neoprene, or CP rubber, such as Hypalon. The newer EPR insulations have improved physical properties that do not require an outer jacket for protection. The following is a guide for the most commonly used 600-V cables:

1. *EPR with or without jacket:* type RHW for 75°C maximum operating temperature in dry or wet locations; and RHH for 90°C in dry locations only
2. *XLPE without jacket:* type XHHW for 75°C maximum operating temperature in wet locations and 90°C in dry locations

3. *Polyvinyl chloride-insulated, nylon jacketed:* type THWN for 75°C maximum operating temperature in wet or dry locations
4. *Metal-clad or interlocked armor cable:* type MC; individual insulated conductors are usually type XHHW or RHH/RHW for use in any raceway, in cable tray, as open runs of cable, for direct burial, or as aerial cable on a messenger
5. *Tray cable:* type TC; multiconductor with an overall flame-retardant nonmetallic jacket; cable takes the rating of the insulation selected; for use in cable trays, raceways, or where supported by a messenger wire

Cables in categories 2 and 4 are usually restricted to conduit or duct applications.

Power-Limited Circuit Cables

When the power in the circuit is limited to levels defined in the NEC, Article 725, for remote control, signaling, and power-limited circuits, the wiring method may utilize power-limited circuit cable, or type PLTC (power-limited tray cable). These cables, rated at 300 V, include copper conductors for electrical circuits and thermocouple alloys for thermocouple extension wire.

Medium-Voltage Cables

Type MV, medium-voltage power cables, have solid extruded dielectric insulation and are rated from 2000 to 35,000 V. These single- and multiple-conductor cables are available with nominal voltage ratings of 5, 8, 15, 25, and 35 kV. EPR and XLPE are the usual insulating compound for type MV cables; however, polyethylene and butyl rubber are also used as insulators. Type MV cables may be installed in raceways in wet or dry locations. The cable must be specifically approved for installation in cable tray, direct burial, exposure to sunlight, or for messenger-supported wiring.

9.1.3 Cable Ratings

Voltage Rating

The selection of the cable insulation (voltage) rating is made on the basis of the phase-to-phase voltage of the system in which the cable is to be applied, and the general system category depending on whether the system is grounded or ungrounded, and the time in which a ground fault on the system is cleared by a protective device. Consequently, 100% voltage-rated cables are applicable to grounded systems provided with protection that will clear a ground fault within 1 minute. 133% rated cables are required on ungrounded systems where the clearing time of the 100% level category cannot be met, yet there is adequate assurance that the faulted section will be cleared within 1 hour. Insulation rated at 173% voltage level is needed on systems where the time required to deenergize a grounded section is indefinite.

Conductor Size

The selection of conductor size is based on the following criteria:

Load Current. The selection of a cable size based on its thermal heating, both from the load current and from mutual heating from nearby cables, is usually considered first. The NEC ampacity tables for low- and medium-voltage cables must be used where the NEC has been adopted. These are derived from IEEE S-135. All ampacity tables show the minimum-size conductor required, but conservative engineering practice, future load growth considerations, voltage drop, and short-circuit heating may make the selection of larger conductors necessary. Higher ambient temperatures require derating of the cables in accordance with the formulas or factors in the ampacity tables.

Conductor sizes over 500 to 750 kcmil may necessitate paralleling two or more smaller cables, because the current-carrying capacity per circular mil of conductor decreases for ac circuits due to the skin effect, proximity effect, and other losses. Where cables are paralleled to increase ampacity and the line overload device will not protect individual cables, consideration should be given to the use of individual in-line limiters at both ends of each cable.

As a guide for conductor selection, Tables 9.2 and 9.3 show the conductor sizes for 75°C rated insulations, for various loads for cables in air, tray, exposed conduit (Table 9.2), and underground ducts (Table 9.3). The load factor in underground runs takes into account the heat capacity of the duct bank and surrounding soil, which responds to the average heat losses. The definition of load factor is the ratio of the average load to the peak load. The peak load is usually the average of a 1/2- to 1-h period of the maximum loading that occurs during the day.

Emergency Overload (Medium- and High-Voltage Cable). As a practical guide, the IPCEA has established maximum emergency-overload temperatures for various types of insulation. Operation at these emergency-overload temperatures should not exceed 100 h per year and such 100-h overload periods should not exceed five during the life of the cable. Table 9.4 gives uprating factors for short-time overloads for various types of insulated cables.

Voltage Drop. The NEC recommends that the steady-state voltage drop in power, heating, or lighting feeders be no more than 3%

Table 9.2 Conductor Requirements (AWG or kcmil) for Indicated Loadings: Above Ground Ratings, 1–15 kV, Conductor Temperature 75°C, Indoor Air Ambient Temperature 40°C

Load Current 50–100% Load Factor (amperes/conductor)	3 Single-Conductor Cables in Air*		3 Single-Conductor Cables in Steel Conduit		Cables in Open Tray or Ladder
	Cu	Al	Cu	Al	
50	8	6	6	6	For ampacities of cables in trays with various tray loadings, refer to tables shown in IPCEA P-54-440-1972
100	4	2	2	1	
150	1	2/0	1/0	3/0	
200	2/0	4/0	3/0	250	
250	4/0	250	250	350	
300	250	350	350	500	
400	350	500	500	750	
500	500	750	750	1000	

From IPCEA S-135-1-1962 and S-135-2-1962.

*Conductor sizes are suitable for outdoor installation on messengers.

and the total drop, including feeders and branch circuits, be no more than 5% overall.

Fault Current. Under short-circuit conditions the temperature of the conductor rises rapidly, then cools off slowly after the short-circuit condition is removed. Failure to check the conductor size for short-circuit heating sould result in severe damage to cable insulation. In addition to the thermal stresses, there are mechanical stresses set up in the cable through expansion upon heating. The minimum conductor size requirements for various rms short-circuit currents and clearing times are shown in Table 9.5.

9.1.4 Cable Specification

Cable specifications generally start with the conductor and progress rapidly through the insulation and coverings. The following items can serve as a checklist for preparing a cable specification:

1. Number of conductors in cable, and phase identification required.
2. Conductor size (AWG, kcmil) and material
3. Insulation type (rubber, PVC, polyethylene, EPR, etc.)
4. Voltage rating
5. Shielding system
6. Outer finishes
7. Installation (cable tray, direct burial, wet location, exposure to sunlight or oil, etc.)
8. Applicable UL listing
9. Test voltage

Table 9.3 Conductor Requirements (AWG or kcmil) for Indicated Loadings: Below-Ground Ratings, 1–15 kV, Conductor Temperature 75°C, Earth Ambient Temperature 20°C; Three Single Conductor Cables per Duct, Rubber or Thermoplastic Insulated, Underground, Earth Resistivity RHO-90

Load Current (amperes/conductor)	1 Loaded Duct				3 Loaded Ducts				6 Loaded Ducts			
	75% Load Factor		100 % Load Factor		75% Load Factor		100% Load Factor		75% Load Factor		100% Load Factor	
	Cu	Al	Cu	Al	Cu	Al	Cu	Al	Cu	Al	Cu	Al
50	8	6	8	6	8	6	8	6	6	6	6	4
100	4	2	4	2	2	1	2	1	2	1/0	1	2/0
150	1	2/0	1	2/0	1/0	3/0	2/0	4/0	2/0	4/0	3/0	250
200	2/0	4/0	2/0	4/0	3/0	250	4/0	350	4/0	350	350	500
250	4/0	350	4/0	350	250	350	350	500	350	750	500	1000
300	250	350	350	500	350	500	500	750	500	750	750	1000
350	350	500	350	750	500	750	750	1000	750	1000	1000	
400	500	750	500	750	750	1000	750		1000			
450	500	750	750	1000	750	1000	1000					
500	750	1000	750	1000	1000							

From IPCEA S-135-1-1962 and S-135-2-1962.

NOTE: All circuits assumed equally loaded. Refer other cases to cable manufacturer.

Table 9.4 Uprating for Short-Time Overloads

Insulation Type	Voltage Class (kV)	Conductor Operating Temperature (°C)	Conductor Overload Temperature (°C)	Uprating Factors for Ambient Temperature 20°C Cu	20°C Al	30°C Cu	30°C Al	40°C Cu	40°C Al	50°C Cu	50°C Al
Paper (solid type)	9	95	115	1.09	1.09	1.11	1.11	1.13	1.13	1.17	1.17
	29	90	110	1.10	1.10	1.12	1.12	1.15	1.15	1.19	1.19
	49	80	100	1.12	1.12	1.15	1.15	1.19	1.19	1.25	1.25
	69	65	80	1.13	1.13	1.17	1.17	1.23	1.23	1.38	1.38
Varnished cambric	5	85	100	1.09	1.08	1.10	1.10	1.13	1.13	1.17	1.17
	15	77	85	1.05	1.05	1.07	1.07	1.09	1.09	1.13	1.13
	28	70	72								
Polyethylene (natural)†	35	75	95	1.13	1.13	1.17	1.17	1.22	1.22	1.30	1.30
SBR rubber	0.6	75	95	1.13	1.13	1.17	1.17	1.22	1.22	1.30	1.30
	5	90	105	1.08	1.08	1.09	1.09	1.11	1.11	1.14	1.14
Butyl RHH	15	85	100	1.09	1.08	1.10	1.10	1.13	1.13	1.17	1.17
	35	80	95	1.09	1.09	1.11	1.11	1.14	1.14	1.20	1.20
Oil-base rubber	35	70	85	1.11	1.11	1.14	1.14	1.20	1.20	1.29	1.29
Polyethylene (cross-linked)†	35	90	130	1.18	1.18	1.22	1.22	1.26	1.26	1.33	1.33
Silicone rubber	5	125	150	1.08	1.08	1.09	1.09	1.10	1.10	1.12	1.11
EPR rubber†	35	90	130	1.18	1.18	1.22	1.22	1.26	1.26	1.33	1.33
Chlorosulfonated polyethylene‡	0.6	75	95	1.13	1.13	1.17	1.17	1.22	1.22	1.30	1.30
Polyvinyl chloride	0.6	60	85	1.22	1.22	1.30	1.30	1.44	1.44	1.80	1.79
	0.6	75	95	1.13	1.13	1.17	1.17	1.22	1.22	1.30	1.30

*To be applied to normal rating determined for such installation conditions.
†Cables are available in 69 kV and higher ratings.
‡For example, Hypalon.

Table 9.5 Minimum Conductor Sizes (AWG or kcmil) for Indicated Fault Current and Clearing Times

Minimum Conductor Sizes, in AWG or kcmil, for Indicated Fault Current and Clearing Times

Total RMS Current (amperes)	Polyethylene and Polyvinyl Chloride, 75–150°C				Oil Base and SBR, 75–200°C				Cross-Linked Polyethylene and EPR, 90–250°C			
	1/2 Cycle (0.0083 s)		10 Cycles (0.166 s)		1/2 Cycle (0.0083 s)		10 Cycles (0.166 s)		1/2 Cycle (0.0083 s)		10 Cycles (0.166 s)	
	Cu	Al	Cu	Al	Cu	Al	Cu	Al	Cu	Al	Cu	Al
5000	10	8	4	2	10	8	4	3	12	10	4	3
15 000	6	4	2/0	4/0	6	4	1/0	3/0	6	4	1	3/0
25 000	3	2	4/0	350	4	2	3/0	250	4	3	3/0	250
50 000	1/0	2/0	400	700	1	2/0	350	500	2	1/0	300	500
75 000	2/0	4/0	600	1000	1/0	3/0	500	750	1/0	3/0	500	700
100 000	4/0	300	800	1250	3/0	250	700	1000	2/0	4/0	600	1000

In addition, the total number of lineal feet of conductors required, the quantity desired shipped in one length, the pulling eyes, and whether it is desired to have several single-conductor cables paralleled on a reel should also be specified.

9.1.5 Installation

There are many different ways to install power distribution cables in industrial plants. Design engineers must select the method most suitable for each application. Each method will transmit power with a unique degree of reliability, safety, economy, and quality for any specific conditions. These conditions include the quantity and characteristics of the power being transmitted, the distance of transmission, and the degree of exposure to adverse mechanical and environmental conditions. The following represent the most common modes.

Open Wire

The open wire method was used extensively in the past. It has been replaced in most applications, but it is still used for primary power distribution over large areas when conditions are suitable. Open wire construction consists of uninsulated conductors on insulators mounted on poles or structures. The attractive features of this method are its low initial cost and easy access for repair work. On the other hand, the uninsulated conductors are a safe hazard and highly susceptible to damage and lightning.

Aerial Cable

Aerial cable is usually limited to incoming service or to distribution between buildings in commercial areas. The greatest usage is in replacing open wiring where it provides greater safety and reliability and requires less space. Aerial cables may be either self-supporting or messenger supported. They may be attached to pole lines or structures. Self-supporting cable is suitable for only relatively short spans. Messenger-supported cable can span large distances. The supporting messenger provides high strength to withstand climatic rigors or mechanical shock.

Direct Attachment

Direct attachment is a low-cost method where adequate support surfaces are available between the source and the load. Its use in commercial buildings is usually limited to low-energy control and telephone circuits. This method employs multiconductor cable attached to surfaces such as structural beams and columns. A cable with metallic covering should be used where exposed to adverse conditions.

For architectural reasons, it is usually limited to service areas, hung ceilings, and electric shafts.

Cable Trays

A continuous rigid cable support is a unit or assembly of units or sections and associated fittings made of metal or other noncombustible material forming a continuous rigid structure used to support cables. These are commonly called cable trays, including ladders, troughs, and channels. They are becoming increasingly popular in commercial and industrial electrical systems because of low installed cost, flexibility, accessibility for repair, or addition of cables, and space saving. Continuous rigid cable supports are available in a number of types and materials. They are a support and not an enclosure or raceway, and therefore may only be used with wiring methods that meet the approval of the NEC and local codes. Covers, either ventilated or nonventilated, may be used to provide additional mechanical protection or electrical shielding for communication circuits. Initial planning of this method of wiring should consider occupancy requirements per NEC and additional space for future expansion. Figure 9.4 shows two typical 12-inch-wide cable trays used to carry 208/120-V feeder cables from a unit substation circuit breaker cubicle to an overhead bus duct.

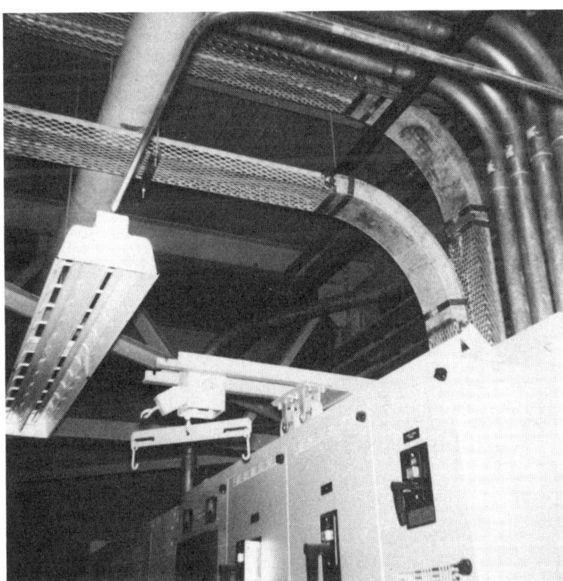

Figure 9.4 Cable tray installed in an industrial plant.

Raceways

Raceway is applied to all types of enclosures or housing providing space, support, and mechanical protection for electric conductors distributing power or control between various units of electric apparatus and equipment. An equally important function of a raceway is to protect life and property from hazards during normal and abnormal conditions. Raceway systems may consist of conduits, EMT, underfloor and cellular floor raceways, wireways, surface metal raceways, and busways. Design engineers should consider advantages and shortcomings of each type and make sure that basic requirements of safety and protection are met. More detailed discussions of some of the raceway systems follow.

Conduit Systems. These are the most common and the most varied of all raceways. Metal conduits are available in galvanized and enameled steel, aluminum, wrought iron, and silicone bronze, and in two wall thicknesses: heavy wall (rigid) and thin wall (EMT). Flexible and liquid-tight metal conduits are also specified for limited applications. Nonmetallic conduits, which may be of plastic, vitreous, fiber, or concrete construction, are generally used underground either directly in the soil or encased in a surrounding envelope of concrete.

Steel or aluminum rigid conduits that can provide maximum mechanical protection for the conductors and also permit wire replacements are commonly installed underground, indoors, or outside in wet or dry locations. Plastic or plastic-coated metallic conduit may be used in extremely corrosive areas. Silicone-bronze conduit, because of its cost, is used only in the most corrosive environments.

Metal conduit raceway systems supply the continuous electrical conductivity, low impedance, and low resistance necessary for grounding of fault currents and to ensure operation of protective relays and circuit interrupters. Rigid conduit sizes range from 1/2 to 6 in., with corresponding fittings, couplings, elbows, nipples, and box openings. Tables for selection of conduit sizes and loading are found in the NEC and many reference books. The NEC specifies, in terms of wire size and percent area fill, the allowable number of conductors per raceway. Flexibility can be designed into raceway systems by allowing spare and oversize conduits for future use. Figure 9.5 shows three 4-in. underground conduits, each carrying 3-1/C 15-kV feeder cables with four 4-in. spare conduits in a plant cable vault.

Electrical Metallic Tubing. When used within the limitations of the NEC, it provides a low-cost raceway for distribution systems below 600 V. However, it lacks the ground-return path and is less adequate than rigid conduits where exposed to hazardous or corrosive

Figure 9.5 Underground conduit systems with cable risers.

surroundings. Nevertheless, it is adequate for many building applications involving branch and feeder circuits.

Flexible Metal Conduits. These are generally used for connection to motors, fans, pumps, and similar equipment where vibration may be present. Flexible conduit may have an inadequate ground return path for power installation. Therefore, a separate internal ground wire is needed to meet NEC requirements.

Underfloor Raceways. These include a wide variety of components to provide flexibility and to meet most requirements of building occupants. Metal ducts, junction boxes, and outlets are electrically continuous and can readily be bonded to structural members.

Wireways. Wireways and auxiliary sheet-metal gutters differ from other raceway systems in that their application is limited to exposed, dry, nonhazardous locations. Manufacturers of wireway systems offer complete lines of sizes and components necessary for the varying conditions found in new or existing buildings. Surface raceways are limited to a voltage of 300 V or less between conductors, unless made of heavier metal sections to improve safety features. The added convenience of multioutlet assemblies in work areas is a major attraction for this type of raceway.

Cable Bus

Cable bus is used to transmit large amounts of power over relatively short distances. It employs insulated conductors supported at maintained spacings by some form of nonmetallic spacer blocks. Cable bus is furnished either as components to field assembly or as completely assembled sections. Cable bus should be installed only for exposed work.

Underground Ducts

These are used when overhead conduits are undesirable for physical or aesthetic reasons, or where the advantages of an underground system are obvious. Underground ducts use rigid steel, plastic, or fiber conduits either encased in concrete or directly buried.

Direct Burial

Cables may be buried directly in the ground where permitted by code when the need for future maintenance along the cable run is not anticipated. The cable used must be suitable for this purpose. The current-carrying capacity may be greater than that of cable in ducts. However, its use might be limited to incoming services for commercial buildings, to circuits between buildings, or to circuits for remote areas.

Hazardous Locations

Wire and cable installed in locations where fire or explosion hazards may exist must comply with Articles 500 to 517 of the NEC. The authorized wiring methods dependent on the class and division of the specified area (see Table 9.6). The wiring method must be approved for the class and division, but is not dependent on the group, which defines the hazardous substance.

Equipment and associated wiring approved as intrinsically safe may be installed in any hazardous location for which it is approved, and the provisions of Article 500-517 of the NEC need not apply to such installation. However, the installation must prevent the passage of gases or vapors from one area to another. Seals must be provided in the wiring system to prevent the passage of the hazardous atmosphere along the wiring system from one division to the other or from a division to a nonhazardous location. The sealing requirements are defined in Articles 501 to 503 of the NEC.

Table 9.6 Wiring Methods for Hazardous Locations

Wiring Method	Class I Division 1	Class I Division 2	Class II Division 1	Class II Division 2	Class III Division 1 or 2
Threaded rigid metal conduit	X	X	X	X	X
Threaded steel intermediate metal conduit	X	X	X	X	X
Rigid metal conduit				X	X
Intermediate metal conduit				X	X
Electrical metallic tubing				X	X
Rigid nonmetallic conduit					X
Type MI mineral insulated cable	X	X	X	X	X
Type MC metal-clad cable		X		X	X
Type SNM shielded nonmetallic cable		X		X	X
Type MV medium-voltage cable		X			
Type TC power and control tray cable		X			
Type PLTC power-limited tray cable		X			
Enclosed gasketed busways or wireways		X			
Dusttight wireways				X	X

9.1.6 Cable Testing

Cable Testing—For and Against

Testing, particularly of elastomeric and plastic insulations, is a useful method of checking the ability of a reasonable future period. The failure to pass a test will cause in-test breakdown of the cable or indicate the need for its immediate replacement. Whether or not to test cables routinely is a decision for each user to make. Following are the arguments for testing:

1. If testing is done properly and good records are kept, bad or marginal cable can be identified and replacement can be planned.
2. Plants with radial feeders can be reasonably certain of having reliable feeder cables, thus eliminating the stock for spare cables.
3. Cable can be replaced under most favorable conditions.

Arguments against testing are as follows:

1. Bad or marginal cable might be usable for sometime. Testing might result in the cable's failure.
2. When there are dual feeders, downtime is incurred because cable failure will be minimal.
3. Much record keeping and work scheduling can be expensive.
4. Testing may not be useful in detecting possible failure from moisture-induced tracking across termination surfaces since this develops primarily during periods of percipitation, condensation, or leakage failure of the enclosure or housing.

Cable Testing—AC or DC

Cable insulation can, without damage, sustain application of dc potential equal to the system basic impulse insulation level for very long periods. In contrast, most cable insulations will sustain degradation from an overpotential proportional to a high power of overvoltage to time of application. Hence it is preferred to utilize dc for any testing that will be repetitive. The manufacturers use ac for an original factory test.

Factory Tests and Field Tests

Factory Tests. All cables are tested by the manufacturer before shipment, normally with ac for a 5-minute period. Unshielded cable is immersed in water (ground) for this test; shielded cable is tested using the shield as the ground return. Test voltages are specified by the manufacturer, by the applicable specification of the ICEA, or by other specifications, such as those of the Association of Edison Illuminating Companies (AEIC). In addition, a test may be made using dc of two to three times the rms value used in the ac test. On cables rated 3000 V and above, corona tests may also be made.

Field Tests. Test voltages and intervals require coordination to attain suitable performance. Tables 9.7 and 9.8 show dc test voltages for pre-1968 and for 1968 and later cables, respectively, as recommended by the ICEA. The AEIC has specified test values for 1968 and later cables that are approximately 20% higher than the ICEA values. The ANSI/IEEE Std 400-1980 specifies much higher voltages than either ICEA or AEIC, which is shown in Table 9.9. The intention here is to reduce cable failures during operation by overstressing the cables during shutdown testing and causing weak cables to fail at that time.

Table 9.7 ICEA Specified DC Cable Test Voltages (Pre 1968 Cable)

Insulation Type	Grounding	Maintenance Test Rated Cable Voltage 5 kV	15 kV	25 kV	35 kV
Elastomeric: butyl, oil base, EPR	Grounded	27	47	—	—
	Ungrounded	—	67	—	—
Polyethylene, including cross-linked polyethylene	Grounded	22	40	67	88
	Ungrounded	—	52	—	—

Table 9.8 ICEA Specified DC Cable Test Voltages (1968 and Later Cable)*

Insulation Type	Insulation Level (%)	Rated Cable Voltage 5 kV 1	5 kV 2	15 kV 1	15 kV 2	25 kV 1	25 kV 2	35 kV 1	35 kV 2
Elastomeric: butyl and oil base	100	25	19	55	41	80	60	—	—
	133	25	19	65	49	—	—	—	—
Elastomeric: EPR	100	25	19	55	41	80	60	100	75
	133	25	19	65	49	100	75	—	—
Polyethylene, including cross-linked polyethylene	100	25	19	55	41	80	60	100	75
	133	25	19	65	49	100	75	—	—

NOTE: Columns 1 — Installation tests, made after installation, before service; columns 2 — maintenance tests, made after cable has been in service.

* These test values are lower than for pre-1968 cables because the insulation is thinner. Hence the ac test voltage is lower. The dc test voltage is specified as three times the ac test voltage, so it is also lower than for older cables.

Table 9.9 IEEE 400-1980 Specified DC Cable Test Voltages

L-L System Voltage (kV)	BIL (kV)	Test Voltage (kV) 100% Insulation Level	133% Insulation Level
2.5	60	40	50
5	75	50	65
8.7	95	65	85
15	110	75	100
23	150	105	140
28	170	120	
34.5	200	140	

NOTE: These test voltages should not be used without the cable manufacturers' concurrence as the cable warranty will be voided.

Testing Procedures

Cables to be tested should have their ends free of equipment and clear from ground. All conductors not under test should be grounded. In field testing, the leakage current of the cable system should be watched closely and recorded for signs of approaching failure. The test voltage may be raised continuously and slowly from zero to the maximum value, or it may be raised in steps, pausing for 1 minute or more at each step. Potential differences between steps are of the order of the ac rms rated voltage of the cable. As the voltage is raised, current will flow at a relatively high rate to charge the capacitance, to supply the dielectric absorption characteristics of the cable, and to supply the leakage current. The capacitance charging current subsides within a second or so, the absorption current subsides slowly and would continue to decrease for 10 minutes or so, finally leaving only the leakage current flowing.

At each step where the calculated leakage resistance decreased markedly (say to 50% of that of the next-lower voltage level), the cable could be near failure. The test should be discontinued should the engineer desire to retain the cable in service until a replacement can be arranged. Normally, the ratio of current after 1 minute to the current after 5 minutes of maximum voltage on good cable will be between 1.25 and 2. Anything less than 1.0 is to be considered a failure.

After completion of the 5-minute maximum test voltage step, the supply voltage control dial should be returned to zero and the charge in the cable allowed to drain off through leakage of the test set and voltmeter circuits. After the remaining potential drops below 10% of the original value, the cable conductor may be solidly grounded. All conductors should be left grounded when not on test, during the testing of other conductors, and for at least 30 minutes after the removal of a dc test potential. Before proceeding with the test, the cable should be disconnected from its attached switching equipment; in particular, lightning arresters, potential transformers, and capacitors must be disconnected. Otherwise, only lower test potentials are to be used.

9.1.7 Locating Cable Faults—Equipment and Methods

In an industrial plant, a wide variety of cable faults can occur. It may be in a communication circuit or in a power circuit, in either the low-, medium-, or high-voltage class. Regardless of the class of equipment involved or the type of fault, one common problem is to find the location of the fault as fast as possible so that repairs can be made to avoid prolonged loss of production. A wide variety of commercially available equipment and a number of different approaches can be used to locate cable faults. The approach used depends on many factors:

1. Nature of the fault
2. Type and voltage rating of cable installation
3. Value of rapid location of faults
4. Frequency of faults
5. Experience and capability of available personnel

The following are the most commonly used methods for locating a cable fault:

1. *Physical evidence.* Observation of a flash, sound, or smoke accompanying the discharge of current through the faulted insulation will help to locate a fault. This is more probable with an overhead circuit than with underground ducts.

2. *Conductor resistance measurement.* This method consists of measuring the resistance of the conductor from the test location to the point of fault by using either the Varley loop or the Murray loop test (Figure 9.6). Once the resistance of the conductor to the point

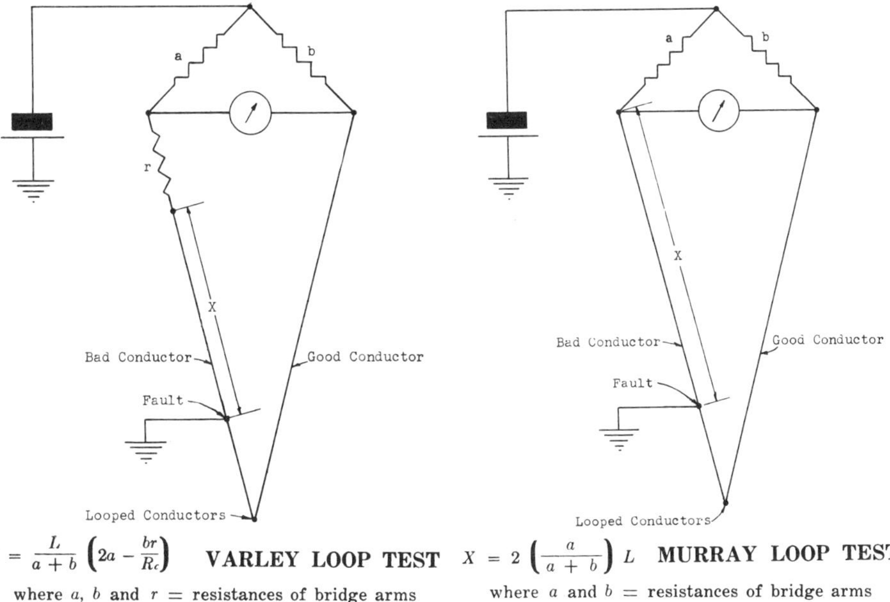

$$X = \frac{L}{a+b}\left(2a - \frac{br}{R_c}\right)$$ **VARLEY LOOP TEST**

where a, b and r = resistances of bridge arms
L = length of cable in feet
X = distance to fault in feet
R_c = resistance of good conductor

$$X = 2\left(\frac{a}{a+b}\right)L$$ **MURRAY LOOP TEST**

where a and b = resistances of bridge arms
L = length of cable in feet
X = distance to fault in feet

Figure 9.6 Varley loop and Murray loop test circuit.

of fault has been measured, it can be translated into distance by using handbook values of resistance per unit length of the size and type of conductor involved, with temperature correction if required. For distribution systems using cables insulated with organic materials, faults of relatively low resistance are normally encountered. The conductor resistance measurement method is most applicable to such systems.

3. *Megohmmeter test.* When the fault resistance is sufficiently low that it can reliably be detected with a megohmmeter, the cable can be sectionalized and each section tested to determine which contains the fault. But this method may involve considerable time and expense and might result in additional splices.

4. *Capacitor discharge.* This method consists of applying a high-voltage high-current impulse to the faulted cable. A high-voltage capacitor is charged by a source of relatively low current capacity, such as that used for high-potential testing. The capacitor is then discharged across an air gap into the cable. Repeated discharging of the capacitor provides periodic pulsing of the faulted cable. Where the cable is accessible, the fault may be located simply by sound. Where the cable is not accessible, such as in duct, the discharge at the fault may not be audible. In such cases, detectors will be required. Detectors may be of the magnetic or the acoustic type. The impulse method appears to be most practical and most commonly used where faults of relatively high resistance are anticipated.

5. *Tone signal.* A fixed-frequency signal, generally in the audio-frequency range, is imposed on the faulted cable. The cable route is then traced by means of a detector, which consists of a pickup coil, receiver, and head set or visual display, to the point where the signal leaves the conductor and enters the ground-return path. This type of equipment may be used on energized ungrounded circuits, usually in the low-voltage field (below 600 V).

Selection Guide

The discussions above cover some of the methods available to the industrial plant operators for locating cable faults. Some require no equipment, others require special equipment and experienced operators. To determine which approach is most practical for a particular plant, the size of the plant and the amount of circuit redundancy should be considered. Equipment that requires considerable experience and operator interpretation for results may be satisfactory for a plant with frequent cable faults, but ineffective for another plant. The final decision as to which method to use also depend on an evaluation of the particular circumstances of the plant in question.

9.2 CONNECTORS AND TERMINATIONS

9.2.1 Connectors

Types of Connectors

Connectors are generally classified in two types: thermal and pressure. Thermal connectors include those involving the application of heat to make soldered, silver soldered, brazed, welded, or cast-on terminals. Soldered connections have been used with copper conductors for many years. However, soldered joints are not commonly used with aluminum. Shielded arc welding of aluminum terminals to aluminum cable makes a satisfactory termination. Torch brazing and silver soldering of copper cable connections is used for underground connections with bare conductors in grounding mats. Thermite welding is also used to connect bare cable for ground mats.

Pressure connectors are available in two different types. The mechanical type may be defined as one in which the pressure to attach the connector to the conductor is by integral screw, cone, or other mechanical parts. The bolt diameter and number of bolts are selected to produce the clamping and contact pressures required for the most economical design. The compression type is one in which the pressure to attach the connector to the conductor is applied externally, changing the size and shape of the connector. It is basically a tube with wall thickness designed to carry the current and to withstand insulation stresses. A joint is made by compressing the conductor to the tube into another shape by means of a specially designed die and tool. The final shape may be indented, cup, hexagon, circular, or oval.

Connectors for Aluminum

While an aluminum conductor performs the same function as copper, it cannot be handled in the same manner. Failure to understand the special properties of aluminum conductors can lead to weak connections, faulty terminations, and even destructive fires. Preparation of the conductor surface is a most important step in proper installation. When raw aluminum is exposed to air, an oxide film forms on the conductor. This high-resistance surface film must be removed to ensure a high-quality, low-resistance connection. It can be done in two ways: the best way to remove an aluminum oxide film mechanically is to abrade the outer strands of the conductor with a wire brush. Brushing should be done through a paste joint compound to prevent reoxidation. It is imperative that only connectors compatible with aluminum conductor be used.

The tendency for aluminum to flow or creep under high unit pressure, and the fact that it expands at a greater rate than copper when heated, created some problems in the early days. Installation procedures today, with all necessary precautions and the special connectors and tools used, has proved to be free of early problems. The most satisfactory connectors are specifically designed for aluminum conductors to prevent any possible trouble from creep, the presence of oxide film, and differences in the coefficients of expansion between aluminum and other metals. Termination of a cable is not complete until the connector is bolted to the bus bar or the terminal plate of the equipment. For heavy-duty service, where the temperature rise will exceed 30°C, it is good practice to use a Belleville, or compression spring washer. A properly designed spring washer will perform two functions. It eliminated the need for using a torque wrench, and will follow creep or flow of dissimilar materials.

Aluminum building wire can be used safely in manufacturing plants under heavy loads and frequent on-off cycling if proper installation procedures are followed:

1. When aluminum wires of No. 8 AWG or larger are to be terminated, spliced, or taped, including connections to panelboards, circuit breakers, and related equipment, the connectors should be of tool-applied compression type. Connectors should be made of aluminum with wire barrels prefilled with oxide-inhibiting compound.
2. Installing tools and dies of the hexagonal or circumferential type, made by the connector manufacturer, should be used for the installation.
3. Terminal lugs with bolting pads should be tin-plated for low contact resistance.
4. Connectors must meet the performance requirements of UL Bulletin 486 (ANSI/UL 486B-1982), except heating or current cycling tests is to be for a minimum of 500 cycles.
5. Compression washers must be used where a temperature rise will exceed 30°C.

Connectors for Various Voltage Cables

Standard mechanical or compression connectors are recommended for all primary voltages provided that the bus is uninsulated. Welded connectors may also be used for conductors sized in circular mils. Up to 600 V, standard connector designs present no problem for insulated or uninsulated conductors. Standard compression connectors are recommended for use on insulated conductors up to 5 kV. Above 5 kV, stress considerations make it desirable to use tapered-end compression connectors or semiconducting tape construction to give the same effect.

9.2.2 Terminations

Function of Terminations

A termination for an insulated power cable must provide certain basic electrical and mechanical functions, which are:

1. To connect the insulated cable conductor to provide a current path
2. To physically protect and support the end of the cable conductor, insulation, shielding system, and overall jacket, sheath, or armor of the cable
3. To effectively control electrical gradients to provide both internal and external dielectric strength to meet the desired insulation level for the cable system

The importance of proper termination is to provide a means of reducing and controlling the stresses within the working limits of the cable insulation and materials used to make up the terminating device itself.

Types of Terminations

In the broad sense of the word "termination," the following are the identifiable types:

1. Taped terminations
2. Armor terminations
3. Potheads
4. Preassembled terminations

This is only a partial list of the many types of terminations in use today, and it is generally agreed that variations within each group listed are impractical to define. Terminations for cables rated 600 V and less generally consist of a lug and tape. Tape is applied over the lower portion of the barrel of the lug and down onto the cable insulation. Terminal connections to buses inside metal-enclosed equipment are usually left untaped. For cables rated over 600 V, the termination requirements vary with the type of cable, its construction, voltage rating, and requirements for installation. For outdoor installation, the termination may be required to perform its intended function while partially or fully immersed in a liquid or gaseous dielectric, such as oil and nitrogen, in order to withstand exposure of the atmospheric contaminants.

The shield system confines the dielectric field to the cable insulation, resulting in a symmetrical radial stress distribution within the cable insulation. However, the unit stress within the wall of insulation is nonlinear, being greater near the conductor and progressively decreasing to lower values near the cable shielding system (see Figure 9.3). At the point of termination, the cutback of the cable shielding to provide necessary creepage distance between the two electrodes (conductor and shielding) introduces a longitudinal stress over the surface of the exposed cable insulation. The resultant combination of radial and longitudinal stress at the termination of cable ends results in the minimum dielectric strength of the cable system. The most common method of reducing these stresses is gradually to increase the total thickness of insulation at the termination by adding insulating tapes in the form of a cone. This construction is commonly referred to as a stress-relief cone and is illustrated in Figure 9.7. The following discussion pertains to several types of terminations in common use today for medium- and high-voltage cables.

Taped Terminations. Taped terminations may be used either indoors or out doors and on shielded and nonshielded cables. In general, it is used at 15 kV and below. Creepage of 1 in. per kilovolt of nominal system voltage is used for indoor applications and from 1 to 2 in. or more per kilovolt for outdoor installations. Additional creepage may be obtained by using a rain hood of neoprene, rubber, plastic, or porcelain for outdoor installations. On upright terminations, the rain hood is usually placed directly over the stress-relief cone, and its primary function is to keep some portion of the cable insulation along the creepage path dry at all times. Where extra rain hoods are needed, potheads or other types of terminating device should be used instead. Figure 9.8 shows a 15-kV taped termination.

Armor Terminations. Cables with a steel, aluminum, or copper metallic jacket of helical continuous weld or interlocking covering require in addition to a taped terminations, an arrangement to serve and ground the armor. Fittings for this purpose are called armor terminators. They are sized to fit the cable armor and designed for use on the cable alone, with brackets, or with locking nuts or adapters for application to other pieces of equipment.

Potheads. A pothead is a hermetically sealed device used to enclose and protect a cable end. It consists of a metallic body with one or more procelain insulators. The body is arranged to accept a variety of optional cable entrance sealing fittings, while the procelains, in turn, are designed to accommodate a number of optional cable conductors and aerial connections. The assembled unit is filled with an insulating compound, such as asphaltic-based materials, resins, and oils.

Potheads can be subdivided into a number of groups or types, related to cable system requirements and construction of the device. Most industrial power cable systems are of the nonpressurized type,

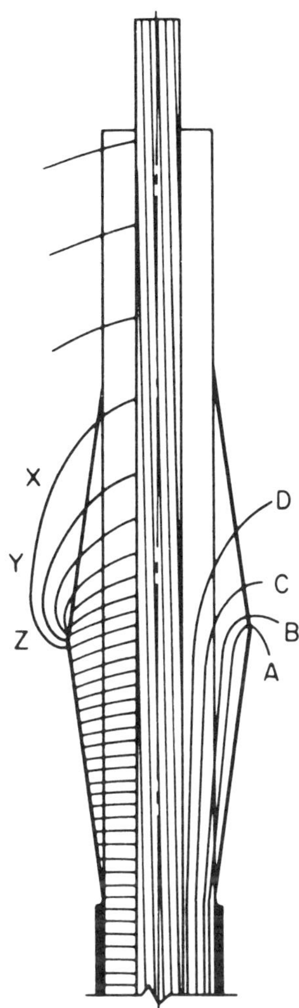

X, Y, Z Electric stress lines
A, B, C, D Equipotential lines

Figure 9.7 Stress-relief cone.

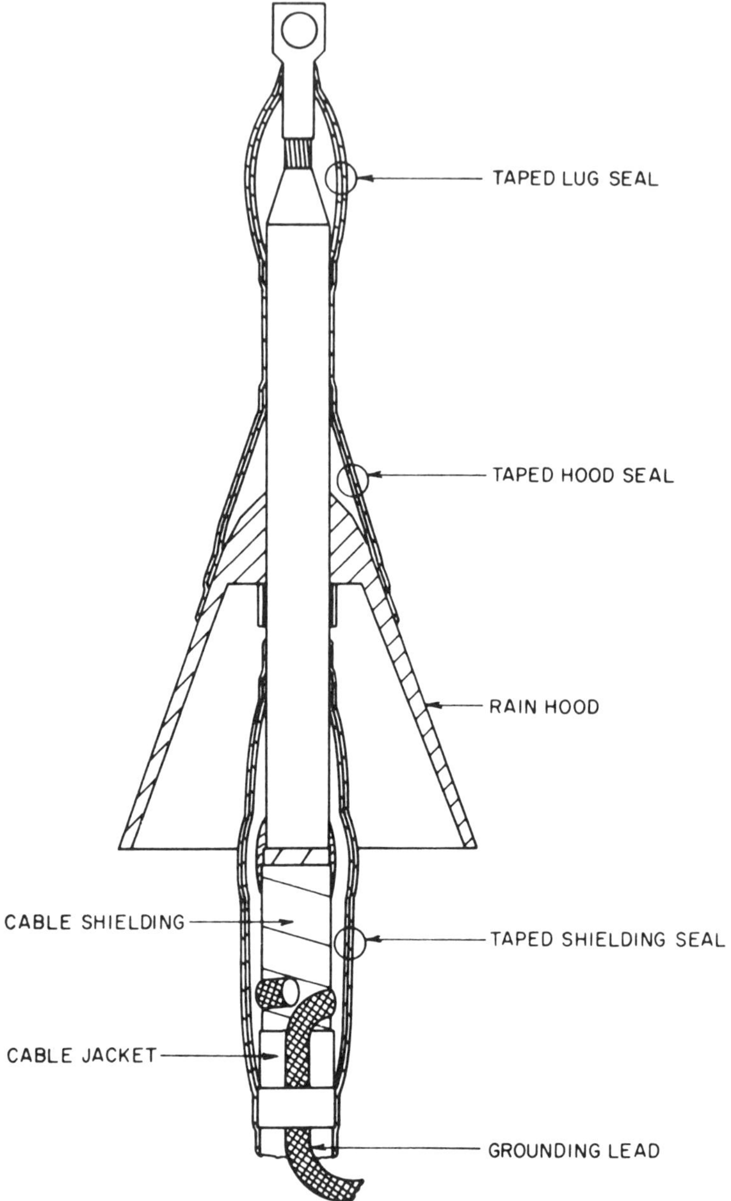

Figure 9.8 15 kV termination.

using solid dielectric-insulated cables. The two most commonly used pothead types are the capnut and the solder-seal. The capnut pothead is made up of cast-metal parts with gasketed joints between metal parts and the procelains. The metal parts for solder-seal potheads are copper spinnings which are solder bonded to the procelain insulator. Figure 9.9 shows a typical three-conductor capnut pothead.

Preassembled Terminations. In recent years, two general types of preassembled terminals have become available. One employs the use of elastomeric materials direct to the cable end. The other type consists of a metal-porcelain housing filled with a gelatin-like substance designed to be partially displaced as the terminator is installed on the cable. They can be installed by a less skilled worker, yet offer a high degree of consistency to the overall quality. They are available in ratings of 15 kV and above for most types of application. When terminating shielded cables 15-kV units can be used on a 5-kV system. Figure 9.10 shows a preassembled terminator, out-door type, with track-resistant porcelain petticoats.

Dead-Front Assemblies. These are two-part devices used in conjunction with high-voltage electrical apparatus. A bushing assembly is attached to the high-voltage apparatus (transformer, switch, or fusing devices, etc.), and a molded plug-in connector is used to terminate the insulated cable and connect the cable system to the bushing. The dead-front feature is obtained by fully shielding the plug-in connector assembly. Two types of dead-front connectors for 15 and 25 kV are available: one load break and the other non-load break.

9.2.3 Splicing Devices and Techniques

In a splice, the highest stresses are around the conductor and connector area. Splicing design must recognize this fundamental concept and provide means to control these stresses to values within the working limits of the materials used to make up the splice. The connectors used to join the cable conductors together must be capable of carrying full rated load, emergency overload, and fault currents without overheating as well as being mechanically strong in order to prevent accidental conductor pull-out or separation. The splice housing or protective cover should provide adequate protection to the splice.

Types of Splices

600 V and Below. Insulated connectors are used where several relatively large cables must be joined together. These terminators, called moles or crabs, are basically insulated buses with provision for making a number of tap connections that can easily be taped or

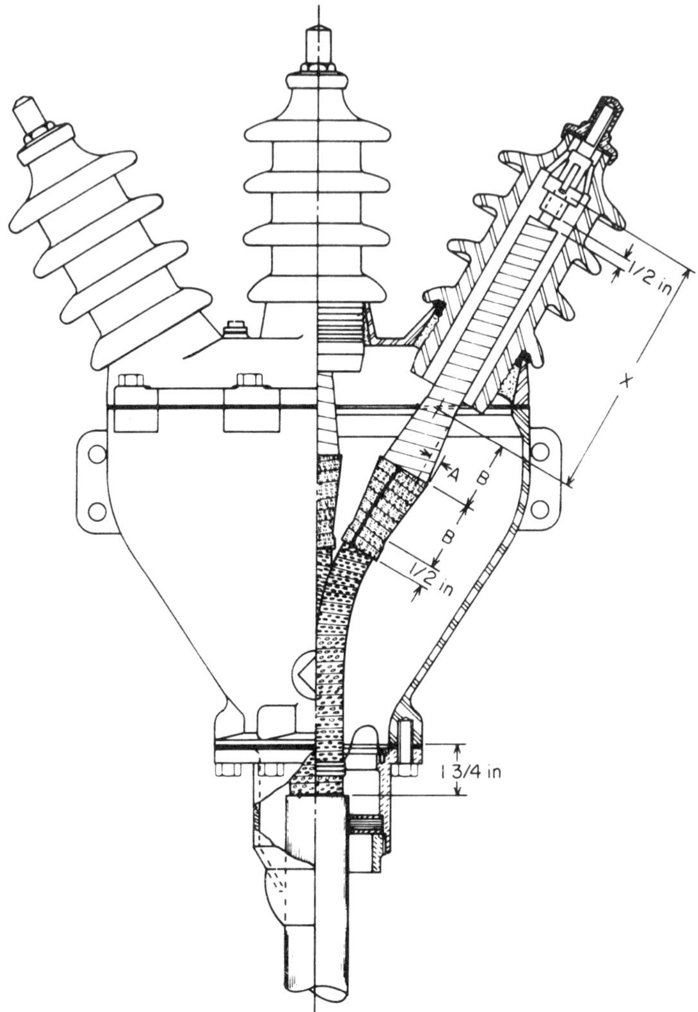

Figure 9.9 Typical 3 conductor capnut type pothead.

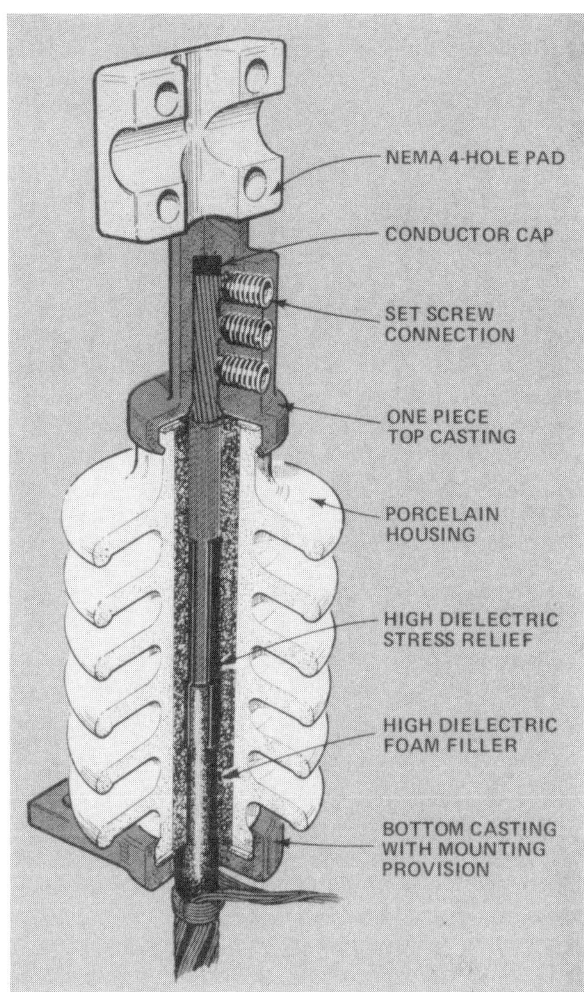

Figure 9.10 Preassembled terminations—outdoor type with track-resistant porcelain petticoats.

covered with an insulating sleeve. One widely used type is a pre-insulated multiple outlet joint in which the cable connections are made by standard compression tooling, which indents the conductor to the tubular cable sockets.

Over 600 V. Splicing of unshielded cable consists of assembling a connector, usually soldered or pressed onto the cable conductors, and applying insulating tapes to build up an insulation wall to a thickness of 1½ to 2 times that of the factory-applied insulation on the cable. Splices or solid dielectric cables are made with uncured tapes, which will fuse together after application and provide waterproof assembly.

Taped Splices. For shielded cables, taped splices have been used successfully for many years. Basic considerations are the same as for unshielded cables. For application of the splice, consideration should be given to such details as providing a moisture seal, thermal stability of tapes, and ease of handling on wye- or tee-type splices. Cables with solid insulation are tapered and those with tape-type insulation are stepped to provide a general transition between conductor-connector diameter and cable insulation diameter prior to the application of insulating tapes. The splice should not be overinsulated since this could restrict heat dissipation at the splice area and risk splice failure. Figure 9.11 shows a typical taped splice in a shielded cable.

Preassembled Splices. There are several types of factory-made splices. The most elementary is an elastomeric unit consisting of a molded housing sized to fit the cable involved, a connector for joining the conductors, and tape seals for sealing the ends of the molded housing to the cable jacket. They are available in two-way, three-way tee type, and multiple configurations for application up to 35 kV. The preassembled splice provides a waterproof seal to the cable jacket and is suitable for submersible, direct-burial, and other applications where the splice housing must provide protection for the splice to the same degree that the cable jacket provides protection to the cable insulation and shielding system. The advantage of these preassembled splices is the reduction in time needed to complete the splice after cable end preparation.

9.3 BUSWAY

Busway was introduced in the late 1920s. It has grown to become an integral part of the low-voltage distribution system for industrial plants at 600 V and below. Busways are particularly convenient to use when numerous current taps are to be made. Plugs with circuit

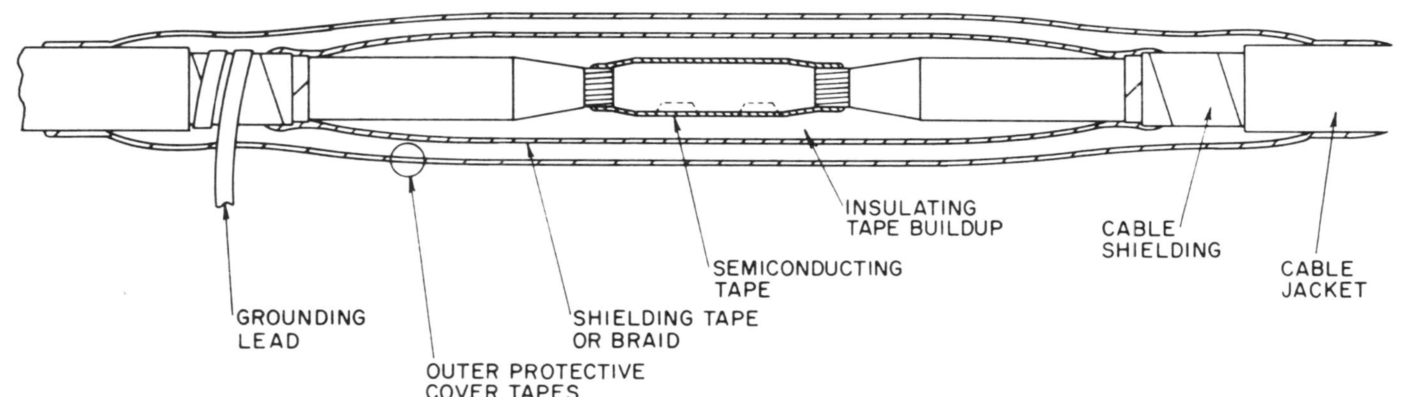

Figure 9.11 Typical taped splice on shielded cable.

breakers or fusible switches can be installed and wired without de-energizing the busway. Busways may be dismantled and reinstalled in whole or part to accommodate changes in the distribution system layout.

9.3.1 Busway Construction and Standards

Originally, a busway consisted of bare copper conductors supported on insulators mounted within a nonventilated steel housing. This type of construction was adequate for current ratings of 225 to 600 A. As the use of busway expanded, with increased loads requiring higher current ratings, the housing was ventilated to provide better cooling at higher capacities. The bus bars were covered with insulation to permit closer spacing of bars of opposite polarity to achieve lower reactance and voltage drop. Busways are available with either copper or aluminum conductors. For equal current-carrying ability, aluminum is lighter in weight and less costly. Busway is usually made in 10-ft sections. Feed and tap fittings to other electric equipment, such as switchboards, transformers, motor control centers, and so on, are also essential components of the busway system. Busways are designed to conform to the following standards:

1. NEC Article 364
2. ANSI/UL 857-1981
3. NEMA BU1-1983

Standards 2 and 3 are primarily manufacturing and testing standards. The NEMA standard is generally an extension of the UL standard to areas that UL does not cover. The NEC is the most important standard for busway installation. State and local electrical codes may have specific requirements over and above ANSI/UL 857-1981 and NEC. Appropriate code authorities and manufacturers should be contacted to ensure that requirements are met.

9.3.2 Types of Busway

Busways are available in the following general types: (1) feeder busway for low-impedance distribution of power; (2) plug-in busway for convenient connection or rearrangement of loads; (3) lighting busway to provide power and mechanical support to fluorescent, high-intensity discharge, and incandescent fixtures; and, (4) trolley busway for mobile power tap-offs to electric hoists, cranes, portable tools, and so on. More detailed discussion of each of these above types follows.

Feeder Busway

A feeder busway is used advantageously to distribute a large amount of power by virtue of its low impedance. Industrial plants use feeder busway from the service equipment to supply large loads directly and to supply smaller current ratings of feeder and plug-in busway, which in turn supply loads through plug-in units.

Current ratings: 600 to 5000 A, 600 V ac; one-phase and three-phase service with 50 to 100% neutral conductor
Short-circuit current ratings: 50,000 to 200,000 A, symmetrical rms
Voltage drop: 1 to 3 V per 100 ft, line to line
Construction: Indoor or weatherproof

Plug-in Busway

A plug-in busway is used in industrial plants as an overhead system to supply power to utilization equipment. Plug-in devices can be placed on the busway near the loads that they supply. Plug-in devices include fusible switches, circuit breakers, static voltage protectors, ground indicators, combination starters, lighting contactors, and capacitor plugs.

Current ratings: 100 to 4000 A; neutral bar 25 to 100% of the phase bar capacity
Short-circuit current ratings: 15,000 to 150,000 A, symmetrical rms
Voltage drop: 1 to 3 V per 100 ft, line to line, for evenly distributed loading

Figure 9.12 shows the installation of a straight-run plug-in busway with individual plug-in circuit breakers in position for power takeoffs.

Lighting Busway

A lighting busway provides power to lighting fixtures and also serves as mechanical support for the fixture. Auxiliary supporting means (strength beams) are available at maximum intervals of 16 ft. Fluorescent fixtures may be suspended from the busway. Lighting busway may be recessed in or surface mounted to a dropped ceiling.

Current rating: Maximum 60 A, 300 V to ground, two, three, or four conductors, 208/120 V or 480/277 V,

Trolley Busway

A trolley busway is constructed to receive stationary or movable take-off devices. It is used to supply power to a motor or a portable tool

Figure 9.12 Plug-in busway system.

moving with a production line, or where operators move back and forth over 10 to 20 ft to perform their specific operations.

9.3.3 Applications

The following important considerations should be made in order to carry out proper application of busways:

Current-Carrying Capacity

Busway should be rated on a temperature rise basis to provide safe operation, long life, and reliable service. The UL requirement for temperature rise (55°C) should be used to specify the maximum temperature rise permitted. Larger cross-sectional areas can be used to provide lower voltage drop and temperature rise.

Short-Circuit Current Ratings

The short-circuit current can generate considerable electromagnetic force. Short-circuit current ratings are generally assigned and tested in accordance with NEMA BU1-1983. They are dependent on many factors, such as bus bar centerline spacing, size, and strength of bus bars and mechanical supports. Since the ratings are different for each bus bar design, the manufacturer should be consulted for specific ratings. These ratings should include the ability of the ground-return path (housing and ground bar if provided) to carry the rated short-cricuit current.

Voltage Drop

Three-phase voltage drops may be determined with reasonable accuracy by the use of Tables 9.10 and 9.11. The voltage drops given in the tables are three-phase, line to line, per 100 ft, at rated load on a concentrated loading basis for feeder, plug-in, and trolley busways. Lighting busway values are single-phase, distributed loading. For other loading and distances, use the following formula:

Table 9.10 Voltage-Drop Values for Three-Phase Busways with Copper Bus Bars in Volts per 100 ft, Line to Line, at Rated Current with Entire Load at End*

Current Rating (amperes)	Load Power Factor (Percent, lagging)									
	20	30	40	50	60	70	80	90	95	100
Totally Enclosed Feeder Busway										
600	2.28	2.51	2.73	2.93	3.09	3.23	3.31	3.31	3.23	2.83
800	1.75	1.93	2.08	2.23	2.35	2.44	2.49	2.48	2.42	2.10
1000	1.51	1.81	2.11	2.39	2.66	2.92	3.15	3.33	3.39	3.29
1350	1.60	1.87	2.13	2.37	2.60	2.80	2.98	3.11	3.13	2.96
1600	1.90	2.10	2.27	2.43	2.56	2.67	2.73	2.72	2.66	2.31
2000	1.82	2.00	2.16	2.30	2.43	2.52	2.57	2.55	2.49	2.15
2500	1.75	1.91	2.06	2.18	2.29	2.36	2.40	2.37	2.30	1.96
3000	1.96	2.14	2.30	2.43	2.55	2.63	2.67	2.63	2.55	2.17
4000	1.84	2.01	2.16	2.29	2.40	2.49	2.53	2.49	2.42	2.07
5000	1.67	1.83	1.98	2.11	2.22	2.30	2.35	2.33	2.27	1.96
Totally Enclosed Plug-In Busway										
225	1.92	2.08	2.22	2.36	2.46	2.54	2.56	2.52	2.42	2.04
400	2.26	2.40	2.52	2.60	2.66	2.70	2.66	2.54	2.40	1.90
600	4.91	5.03	5.10	5.11	5.04	4.89	4.62	4.11	3.67	2.38
800	5.75	5.91	6.00	6.02	5.96	5.80	5.50	4.92	4.42	2.92
1000	4.77	4.91	4.98	5.02	4.98	4.84	4.60	4.12	3.70	2.46
1350	3.72	3.84	3.92	3.94	3.94	3.84	3.68	3.32	3.01	2.06
1600	3.58	3.70	3.78	3.82	3.80	3.72	3.54	3.22	2.92	2.00
2000	4.67	4.79	4.86	4.86	4.82	4.68	4.42	3.94	3.52	2.30
2500	4.08	4.20	4.26	4.30	4.26	4.14	3.94	3.54	3.18	2.12
3000	3.76	3.87	3.92	3.94	3.90	3.80	3.60	3.24	2.90	1.92
4000	4.64	4.74	4.80	4.79	4.73	4.57	4.30	3.81	3.38	2.15
5000	3.66	3.75	3.78	3.78	3.78	3.62	3.40	3.02	2.70	1.76
Lighting, Single Phase, Distributed Loading										
30	0.84	1.11	1.38	1.65	1.89	2.13	2.40	2.51	2.20	2.75
60	1.08	1.38	1.62	1.98	2.22	2.46	2.70	2.88	3.00	3.00
Trolley										
100	1.16	1.38	1.56	1.74	1.90	2.06	2.20	2.28	2.30	2.18

NOTE: Voltage-drop values are based on bus bar resistance at 75°C (room ambient temperature 25°C plus average conductor temperature at full load of 50°C rise).

*Divide values by 2 for distributed loading.

Table 9.11 Voltage-Drop Values for Three-Phase Busways with Aluminum Bus Bars, in Volts per 100 ft, Line to Line, at Rated Current with Entire Load at End*

Current Rating (amperes)	Load Power Factor (Percent, lagging) 20	30	40	50	60	70	80	90	95	100
Totally Enclosed Feeder Busway										
600	1.64	1.93	2.21	2.48	2.73	2.96	3.16	3.30	3.34	3.17
800	1.69	1.95	2.21	2.44	2.66	2.86	3.03	3.14	3.15	2.94
1000	1.51	1.81	2.11	2.39	2.66	2.92	3.15	3.33	3.39	3.29
1350	1.60	1.87	2.13	2.37	2.60	2.80	2.98	3.11	3.13	2.96
1600	1.70	1.97	2.22	2.45	2.67	2.87	3.04	3.14	3.15	2.94
2000	1.57	1.81	2.03	2.23	2.42	2.59	2.73	2.81	2.81	2.60
2500	1.56	1.78	1.98	2.18	2.35	2.51	2.63	2.70	2.69	2.48
3000	1.64	1.94	2.14	2.37	2.58	2.78	2.94	3.04	3.05	2.85
4000	1.60	1.83	2.04	2.24	2.42	2.59	2.71	2.79	2.78	2.56
Totally Enclosed Plug-In Busway										
100	2.05	2.63	3.20	3.76	4.30	4.83	5.33	5.79	5.98	6.01
225	1.94	2.22	2.49	2.73	2.96	3.15	3.31	3.41	3.40	3.13
400	3.47	3.66	3.81	3.92	3.99	3.99	3.92	3.69	3.45	2.64
600	4.62	4.89	5.12	5.30	5.41	5.45	5.37	5.10	4.80	3.76
800	4.09	4.34	4.54	4.70	4.81	4.84	4.78	4.54	4.28	3.36
1000	3.22	3.43	3.61	3.75	3.85	3.89	3.86	3.70	3.50	2.79
1350	2.92	3.10	3.12	3.36	3.44	3.48	3.44	3.28	3.08	2.44
1600	3.98	4.20	4.38	4.51	4.59	4.61	4.52	4.27	3.99	3.07
2000	3.48	3.68	3.85	3.99	4.07	4.09	4.04	3.83	3.60	2.81
2500	2.83	3.00	3.13	3.24	3.30	3.32	3.27	3.10	2.92	2.27
3000	3.68	3.85	3.99	4.09	4.14	4.12	4.01	3.74	3.47	2.60
4000	3.11	3.27	3.40	3.50	3.55	3.55	3.47	3.26	3.04	2.31

NOTE: Voltage-drop values are based on bus bar resistance at 75°C (room ambient temperature 25°C plus average conductor temperature at full load of 50°C rise).

*Divide values by 2 for distributed loading.

$$\text{VD} = \text{table VD} \times \frac{\text{actual load}}{\text{rated load}} \times \frac{\text{actual distance in ft}}{100 \text{ ft}}$$

The voltage drop for a single-phase load connected to a three-phase busway is 15.5% higher than the value shown in the tables. Table 9.12 shows typical values of resistance and reactance for both copper and aluminum busways. Resistance value is shown at normal room temperature (25°C). This is the value to be used in calculating the short-circuit current available in systems since short circuits can occur when busway is either at light load or initially energized. To calculate voltage drop when fully loaded (75°C), the resistance of copper or aluminum should be multiplied by 1.19.

Resistance Welding

Electrical resistance welding is a means of joining two or more metals together by the use of heat and pressure. The heat is generated by an electric current flowing across the intended joint location and

Table 9.12 Typical Values of Resistance and Reactance for Copper and Aluminum Busways

ALUMINUM CONDUCTORS

Rating	10^{-4} ohms Per Foot				Percent Power Factor
	Leg Resistance	Leg Reactance	Leg Impedance	Line-to-Line Impedance	
225A	46.15	18.2	49.6	85.95	93
400A	40.0	15.8	43.0	74.5	93
600A	30.1	12.8	32.7	56.5	92
800A	24.0	8.5	25.4	44.0	91
1000A	18.0	6.5	19.1	33.1	94
1200A	14.75	5.5	15.7	27.2	94
1350A	12.25	4.8	13.2	22.82	94
1600A	10.63	4.41	11.93	20.35	90
2000A	8.25	3.6	9.00	15.5	92
2500A	6.37	2.8	6.96	12.1	92
3000A	5.43	2.55	6.00	10.4	90
4000A	3.81	2.12	4.36	7.55	87

COPPER CONDUCTORS

Rating	10^{-4} ohms Per Foot				Percent Power Factor
	Leg Resistance	Leg Reactance	Leg Impedance	Line-to-Line Impedance	
225A	69.8	45.1	83.1	143.9	84
400A	35.2	22.7	41.9	72.6	84
600A	23.6	15.25	28.1	48.6	84
800A	17.75	12.03	21.4	37.1	83
1000A	13.65	8.64	16.17	28.0	85
1200A	9.75	7.35	12.21	21.35	80
1350A	9.25	7.10	11.68	20.25	80
1600A	8.07	6.37	10.28	17.8	79
2000A	7.10	5.50	8.96	15.55	79
2500A	5.14	3.70	6.32	10.95	81
3000A	4.18	2.87	5.07	8.79	81
4000A	3.00	2.15	3.64	6.32	81
5000A	2.35	1.96	3.06	5.30	77

the pressure is applied by the two electrodes that carry the current, using hydraulic, mechanical, or other means.

The busway distribution system for a resistance welder installation must meet two requirements: first, it must provide sufficient current-carying capacity to avoid overheating the busway; and second, it must not allow the permissible voltage drop to be exceeded. Due to the intermittent character of resistance welder loads, the voltage drop requirement is most difficult to meet. Both requirements must be carefully determined by separate calculations.

The operation of resistance welders may be considered as either constant or varying. Constant operation means that the actual primary current during weld and the duty cycle are known and do not vary. In varying operation the duty cycle and type and thickness of material being welded will not be constant; reasonable assumptions must be made for these varying quantities.

Current-Carrying Capacity. To determine the current carrying capacity required, it is necessary to convert the intermittent welder loads to an equivalent continuous load or effective kVA. If the during-weld kVA demand and the duty cycle for a welder are known, the effective kVA can be obtained by multiplying the during-weld kVA demand by the square root of the duty cycle. The multipliers for various duty cycles are listed in Table 9.13.

Table 9.13 Multipliers for Various Duty Cycles

Multiplier	Duty cycle (%)
0.71	50
0.63	40
0.55	30
0.50	25
0.45	20
0.39	15
0.32	10
0.27	7.5
0.22	5.0

If both the during-weld kVA and the duty cycle are unknown, the effective kVA can be assumed to be 70% of the nameplate kVA rating for seam and automatic welders and 50% of the nameplate kVA rating for manually operated welders other than seam. Nameplate kVA rating is defined as the maximum load that can be imposed on the welding machine transformer at a 50% duty cycle.

The total effective kVA of a group of welders is equal to the effective kVA of the largest welder plus 60% of the sum of the effective kVA of the remaining welders. Once the effective kVA has been determined, the current-carrying capacity can be calculated by using the following formulas:

1. For single-phase distribution systems:

$$\text{current-carrying requirement} = \frac{\text{total effective kVA} \times 1000}{\text{system voltage}}$$

2. For three-phase distribution systems:

$$\text{current-carrying requirement} = \frac{\text{total effective kVA} \times 1000}{\sqrt{3}\ \text{system voltage}}$$

Voltage Drop. To assure consistently good welds, the voltage drop in a distribution system should be limited to 10%. This limit includes voltage drop in the primary distribution system, the distribution transformers, and the secondary distribution system. The voltage drop in the primary distribution system can be obtained from the power company provided that the maximum kVA demand and the power factor of the largest welder is provided them. The voltage drop in the distribution transformer can be calculated from the formula

$$\text{voltage drop (\%)} = \frac{\text{during weld kVA} \times \text{transformer impedance in \%}}{\text{transformer kVA rating}}$$

It is general practice to permit 2% voltage drop in the primary distribution system, 5% in the distribution transformer, and 3% in the secondary distribution system. If the during-weld kVA is unknown, it can be assumed to be about 4 times the nameplate kVA rating for large projection and butt welders, and $2\frac{1}{2}$ times for other types.

9.3.4 Installation

Layout

Busway must be tailored to the building in which it is installed. The initial step is to identify and locate the building structure and other equipment that is in the busway route. It is found that to limit the busway installations to a minimum number of current ratings and to maintain as many 10-ft lengths as possible. This permits the reuse of busway components to maximum advantage when relocation of the busway may become necessary at a future date.

Precautions for Installation

Study manufacturers' drawings carefully if supplied. Check all components and identify them properly, and identify the defective pieces promptly to save the time and cost of installation. Finally, preposition hanger supports to get ready for actual installation of the busway components.

Field Testing

The installed busway should be tested electrically prior to being energized. Phasing and continuity tests are important. Also tests should be done with a megohmmeter or high-potential tester to determine that there is no excessive leakage path between phases and ground.

9.3.5 Busways over 600 V

Busways over 600 V are referred to as metal-enclosed bus and consist of three types: isolated phase, segregated phase, and nonsegregated phase. Industrial plants outside power generation areas normally use nonsegregated phase for connection of transformers and switchgear and interconnection of switchgear lineups. It is rarely used to feed individual loads. Metal-enclosed bus was first covered in the 1975 NEC. The NEC requires that the nameplate of this type of busway specify its rated voltage, continuous current, frequency, 60-Hz withstand voltage, and momentary current as well as that it be constructed and tested in accordance with ANSI/IEEE C37.20-1969 (R1981).

BIBLIOGRAPHY

AEIC CS5-1982, Specifications for Thermoplastic and Crosslinked Polyethylene Insulated Shielded Power Cables Rated 5-46 kV.

AEIC CS6-1982, Specifications for Ethylene Propylene Rubber Insulated Shielded Power Cables Rated 5–69 kV.

ANSI/IEEE C37.20-1969 (R1981), IEEE Standard for Switchgear Assemblies Including Metal-Enclosed Bus.

ANSI/IEEE Standard 141-1986, Recommended Practice for Electric Power Distribution for Industrial Plants.

ANSI/IEEE Standard 241-1974, Recommended Practice for Electric Power Systems in Commercial Buildings.

ANSI/IEEE Standard 400-1980, IEEE Guide for Making High-Direct-Voltage Tests on Power Cable Systems in the Field.

ANSI/NFPA 70-1987, National Electrical Code.

ANSI/UL 486A-1982, Safety Standard for Wire Connectors and Soldering Lugs for Use with Copper Conductors.

ANSI/UL 486B-1982, Safety Standard for Wire Connectors for Use with Aluminum Conductors.

IEEE S-135, IEEE/IPCEA (ICEA) Power Cable Ampacities (SHO7096).

IEEE Standard 48-1975, IEEE Test Procedures and Requirements for High-Voltage AC Cable Terminations.

IPECA S-61-402/NEMA WC5-1973, Thermoplastic-Insulated Wire and Cable for Transmission and Distribution of Electrical Energy, Revision 11, Dec. 1984.

Nelson, Roy A., Wire and Cable Design and Applications for Industrial Plants, IEEE-IAS Industrial and Commercial Power Systems Conference Record, Oct. 8, 1964.

NEMA BU1-1983, Busways.

NEMA BU1.1-1986, Instructions for Safe Handling, Installation, Operation, and Maintenance of Busway and Associated Fitting Rated 600 Volts or Less.

Nestor, A. Thomas, High-Potential Testing of Medium-Voltage Cable, *Plant Engineering*, Apr. 16, 1981, pp. 161–165.

Palko, Ed, Splicing and Terminating Medium-Voltage Power Cable, *Plant Engineering*, Apr. 12, 1984, pp. 62–69.

UL 1072-1986, Medium-Voltage Power Cables.

10

Power Distribution for Computers

10.1 POWER PROBLEMS

Electric Power is the energy source that makes our lives more enjoyable and our daily tasks easier to accomplish. If one has been caught in any of the famous balckouts of recent times, he or she will realize just how dependent we are on electricity to heat and cool homes and offices, to telecopy to business associates, or to run equipment from typewriters to computers. Ac electric power makes it all work. What we take for granted as a readily available, constant source of power is really a highly complex system subject to variations and fluctuations during the transmission process.

Theoretically, all power delivered to the consumer is "clean", meaning that fluctuations of no more than ±10% of the rated voltage are allowed. In reality, electric power gets "dirty", resulting in typical line problems of spikes, surges, faults, brownouts, over- and undervoltages, blackouts, and noise.

10.1.1 Terminologies of Power Disturbances and Their Solutions

Voltage Transients (Spikes)

Voltage transients are brief, high-frequency spikes that appear on the 60-Hz voltage wave. These spikes are typically of microseconds' duration and can be many times the amplitude of the fundamental wave. The typical industrial plants experience hundreds of thousands of these voltage surges a year. The amplitude, duration, and frequency of occurrence determine whether they do cause a problem. The effect of transients on sensitive equipment is to cause erratic and erroneous operations by introducing spurious command signals or negating valid command signals, permanent memory loss or program damage, and damage to components.

Common causes of voltage transients are lightning strikes, switching operations, arcing faults, static discharges, and the firing of SCRs or TRIACs. Anything that interrupts power or draws an arc on the power system causes a transient of some magnitude.

Dedicated lines can alleviate the problems of transients originating in the plant, but they can do nothing about transients originating upstream of the dedicated line's point of connection. The most effective way to deal with voltage transients is to use a transient suppressor. How sophisticated a suppressor will be needed depends on the severity of the transients and the tolerance of the equipment being protected.

Voltage Instability

Voltage instability is far more common than consistent high or low voltage. Consistent high or low voltage can be overcome by using a buck-boost transformer or a transformer tap changer. Voltage can fluctuate with the day of the week or with the season, sometimes dropping 15% or more during summertime air-conditioning brownouts. Voltage can drop to intolerably low levels when large motors are started up within the plant.

The effect of low voltage on sensitive electronic equipment is sluggish operation or erroneous operation, and memory lapse. Equipment damage can also occur. High voltage can cause memory loss and damage to components. The voltage fluctuation problem can be solved with one of several types of voltage regulators.

Electrical Noise

Noise is airborne power-line pollution induced on the line by electromagnetic radiation or electrostatic coupling, with the power wiring acting as an antenna, secondary of a transformer, or plate of a capacitor. Noise appears in the power system in two forms: common mode and transverse mode (sometimes called normal and differential modes). Common-mode noise appears from line to ground, while transverse-mode noise appears from line to line. Either mode of noise existing on the line can generate the other.

Common causes of electrical noise are broadcast transmission; microwave radiation; corona discharge; electrostatic processes; arc produced by equipment, such as welding machines, ignition systems, and switching devices; arcing faults and sparking commutations of motors and generators; and far-off lightning storms. As opposed to transients, the causes of noise need not be physically connected to the power system; coupling is achieved electromagnetically or electrostatically. Transients are random, high-frequency, high-amplitude spikes of microseconds' duration. Noise, on the other hand, is of lesser amplitude and is characterized by being somewhat repetitive and of long duration—appearing more as a ripple than as a spike superimposed on the fundamental voltage wave.

The effect of noise is to cause spurious, erractic, erroneous operation. It can also result in slow degradation of components. A dedicated line can alleviate noise problems that originate within the plant, as can properly shielded and grounded data-link wiring. Major common-mode noise problems require the installation of an electrostatically shielded isolation transformer.

Noise can be a vexing problem, because in many cases, the equipment to be protected from noise is itself capable of generating noise. Computers and peripheral equipment such as alphanumeric printers can produce noise and "crosstalk" the noise between devices.

Power Interruptions

The effects of power interruptions range from minor inconvenience to lost production to chaos. If power to some types of electronic equipment is severed abruptly, high-voltage transients can develop, causing major component damage. Computers can suffer from loss of irreplaceable data.

A rotary filter [rotary uninterruptible power supply (UPS)] can deal effectively with power interruptions of less than a second, but beyond that, only a full-scale static UPS suffices. A UPS is the most comprehensive and expensive solution that can be applied to a dirty power problem. Figure 10.1 is the waveform representation for various power disturbances discussed above.

Frequency Deviations

Frequency deviations seldom occur in most parts of the U.S. Most utility generating equipment is a part of a national grid that rigidly locks all generators in synchronism. Consequently, power derived from the grid seldom deviates by more than a few tenths of a cycle from the nominal 60 Hz.

The manufacturers' tolerance on 60 Hz equipment ranges from ±0.5 Hz to ±1%. Excessive frequency deviation can affect the operation of computer disk drives and cause improper operation of timing and logic circuits.

Although frequency deviation is a rare problem, simple, relatively inexpensive solutions are not sufficient when it does exist. A full-scale UPS with its inherent frequency inverter is the sure solution.

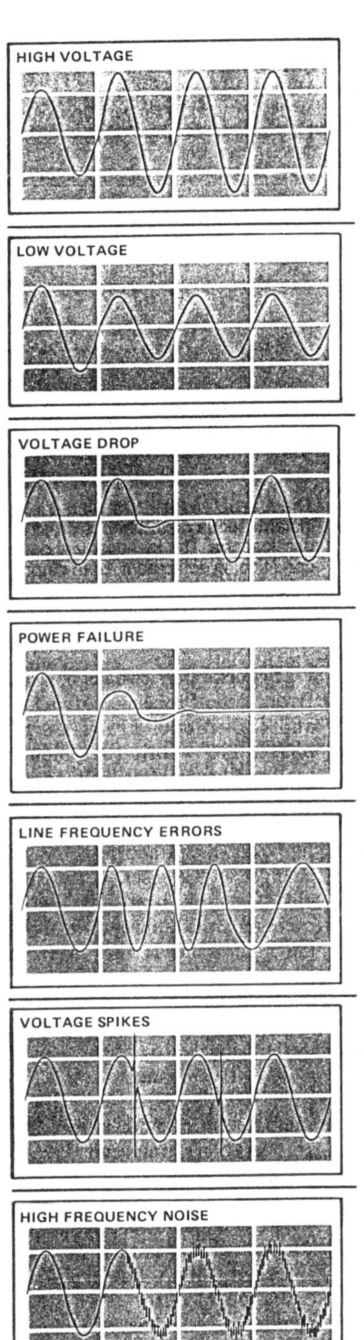

Figure 10.1 Waveform representation for power disturbances.

used. When many analog circuits are present, the shielded cables may be contained within a single outer shield, terminated to ground at both ends. Conduits are well suited for this purpose.

ANSI/IEEE Standard 518-1982 contains very useful guidelines for wiring installation in industrial plants. Programmable controllers (PCs) have installation requirements specified by the manufacturer of the equipment. Such items as cover plates, ground connections, and mounting requirements may be very important to PCs' resistance to electromagnetic interference (EMI); they can often help contain noise generated within the system.

10.2 COMPUTER POWER REQUIREMENTS

Today with the advent of computers and data processors, as well as many other semiconductor-controlled devices, many previously disregarded power disturbances pose a direct threat to the proper operation of these equipment. Based on a past study, Table 10.1 shows various types and characteristics of power-line disturbances which have been discussed previously and are now grouped into three types. Each type is given its causes, threshold level, and duration, respectively.

10.2.1 Industry Standards

There are standards set by the utility that govern the allowable voltage tolerances for power as delivered to the user. The utility power profile is specified in ANSI Standard C84.1. The Computer Business Equipment Manufacturers Association (CBEMA) has standardized on voltage limits within which computers and similar equipment should operate reliably. Voltages outside these limits could cause erratic, unreliable, or incorrect operation. Figure 10.2 shows the ANSI utility power profile with the CBEMA voltage "envelope" superimposed. It indicates that ANSI standards for steady-state voltage tolerances for the utility at the source are ±5% for residential power and ±10% for industrial power. After allowances for voltage drop up to and on the user's premises, CBEMA steady-state tolerances for industrial power are +6% and −13%. These superimposed computer tolerances and utility profile curves make it clear that there are potential problems in the momentary outage area and in the high-voltage transient area in power from the utility, in addition to any user-caused transients. Actual site voltage measurements of incoming utility power using rapid-response instruments (small circles in Figure 10.2) show that the utilities, despite their best efforts, often cannot maintain the specified ANSI voltage tolerances at all times.

Most computer equipment can accept the +6%, −13% tolerance, but some computer manufacturers call for even more stringent voltage

Table 10.1 Various Types and Characteristics of Power-Line Disturbances

Definition	Type I Transient and oscillatory overvoltage	Type II Momentary undervoltage or overvoltage	Type III Outage
Causes	Lightning; power network switching (particularly large capacitors or inductors); operation of on-site loads	Faults on power system; large load changes; utility equipment malfunctions; on-site load changes	Faults on power system; unacceptable load changes; utility or on-site equipment malfunctions
Threshold level*	200 to 400% rated rms voltage or higher (peak instantaneous above or below rated rms)	Below 80 to 85% and above 110% of rated rms voltage	Below 80 to 85% rated rms voltage
Duration	Spikes 0.5 to 200 microsecs wide and oscillatory up to 16.7 millisecs at frequencies of 0.2 to 5 kHz and higher	From 4 to 60 cycles, depending on type of power system and on-site distribution	From 2 to 60 sec if correction is automatic; unlimited if manual
Duration (cycles of 60-Hz wave)	0	0.5	120

* Approximate limits beyond which disturbance is considered harmful to computer.

regulation. Typical transient tolerances permit no dips or surges greater than about ±20% for no longer than 30 ms, with briefer transients capable of creating errors. Most computers cannot tolerate a loss of voltage for longer than about 15 ms.

Transient voltages from utility switching at the user's site can exceed the transient-voltage tolerances, and momentary outages exceeding the 15 ms level can occur from utility power network switching and lightning-strike circuit breaker tripping and reclosing. About the only requirement that rarely causes problems is the frequency tolerance, ±0.5 Hz. Input power variations beyond acceptable limits can cause errors in calculations, output errors, loss of data, unscheduled shutdowns, and even equipment damage. Table 10.2 summarizes the foregoing discussion in an orderly and concise fashion.

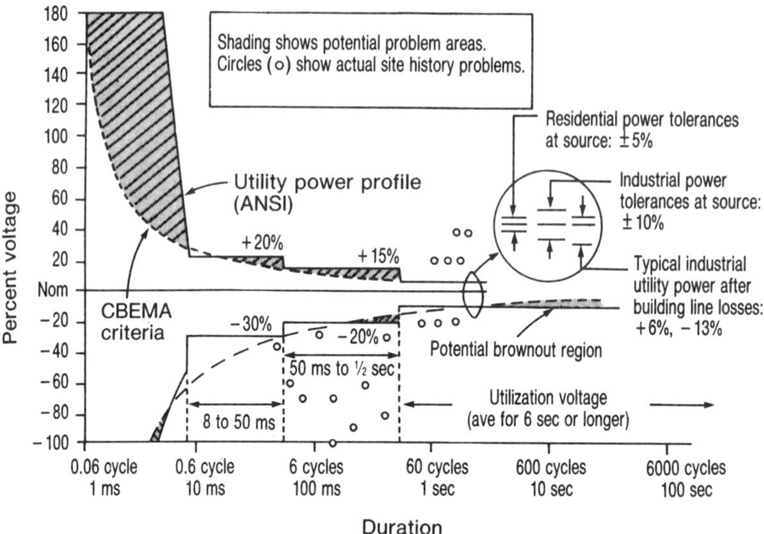

Figure 10.2 ANSI utility power profile superimposed with CBEMA voltage envelopes.

10.2.2 Power Problem Study Data

In early 1974, two engineers of IBM published a paper "Monitoring of Computer Installations for Power Line Disturbances." This study is one of the most extensive records of actual disturbances to power systems supplying computers, encompassing both utility source and on-site-generated voltage excursions, including spikes, oscillatory transients, overvoltages, and undervoltages. The data accumulated represented a climate and geographical cross section of the United States, with representative loads from heavy industry, light industry, office buildings, retail stores, residences, and mixed locations. Utility systems included overhead and underground, single- and dual-feeder distribution. Table 10.3 summarizes the results of this study. It is noted that the study shows that only 37% of the 4507 total transients measured were of the type that an isolation transformer can eliminate, or in other words, at least 63% of all transients pass through the transformer and on to the computer. In contrast, a motor-generator eliminates all types of power transients because the primary and secondary circuits of the system are mechanically, electrically, and magnetically separated. The correction time of most of the many types of voltage-regulating transformers is so slow as to be ineffective in protecting against common sag/surge problems that last only

Table 10.2 Representative Power Quality Attributes

Environmental attribute	Typical environment	Typical acceptable limits for computers and power sources		
		Normal	Critical	Units affected and comments
Line frequency	±0.1%–±3%	±1%	±0.3%	Disk packs, tape, regulators
Rate of frequency change	0.5–20 Hz/s	1.5 Hz/s	0.3 Hz/s	Disk packs
Over and undervoltage	±5%–+6, -13.3%	+5%, -10%	±3%	Unregulated power supplies
Phase imbalance	2%–10%	5% max	3% max	Polyphase rectifiers, motors
Power Source: Tolerance to low power factor	0.85–0.6 lagging	0.8 lagging	Less than 0.6 lagging or 0.9 leading	Indirectly limits power source or requires greater capacity unit with reduced overall efficiency
Tolerance to high steady-state peak current	1.3–1.6 peak/rms	1.0–2.5 peak/rms	Greater than 2.5 peak/rms	1.414 normal; departures cause wave shape distortion.
Harmonics (Voltage)	0–20% total rms	10–20% total 5–10% largest	5% max total 3% largest	Voltage regulators, signal circuits
dc load current capability of power source	Negligible to 5% or more	Less than 0.1% w/exceptions	As low as 0.5%	Half wave rectifier load can saturate some power sources, trip circuits.
Voltage deviation from sine wave	5–50%	5–10%	3–5%	Affects regulators, signal circuits
Voltage modulation	Negligible to 10%	3% max	1% max	Voltage regulators, servo motors
Transient surges/sags	+10%, -15%	+20%, -30%	+5%, -5%	Regulated power, motor torque
Transient impulses	2 to 3 times nominal peak value (0–130% V-s)	Varies; 1,000–1,500 V typical	Varies; 200–500 V typical	Memory, disks, tapes having high data transfer rates, low level data signals
RFI/EMI and "tone bursts" normal and common modes	10 V up to 20 kHz; less at higher freq.	Varies widely 3 V typical	Varies widely 0.3 V typical	Same as above
Ground currents	0–10 A rms + impulse noise current	0.001–0.5 A or more	0.0035 A or less	May trip GFI devices, violate code, introduce noise in signal circuits

Table 10.3 Study of Power-Line Disturbances Summary

Power Problem Type	Total Incidents Measured	NUMBER OF INCIDENTS CORRECTED					
		Regulating Ferro	Transformer SCR	Isolation Transformer	Motor-generator	UPS	UPS with Diesel
Undervoltages	1569	605	737	0	1559	1569	1569
Overvoltages	103	33	34	0	103	103	103
Outages	65	0	0	0	14	57	65
Transients-Common	1676	0	0	1676	1676	1676	1676
Transients-Normal	2831	0	0	0	2831	2831	2831
Total Incidents	6244	638	771	1676	6183	6236	6244
Power Protection Offered		10%	12%	27%	99%	99.8%	100%

milliseconds. Most computer manufacturers specify that any voltage condition out of tolerance for longer than one-fourth of a cycle (4 ms) could result in hardware and software malfunctions.

10.3 POWER CONDITIONERS FOR COMPUTERS

10.3.1 Different Types of Power Conditioners

The survey results as presented in Table 10.3 also indicate the relative effectiveness of the different types of power conditioners to combat a variety of power problems. Selecting the proper solution for a "dirty" power problem requires knowing which components of dirty power are causing the problem—frequency deviation, voltage transients, over-, under-, or fluctuating voltage, electrical noise, or power interruptions. Some types of power conditioners can cope with one or more types of problems, but not others. Minor transient surge problems can sometimes be handled with simple, limited-capacity transient suppressors costing less than $30; multiple, major problems on critical equipment might require a full-scale uninterruptible power supply (UPS), costing tens of thousands of dollars. The optimum solution is a balance between economies and the nature of the problem. It is therefore important to understand clearly the various types of power conditioners that are available on the market today, their advantages and limitations, and their relative cost.

Dedicated Lines

Dedicated line is a circuit run in its own electrically continuous, bonded, properly routed, and grounded metallic raceway from a point well upstream directly to the protected equipment. Dedicated lines are commonly recommended by mainframe computer manufacturers, but their limited capabilities often require that they be supplemented by one or more types of power conditioners.

Advantages. Dedicated lines can shield out noise-producing electromagnetic and electrostatic coupling along its length and bypass power interruptions downstream of their point of connection.

Limitations. It cannot solve problems of improper voltage, transients, noise, and power interruptions originating upstream of its point of connection.

Transient Suppressors

Transient suppressors range from simple, low-cost devices to highly engineered, sophisticated devices. Simple ones are little more than electronic fuses intended to self-destruct in performing their duty.

They clamp, restrict, or divert transient surges by operating on overvoltage, high frequency, or both. There are versions that conduct from line to neutral, line to ground, or neutral to ground, or combinations.

Advantages. Transient suppressors typically supress spikes having a frequency greater than 10 kHz and peaks greater than 250 V. High-quality units effectively cope with virtually any transient voltage problems.

Limitations. They do nothing about frequency deviation, voltage problems, and power interruptions, and even the best-quality units have little effect on electrical noise.

Voltage Regulators

Voltage regulators maintain voltage output within specified limits upon fluctuations in input line voltage. Among those offered are ferroresonant, saturable reactor, total solid-state and tap-changer types in both electromechanical and electronic versions. All tap-changer types regulate in discrete steps rather than in a smooth sweep.

Advantages. Good-quality voltage regulators are energy efficient, respond rapidly to fluctuations in input voltage, and maintain voltage output within extremely narrow limits upon wide fluctuations in input (constant output voltage within 1% when input voltage fluctuates up to 15%).

Limitations. They have no effect on other types of dirty power problems.

Isolation Transformers

Electrostatically shielded isolation transformers are designed to prevent electrical noise on the power line from being passed through to the equipment being protected. These isolation transformers should not be confused with oridinary isolation transformers. At one time, noise isolation transformers were offered only in one-to-one transformation ratios, but they are now available in step-down versions. Metal-foil shielding is used on the primary to prevent noise from being induced in the secondary.

Advantages. Good-quality units are energy efficient and can effectively block common-mode noise, but it is generally ineffective on transverse-mode noise. Hybrid versions, however, have supression devices that enhance transverse-mode attenuation.

Limitations. They are generally ineffective in coping with other types of dirty power.

Hybrid Power Conditioners

Hybrid power conditioners combine two or three of the functions of transient supression, voltage regulation, and noise attenuation in one device.

Advantages. Best-quality units deal effectively with problems of transient surges, voltage fluctuation, and noise on the input power supply, and operate at high energy efficiency.

Limitations. They do nothing about frequency deviation, cannot cope with power interruptions, and can do nothing about transients and noise originating on the output side.

Rotary Filters (or Rotary UPS Systems)

Rotary filters typically have a synchronous motor fed from the utility supply and driving a synchronous generator. The generator output is fed to the protected load. The kinetic energy of the spinning mass provides ride-through of momentary power interruptions. Figure 10.3 shows the arrangement of an uninterruptible synchronous motor-generator set.

Advantages. They provide ride-through of momentary power interruptions on the order of 500 ms, with actual capability varying with the design of the unit. On longer power interruptions, they also provide buffering to prevent the development of transients on their output side. They maintain stable voltage at the load, and deal effectively with noise and transients on the incoming power supply.

Limitations. They cannot cope with power interruptions of more than momentary duration, and they cannot handle frequency deviation.

Static Uninterruptible Power Supplies

Static uninterruptible power supplies (UPSs) are comprised of a rectifier/charger unit, solid-state inverter, battery bank, and appropriate transfer and bypass switching. UPS systems are offered in small units that can fit under a desk through large, unitized systems for floor mounting. Two configurations can be employed for a static UPS: forward transfer or reverse transfer. Figure 10.4 shows these two types of configurations. With a reverse transfer system, the UPS is continually on line; a solid-state transfer switch is typically used to switch to raw utility power if some elements of the UPS fail. In the forward transfer system, the load is continuously served with

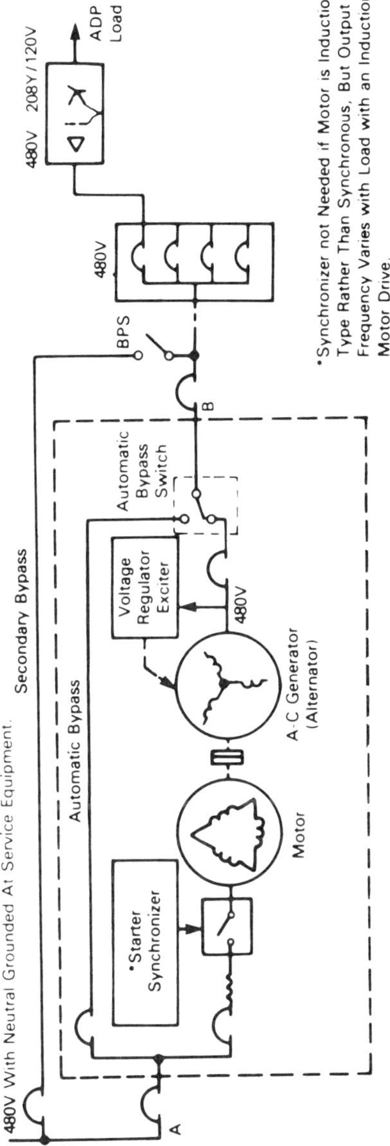

Figure 10.3 Arrangement of an uninterruptible synchronous motor-generator set.

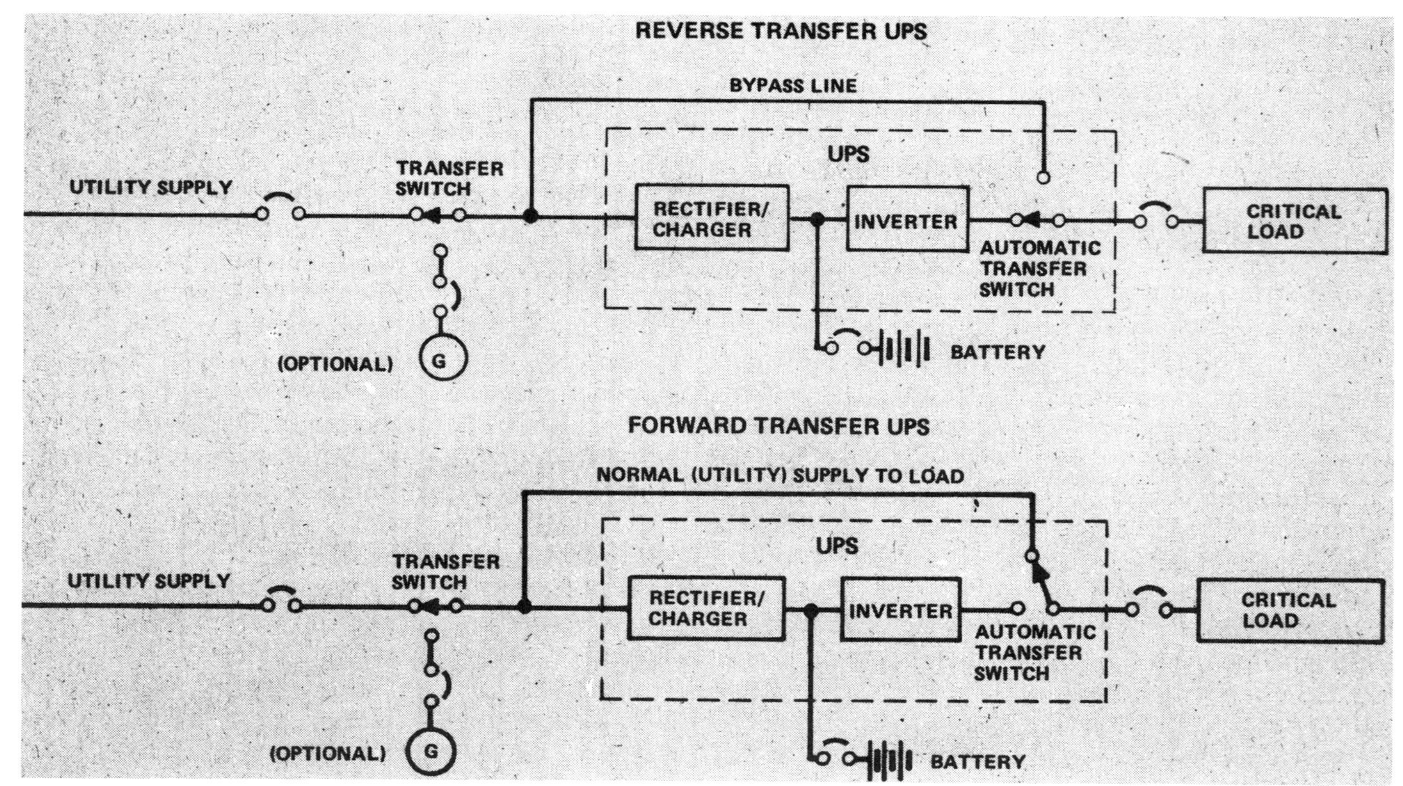

Figure 10.4 Two types of configurations for a static UPS.

utility power, and transferred to the UPS if utility supply is interrupted.

Advantages. A reverse transfer UPS can cope with any and all dirty power problems on the incoming supply—frequency deviation, voltage transients, under-, over-, or fluctuating voltage, electrical noise, and momentary and sustained power outages. A forward transfer system can be modified with controls to transfer the load to UPS for other reasons, such as frequency deviation or unacceptable high or low voltage.

Limitations. Unless supplemented with a standby generator, both forward and reverse transfer systems are limited in operation on loss of the utility supply by the size of the battery bank.

Hybrid Rotary/Static UPS Systems

A hybrid rotary/static UPS system combines the feature of a rotary filter with those of a static UPS and draw on the advantages of both. In normal operations, the rotary filter is on line serving the protected load. Immediately upon sensing a power interruption, a high-speed transfer switch transfers the load to the static system. The spinning momentum of the rotary unit preserves power to the load during the transfer interval. Figure 10.5 shows a schematic for a combined static, battery, and rotary UPS system.

Advantages. This system protects against any and all dirty power problems; the system can be provided with a frequency relay to transfer from the rotary to the static mode if the problem of frequency deviation arises. Hybrid systems have an advantage over static-only systems in that they use fewer elements to provide full uninterruptible protection. This tends to minimize the possibility of electronic components failure.

Limitations. Unless supplemented with standby generation, sustained operation during power outages is limited by battery capacity. Rotating elements require periodic routine maintenance required for all rotating equipment.

10.3.2 Economics of Different Types of Power Conditioners

Figure 10.6 summarizes the efficiency and relative cost of several different systems:

1. UPS/engine generator
2. Redundant UPS

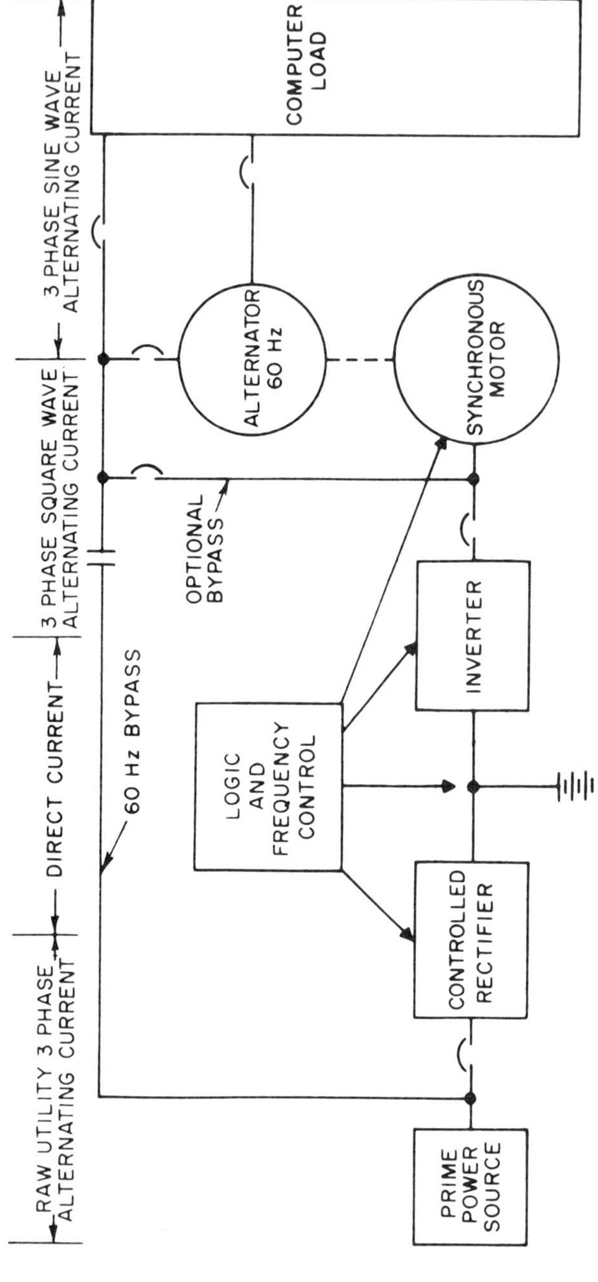

Figure 10.5 A schematic for a combined static, battery, and rotary UPS system.

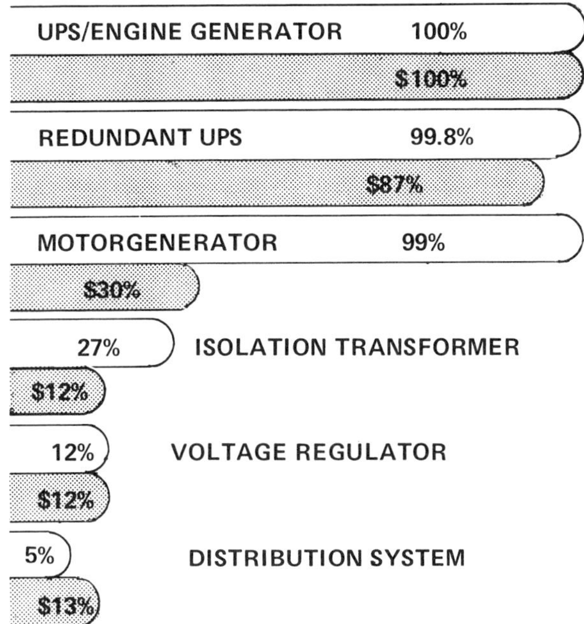

Figure 10.6 Summary of efficiency and relative cost of several different power conditioning systems.

3. Motor-generator
4. Isolation transformer
5. Voltage regulator
6. Distribution system

It is noted that the motor-generator system and UPS system perform the same function by protecting a computer from 99% of all power problems. However, a motor-generator system is much less costly to install and operate. If the computer application justifies 100% protection and cost is not an issue, consideration must be given to a perperly engineered redundant UPS system with backup diesel generators.

10.3.3 Specifications for a UPS System

Developing specifications for an uninterruptible power supply requires careful analysis of variable factors before a system is chosen. Electrical power required by a sensitive load may vary widely, depending on the configuration of the load. The power demand of a computer system, for example, depends on the number of central processing units and the number and types of peripheral devices, such as disk and tape drives.

The task of developing specifications becomes complex because nearly any piece of load equipment can be found in nearly any application. There are characteristics of the connected load that dictate the design, performance limits, and capabilities of the UPS to be specified. The principal factors determined by hardware, which vary from load to load, are: power requirements, power factor, load configuration, number of phases, voltage regulation, phase displacement, frequency stability, slew rate, unbalanced loading, filtering and distortion, inrush handling capabilities, overload capacity, bypass capacity, transfer time, and reverse time. Several of these items which require further explanation are discussed as follows:

Power Requirements

The UPS can be sized either to accommodate the total anticipated critical load growth at the outset or to accommodate the initial critical load, with provisions for expansion. The specific power requirement, the specified kW and kVA, can be determined in one of two ways. The preferred method uses actual load measurements to determine the exact power requirements of each device that will be powered by the UPS. Sometimes it becomes necessary to use the nameplate ratings found on various devices or in the computer installation manual. However, the nameplate ratings taken in conjunction with a diversity factor of 100% can result in a substantially oversized UPS and increased operating costs. On the other hand, if one conscientiously sizes the UPS for actual measured load, it is often to discover other unaccounted-for critical loads after starting up the system. Therefore a modest "pad" is often applied to the final load power requirement calculations as a means of refining process.

Number of Phases

A three-phase load cannot be powered by a single-phase UPS, but a single-phase load can be powered by a three-phase UPS. For computer systems requiring three-phase input, with some of the peripherals wired for 208 V single-phase input, a three-phase UPS is ideal because it provides 208 V three-phase, 208 V single-phase, and 120 V single-phase inputs. A three-phase system is generally more efficient than a single-phase system. Therefore, a three-phase system is always preferred in sizes above 10 kVA.

Phase Displacement

The angle between phases of a three-phase voltage supply should be exactly 120 electrical degrees. The voltage-regulating scheme used by the UPS inverter dictates how well each phase remains fixed in relation to the other phases. The deviation from nominal is often stated at ±1 or 2 degrees for the nominal 120-degree phase displacement. One method regulates the line-to-line voltages of the inverter output, rather than the line-to-neutral voltages. This results in the angle between phases being inherently regulated to 120 degrees.

Slew Rate

Slew rate is defined as the rate of change of frequency, in hertz per second. It is important for installations employing motors, such as the disk drives in a computer system. Because motor speed is proportional to the frequency, a sudden change in frequency will result in a high inrush to the motor. Most critical loads can tolerate a maximum slew rate of 0.5 Hz/s.

Unbalanced Loading

In most installations requiring three-phase power, at least 75% of the loads can actually be single-phase devices. Although the 20% unbalanced capability available with most units may appear adequate, this capability can easily be exceeded. In a 15-kW unit, a mere 1-kW unbalance in one phase represents a 20% total unbalance. Where loading exceeds the specified maximum unbalance, UPS reliability suffers as components are stressed, regulation begins to deteriorate rapidly, and distortion increases.

Filtering Capabilities and Distortion Requirements

Every load has a limited tolerance for noise and harmonic distortion. Total harmonic distortion (THD), which is the measure of the quality of the waveform applied to the load, is calculated by geometrically summing the harmonic voltages present in the waveform and by relating this sum as a percentage of the fundamental voltage of the waveform. Typically, critical loads can withstand 5% THD, where no single harmonic exceeds 3%.

One type of noise that is difficult for the UPS to attenuate is load-produced noise. If the critical load is very noisy or has an extremely nonsinusoidal waveform, the result will be reflected in the UPS output. Special filters may be required to prevent a false trigger or misfiring circuit element or to prevent an adverse effect by reflected noise on other loads connected to the UPS output. UPS designs that already have inherently large amounts of filtering on their output are less likely to need addition of an external filter.

Inrush Handling Capabilities

A surge, lasting from a half-cycle to a few cycles, can produce an effect similar to that of a short circuit. In specifying a UPS for such applications, the short-circuit capabilities of the unit must be sufficient to accommodate such an inrush surge, particularly when the bypass source is unavailable. When the inrush is greater than 125% of the normal current, the UPS will attempt to transfer to the bypass source until the current demands return to acceptable limits. Without a bypass source, the UPS will be forced to shut down. Normally, the bypass source is available and is used to start the heavy loads automatically via the UPS static transfer switch.

Bypass Capability

The bypass source must be sized to power the load, and its output must be maintained within parameters acceptable to the load. When a commercial ac line is used, a line-voltage regulator may be required.

Transfer Time

The UPS configuration and transfer mechanism must be adequate in terms of transfer time. A decision regarding the type of output transfer switch employed—electromechanical, solid-state, or hybrid—depends on the load device's ride-through capability. Some devices can withstand three to six cycles of interruption; others cannot tolerate as little as a half-cycle. Most computer loads need usable voltage continuously during a transfer (make-before-break).

Reserve Time

Most often, the standby battery plant employed by the UPS is sized to provide 15 to 30 min of reserve time, with a standby engine-generator being considered for longer outages.

10.4 POWER DISTRIBUTION DESIGN CONSIDERATIONS FOR COMPUTERS

10.4.1 Reliable, Continuous, and Clean Power

Power system must be reliable, continuous, and "clean" for the vital operations of a modern computer processing center. If "clean" power is the only requirement of the UPS system, a simple ac-to-ac motor-generator (MG) set can satisfy the need. This type of "buffer" MG set was, in fact, used almost exclusively with early computer systems. However, "continuous" power refers to the ability to supply power even during complete interruptions of power. Recloser operation of 15 to 20 cycles' duration can easily be handled by the "buffer" MG set, but longer interruptions require additional protection. This is where more sophisticated UPS systems come in. The static UPS system is perhaps the simplest such UPS system, and, in general, also the most economical to install.

When a power level goes beyond, say, 250 kVA, reliability of the system should be further emphasized. Redundancy is a means of providing increased system reliability. With redundancy in its simplest form, any one inverter or rectifier-charger can be removed from the system without affecting the ability to continue feeding the critical load. Figure 10.7 shows one of many different redundant configurations. It should be noted that it is rarely feasible to consider using operating redundancy with rotating UPS systems.

10.4.2 Grounding Requirements for a Computer System (or Automatic Data Processing Equipment)

Purpose of Grounding

Grounding accomplishes multiple functions, all of which must be considered in the design and installation of an automatic data-processing (ADP) system. Grounding is required both for safety reasons and because of the need for highly sensitive computer circuits to operate reliably.

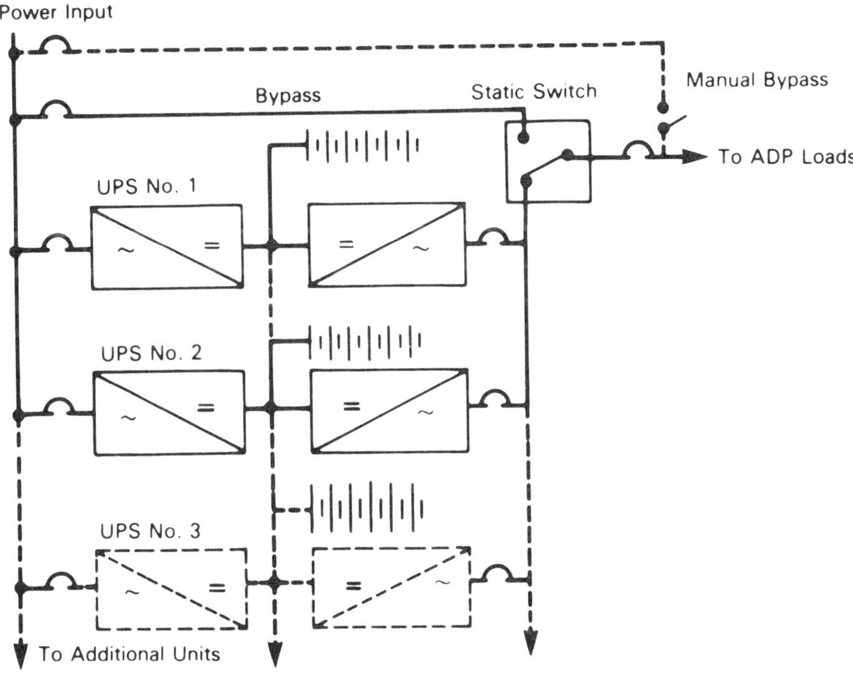

Figure 10.7 Redundant configurations of UPS system.

Low-resistance ground connections to earth are appropriate for lightning and transmission-line ground faults. However, this is not the rationale for applying grounding principles to 120, 240, and 208 V circuits. At these voltages, a system of interconnected or bonded conductors acting as a voltage reference network can equalize voltage differences throughout the network much more effectively than can multiple low-impedance earth contacts. Such an interconnected, bonded network can serve as both a power and a signal reference, regardless of its voltage with respect to earth ground. To avoid shock hazard and to minimize voltage differences between individual reference networks, it is not only accepted practice but mandatory for safety purposes that these networks be connected to earth ground. Since the connection to earth is never expected to carry load or fault current, NEC permits this conductor to be smaller than the equipment ground conductors in the network.

For computer or ADP systems, a copper rod driven into moist soil is not a magic cure-all for grounding problems, nor is it necessarily a requirement. Various functions of ADP system grounding may be summarized as follows:

1. Touch voltage differences must be limited by bonding and grounding to avoid shock hazard.
2. Ground fault current path to power source must have low impedance to enable it to actuate overcurrent protection and disconnect the source.
3. Ground potential differences in the area must be reduced to essentially a constant reference.
4. Grounded conducting enclosures serve as electromagnetic shielding for sensitive circuits.
5. Grounding in compliance with safety code is mandatory.

Techniques of Grounding

When the grounding instructions given by the computer manufacturer appear to conflict with safety code requirements, the manufacturer should be consulted and asked to resolve the problem. Any arbitrary independent departure from these instructions or electrical codes can result in taking the liability for safety hazard or failure of the computer equipment to perform reliably.

The central grounding point shows a "single point ground" in an ADP system (Figure 10.8). This point should be readily identifiable. It should be the point where the interconnected parts of the computer's grounding system are connected to other ground conductors that extend beyond and outside the ADP room.

Within very large systems, there may be subsystems, each with a central grounding point for connection to other central grounding points. However, separate connections to separate external grounds would create unwanted external ground loops. Impulse ground cur-

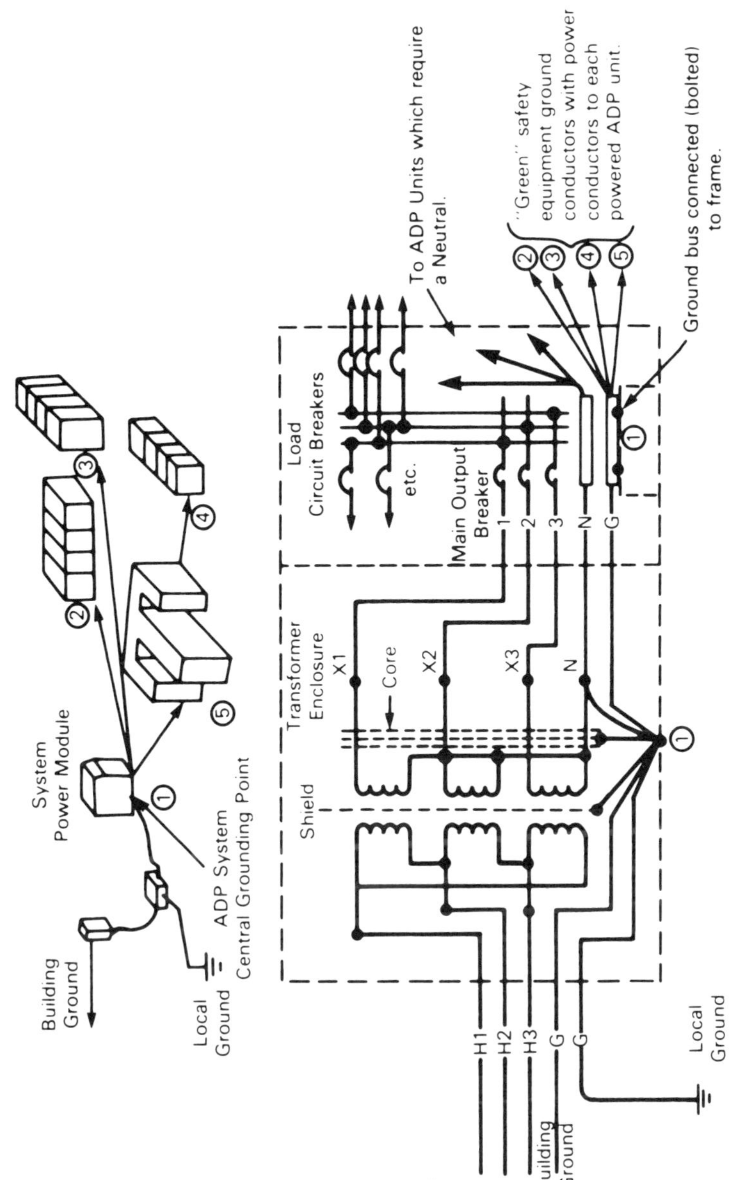

Figure 10.8 Illustration of single point ground system.

rents can find paths in the ground shields and grounded conductors, or signal pairs and coaxial cables. Intercoupling with digital circuits and signal corruption can be the unwanted results.

Future computers with higher-frequency performance circuits will probably require careful shielding of all circuits longer than a few inches. Coaxial conductors and waveguides become appropriate for higher frequencies. However, these are prime candidates for replacement with fiber optic signal transmission, which is already being put into communications service to carry digital signals at 90 to 400 megabits per second. Fiber optics will solve some of the common-mode noise problems that bother today's computer circuits, but other limitations will take their place. The signal energies may continue to decrease as noise is reduced, but there are theoretical limits that will ultimately be approached.

In today's techniques with digital data and control signals, any dc and low-ac frequency (100 kHz or less) signal currents will follow the lowest-resistance paths. At high radio frequencies (above 100 kHz), stray capacitance and electromagnetic coupling become significant circuit paths. At low frequencies where currents follow conductors, single-point grounding is generally preferred. As signal frequencies exceed approximately 10 mHz and greater, the noise currents and voltage signals cannot easily be confined to conductors. In this realm, multipoint grounding becomes necessary if it is to be effective.

If the desired signal must be protected against both high- and low-frequency interference, a solid metallic galvanic grounding connection is needed for a single-point ground, while at high frequencies one can use multiple ground paths via deliberate use of stray or discrete capacitors. A very effective technique is to have multipoint-ground connections to an outer shield over an inner insulated shield that has a single-point ground.

Earth Ground Connection and Isolated Ground

Earth ground connection can be important to reliable performance of electronic circuits in ADP systems, minicomputers, and word processors, but its role is often misunderstood. As a result, much effort and sometimes needless expense is incurred in achieving a "quiet, isolated ground" with a very low ground resistance. The quiet, low-resistance attributes are always desirable, but a misunderstanding of the term "isolated" can lead to dangerous grounding practices that violate safety codes and will not solve noise problems.

Stand-alone word processors, minicomputers, their peripherals, and other electronic office machines are typically powered and grounded solely by their three-prong grounding-type 120-V plugs on their power cords. If a minisystem requires more than one separately powered unit, it is common practice to use a grounding-type duplex receptacle cube tap, or portable receptacle strip provided that the total load does not exceed 80% of the 15- or 20-A circuit protection rating for the circuit.

In many instances, such an arrangement has been used successfully. Unfortunately, there are often reasons why this may be marginally successful or not work at all:

1. The ground pin at the receptacle, the neutral grounding point at the building entrance service equipment, and the transformer power source's secondary output winding ground point are separated and may be grounded at two or three separate locations. This could permit noise voltages to develop between them and appear as common-mode noise.
2. Older buildings may have wiring without an equipment ground conductor and may even lack electrically continuous conduit to serve in its stead. If the receptacle enclosure is grounded to a local water pipe, driven earth electrode, or building structural steel, and there is no provision for an electrical return path such as an equipment ground conductor or continuous conducting path provided by conduit back to the power source grounding point, such an installation could be unsafe.
3. The equipment ground conductor in the receptacle may be connected permanently to the conducting enclosure in which it is mounted. A connection that is integrally built into the receptacle normally creates its ground path. Noise currents originating from a load plugging into an adjacent or nearby receptacle could reach the sensitive equipment via the path.

One solution is to install an "isolated ground" receptacle (sometimes identified by orange color) in which the ground terminal is isolated from the mounting strip (Figure 10.9). An insulated equipment ground conductor is then connected from the grounding terminal of the receptacle in accordance with the NEC Article 250-74, Exception 4, and is passed through one or more panelboards without connecting to their grounding terminals (Article 384-27, Exception 1) for direct connection to the applicable derived system or service grounding terminal.

Contrary to illustrations in some catalogs which show examples of the equipment ground connected to an independent earth electrode without any other connections to the building ground, the equipment grounding conductor from the receptacle with the isolated ground must be connected directly to the neutral grounding point for the building. This is necessary for safety, compliance with code, and for low electrical noise at the computer unit.

Single-Point Versus Multipoint Grounding

This has been discussed in theory. However, grounding practice is seldom like the theory. Many unintentional ground connections not discussed in theory do appear in practice. It is virtually impossible to enclose power, communications, and other conductors in a conduit

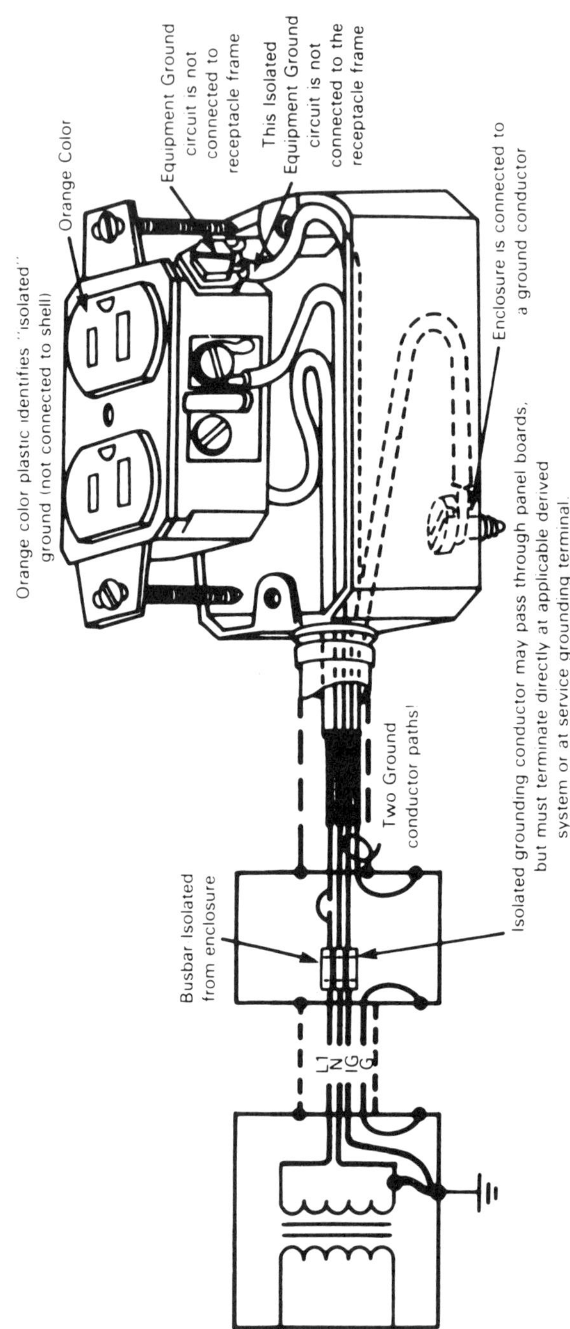

Figure 10.9 Wiring and grounding for an isolated ground receptacle.

and have electrical noise, voltages, and currents on that conduit without coupling to the conductors contained within it. For this reason it is possible to make a great improvement if some control can be exercised over all conducting members that enter the computer room or are run nearby. A single-point entry makes it possible to equalize the voltage differences without having the resulting current conducted throughout the computer room. If a single-point strategy is to be followed, this may be possible within the computer room, but as soon as circuits leave the room, they must be considered to be grounded at multiple points and be sources of good loop noise currents.

10.4.3 Standby Power Systems

As discussed in Section 10.3.1, none of the power conditioning systems can sustain a prolonged power outage unless they are supplemented with a standby generator. Standby generators can range from small units serving a few emergency lights, communications, and security systems to large, comprehensive installations capable of sustaining business as usual. Standby generators are not intended to provide power at lower cost than that from utility. They are offered in a wide variety of prime-mover choices, such as gasoline engine, diesel engine, gas turbine, and dual-fuel diesel/natural gas engine. In general, the gasoline engine unit will have the lowest initial cost, but at a higher maintenance expense than with other types. Diesel engine has lower fuel costs and maintenance expense, and loger life expectancy, and less failure rates. Gas turbine has also the advantage of fuel economy, minimum maintenance, and long life, but higher initial costs.

Static power supplies provide a reliable, though more expensive alternative to engine-generator sets. Static standby power systems provide some advantages not obtainable with engine-generator sets. Where ultrapure, totally uninterruptible power is required, static power supplies provide the only answer in light of today's technology. In addition to providing an uninetrruptible source of power for critical loads such as computers, an uninterruptible power supply (UPS) also acts as a filter, filtering out any undesirable voltage transients on the power system. A static power supply consists of four major elements: a stationary battery to provide necessary power reserve, a charger to maintain the battery at full charge, a solid-state inverter to convert battery dc power to ac, and appropriate switching equipment.

10.4.4 Equipment Interface

In designing a buffer system for powering computers and other sensitive electronic equipment, a number of important interfaces should be carefully considered. One of these is the interface between a standby generator and an UPS.

Because the output impedance of a generator is higher than that of the utility, the standby generator must be oversized so that the UPS current draw does not distort the generator output voltage. Any equipment using SCR ac-to-dc conversion introduces harmonics into the feeder supply source. The voltage distortion in the conductors between generator and UPS can be such that the UPS will no longer recognize the generator as an acceptable power source and transfer the supply to battery backup. As soon as this happens, the output voltage waveform of the generator returns to within specifications. The UPS transfers it back to the generator. Distortion of waveform starts again. The back-and-forth transfers continue until the batteries are drained.

As a good empirical formula, the generator should be oversized in such a way that only 60% of its capacity is required for the UPS. The remaining capacity may be used for lighting, air conditioning, and life safety.

UPS that uses solid-state rectifiers for battery charging represents a difficult load for a standby generator to handle because the current contains substantial harmonic distortion. The THD of an UPS input current should be no more than 10 to 12% for proper generator/UPS operation. The addition of harmonic filter in the UPS can reduce the current distortion to less than 5% THD and also improve the power factor at the generator/UPS interface. Frequency monitoring can be accomplished by equipping the generator with an isochronous governor or its equivalent.

BIBLIOGRAPHY

ANSI/IEEE C62.41-1980, Guide for Surge Voltage in Low Voltage AC Power Circuits.

ANSI/IEEE Standard 518-1982, Guide for the Installation of Electrical Equipment to Minimize Noise Inputs to Controllers from External Sources.

ANSI/NFPA 70-1987, National Electrical Code.

Guideline on Electrical Power for ADP Installations, Federal Information Processing Standards Publication 94, U.S. Department of Commerce, Sept. 21, 1983.

Kesterson, Albert and Maher, Pat, Computer Power: Problems and Solutions, *Electrical Construction and Maintenance*, Dec. 1982, pp. 67–72.

O'Neill, Thomas S., Understanding Uninterruptible Power Supplies, Part 8, *Electrical Construction and Maintenance*, Apr. 1984, pp. 78–81.

Palko, Ed, Monitoring and Analyzing Quality of Electric Power to Electronic Equipment, *Plant Engineering*, Apr. 25, 1985, pp. 44–51.

Palko, Ed, Providing Clean, Stable Power to Sensitive Electronic Equipment, *Plant Engineering*, Mar. 17, 1983, pp. 32–37.

Part II

11

Power Distribution and Illumination

11.1 DISTRIBUTION VOLTAGE PROBLEMS

Chapter 10 has covered all types of power disturbances that may occur on a plant distribution system during its operation. However, the discussions have thus far centered on the effects of such disturbances on a computer's operation and performance. It must be realized that the same disturbances—voltage transients, voltage instability, and power interruptions—will affect lamp life and lumen output as much as they will affect computers.

11.2 EFFECTS OF VOLTAGE VARIATION ON LAMP LIFE AND ITS LUMEN OUTPUT

Whenever the voltage at the terminals of a utilization device varies from the device's nameplate rating, something is sacrificed in either the life or performance of the equipment. This is certainly true for all types of lamps. For instance, with incandescent lamps a 1% deviation from rated voltage will cause a change of 3 to 3½% in lumen output. A 10% reduction in lamp voltage will result in a 30% reduction in lumen output; while with an overvoltage of 10%, the lamp life is reduced to one-half of normal (Figure 11.1a). With fluorescent lamps, a 1% variation in the line voltage will change the lumen output only 1%. Both low and high voltages are undesirable and tend to reduce lamp life and lower lumen maintenance (Figure 11.1b). Mercury lamps (one of the HID lamps) are less sensitive and generally give good performence within ±5% variation in line voltage.

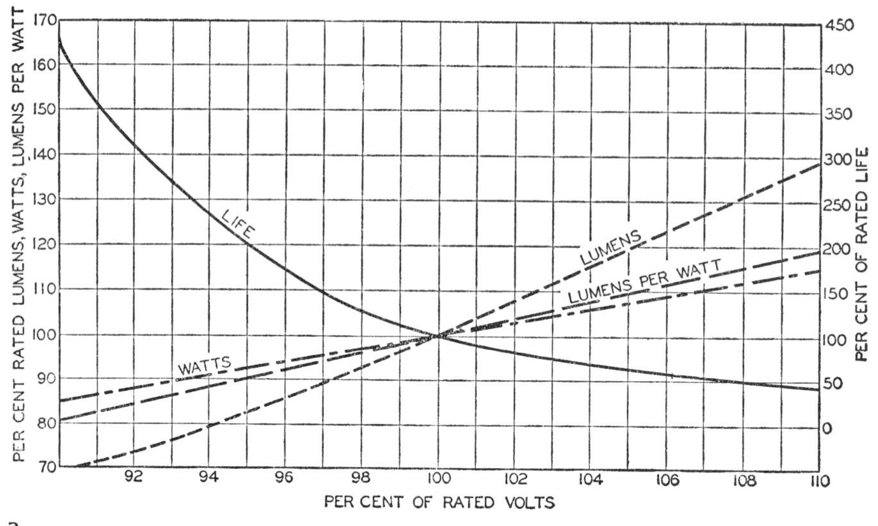

a

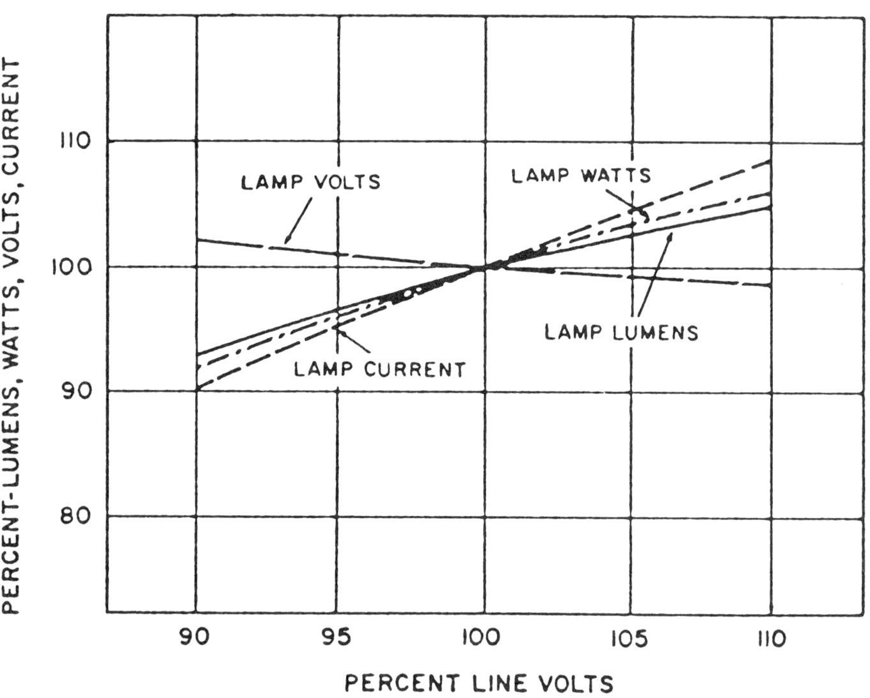

b

Figure 11.1 Lamp performance curves versus voltage variation. (a), incandescent lamps; (b) fluorescent lamps; (c) mercury lamps.

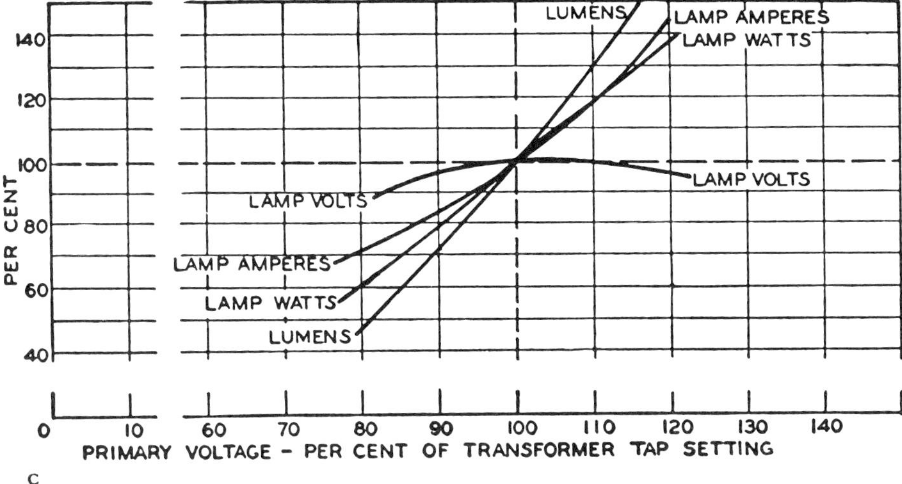

c

Figure 11.1 (Continued).

From the data above, it is evident that good voltage regulation of the plant distribution system is important for better lamp performance. An inherent voltage-regulating device in the form of a load tap changer in the substation transformer can contribute a great deal toward this goal. The initial investment of this type of special equipment can be recovered handsomely from the improved performance of the overall plant power distribution and utilization equipment, including illuminating system components.

11.3 OPERATING VOLTAGE LEVELS FOR ILLUMINATING SYSTEMS

The most common operating voltage level for major types of illuminating systems is 120 V. The 120 V general service filament lamp is considered to be a "standard voltage" lamp. Filament lamps can also operate on a 230 or 250 V system. For fluorescent lamps, most ballasts designed for a 120 V supply can operate over a range of 110 to 125 V. A 277 V ballast can operate over a range of 260 to 290 V. Ballast for mercury lamps are often designed with two primary voltage taps. A connection should be made to the tap which corresponds closely to the supply voltage. The lamps can operate on 115, 230, or 460 V. Fluorescent and mercury lamp ballasts made for high branch-circuit voltage (265 or 277 V) are now used in some industrial plants. Sizable savings can be realized as a result of reduced wiring

Table 11.1 Effect of Varying Voltages on Two Types of Mercury Ballasts and Lamp Performance

Ballast Type	Reactor Ballast			Stabilizer Ballast		
Voltages	Low	High	per cent change	Low	High	per cent change
Line voltage	210	240	14	210	240	14
Lumen output	16,500	22,600	27	21,150	20,600	2
Lamp current (amps)	2.67	3.45	23	3.14	3.24	3.1
Lamp watts	322	437	26	392	403	2.2

and distribution equipment cost. Table 11.1 shows the effect of varying voltages on two types of mercury ballasts and lamps.

11.4 PLANT POWER DISTRIBUTION CONSIDERATIONS FOR LIGHTING LOADS

11.4.1 Based on Lighting-Load Carrying Capacity

Illumination levels have risen in the past years. They have approximately doubled every 10 years. In 1900, illumination levels were about 2 fc; by 1920 they were up to 10; in 1945, 50 fc. Today, 200 fc is fairly common. As the illumination level increases, the requirements for power also increase. As the power requirements go up, the higher voltage level offers greater circuit capacity. In 1900, circuit voltages were in the range 110 to 150 V. About 1910, they were at the level 120/240 V or 120/208 V. About 1946 a new distribution voltage for illumination, 480/277 V, came into being. Footcandle levels are now somewhere around 100. A given circuit is capable of supplying a greater amount of lighting load with higher voltage levels. In the range 100 to 200 fc, the higher voltage of 480/277 V will be required. Above 250 fc, a 480/277 V three-phase four-wire system can be applied. Figure 11.2 shows the trend of illumination levels versus circuit voltages from 1900 to 1960. Figure 11.3 shows the carrying capacity of a 20-A branch circuit at unity power factor at various voltage levels and a variety of systems. It is noted that the 20 A circuit carries up to 12,400 W on a 480/277 V three-phase four-wire system. This represents the high-water mark for the low-voltage distribution. The next step, 600/346 V, presents some technical problems.

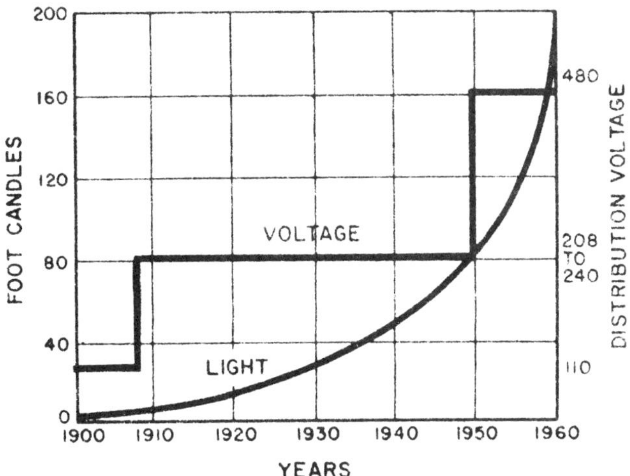

Figure 11.2 Illumination levels versus circuit voltages from 1900 to 1960.

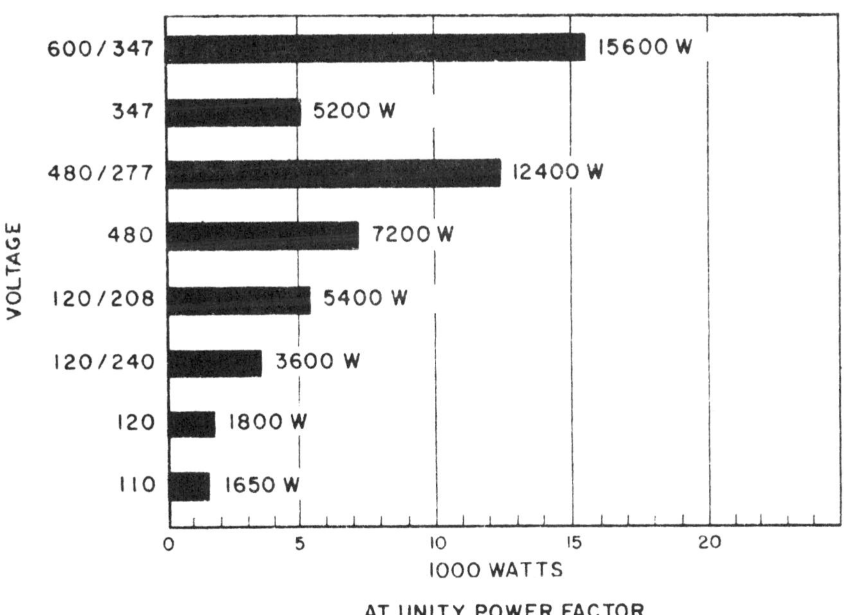

Figure 11.3 Branch circuit carrying capacity on distribution systems.

11.4.2 Based on Service Reliability

Thus far we have realized voltage regulations as one of the most important considerations for power distribution, which can improve illuminating system performance. Service reliability is another important factor in a better illuminating system. Reference should be made to Chapter 4 for various power distributions systems commonly used in industrial plants and their advantages and shortcomings.

In general, a network system has higher reliability than a simple radial system. However, the plant size often determines whether it would be capable of supporting one of the more reliable distribution systems.

11.4.3 Based on Economics

A few years ago, extensive studies were made by Westinghouse engineers on the relative merits of several methods of distributing power to lighting loads. The findings of these studies indicated that a combined distribution system for supplying both power and lighting loads in an industrial plant would be more economical. Table 11.2 shows the practical voltage levels for major illuminating systems.

Table 11.2 Practical Utilization Voltage for Interior Illuminating Systems

Approximate equipment rating (V)	Distribution system	Lamp type[a]		
		Filament	Fluorescent	HID
110–125	120/240 V, 3-wire, 1-phase	*	*	*
110–125	120/208 V, 4-wire, 3-phase	*	*	*
199–216	120/208 V, 3-wire, 3-phase		*	*
220–250	240 V, 2-wire, 1-phase		*	*
220–250	240 V, 3-wire, 3-phase		*	*
254–277	265/460 V, 4-wire, 3-phase		*	*
440–480	460 V, 3-wire, 3-phase			*

[a]Asterisk denotes where practical.

BIBLIOGRAPHY

Chen, Kao, Plant Distribution as Viewed by an Illuminating Engineer, a lecture delivered before the Kansas chapter of IES, May 19, 1966.

Chen, Kao, Industrial and Commercial Distribution Considerations for Improved Lighting Performances, IEEE Industrial and Commercial Power Systems Conference Record, May 1967.

Kahler, W. H., and Bell, R. N., Electric Distribution and Control for Lighting Systems, AIEE Fall General Meeting Proceedings, Oct. 1951.

Lighting Handbook, Westinghouse Electric Corporation, Bloomfield, N.J., Jan. 1976.

12

Illuminating Design Principles

12.1 BASIC CONSIDERATIONS FOR ILLUMINATING DESIGN

12.1.1 Space Function

The function of a space greatly influences the way in which lighting is applied. The same type of visual task may be encountered, regardless of location, in a factory, store, or home. But factors such as economics, appearance, and quality of lighting results desired can influence the lighting design developed for the task. Thus application techniques generally designated as industrial lighting, store lighting, and so on, have developed based on lighting solutions for the types of visual tasks encountered in each type of occupancy. It is necessary to relate the design of a lighting installation to the particular occupancy of the space it is to serve.

12.1.2 Provision of Quality and Quantity of Illumination

Illuminating engineers must know and fully understand the visual sense, and have basic knowledge with regard to selecting relative luminance for the task, its immediate surroundings, and anything else in the peripheral field of view. Research indicates that desirable seeing conditions exist when the luminances of the surroundings and the visual task are relatively uniform and veiling reflections are eliminated or effectively reduced and diminished. Since this condition is not always practical, luminance limitation recommendations will provide a generally satisfactory visual environment. Good practice calls for provision of both quality and quantity of lighting commensurate with the degree of severity of the seeing task. It should be noted that not all visual tasks are in the horizontal plane. Much critical seeing in industry is involved with tasks in a vertical or other nonhorizontal plane. Illuminating engineers must make special provision for luminance distribution and placements to provide task luminance in these nonhorizontal planes.

12.1.3 Selection of Lighting Systems, Sources, Luminaires, and Controls

Illuminating engineers have at their disposal a wide range of types and sizes of light sources, luminaires, and lighting equipment. They should exercise their professional judgment to make choices based on economic analyses and application requirements. General types of lighting are classified as general, local, localized general, supplementary and task-ambient. Luminaires are grouped in five classifications, based on light distribution characteristics, as follows: direct, semi-direct, general diffuse, semi-indirect, and indirect. These classifications are intended to simplify professional discussion relating to lighting techniques as employed for lighting any specific area. Often the local conditions of vibration, ambient temperature, dust, dirt, and color influence light source selection, application, operation, and indirectly the lighting application technique.

Today's interest in lighting energy management dictates more consideration of lighting control. Lighting control systems range from simple photoelectric controls to turn light on and off to sophisticated microprocessor controllers. Lighting control is discussed in more detail in Chapter 18.

12.1.4 Economics

The design of a lighting system is affected by both initial and operating costs. In view of today's high energy cost, considerations of operating cost often outweigh the initial cost. Life-cycle costing is often the best way to measure lighting costs. There is no easy way to predict the exact value of industrial lighting in terms of production, safety, quality control, employee morale, or employee health. Nevertheless, illuminating engineers must balance costs against the attainable results in developing any lighting design, relying to a great extent on experience gained in the solution of comparable problems.

12.1.5 Interior—Exterior Relationship

With increasing widespread circulation and activity at night, the problem of building esthetics extends beyond daytime architecture. The brightness of visible interior surfaces, the pattern of luminaires, and the color of the light source may exert an important influence on the exterior appearance of the building. Also at night, good floodlighting can enhance the factory environment and harvest the advertizing advantage of one's manufactured products. For the subject of floodlighting, more detailed presentations are made in Chapter 16.

12.1.6 Definitions Basic to Illuminating Design

Illumination. Illumination is the density of luminous flux on a surface, expressed in either footcandles (lumens/ft^2) or lux (lx) (lux = 0.0929 fc).

Luminance (or Photometric Brightness). Luminance is the luminous intensity of a surface in a given direction per unit of projected area of the surface, expressed in either candelas per unit area or in lumens per unit area.

Reflectance. Reflectance is the ratio of the light reflected from a surface to that incident upon it. Reflection may be of several types, the most common being specular, diffuse, spread, and mixed.

Glare. Glare is any brightness that causes discomfort, interference with vision, or eye fatigue.

Color Rendering Index (CRI). The CRI is nothing new. In 1964, the CIE (Commission Internationale de l'Eclairage) officially adopted the IES procedure for rating lighting sources and developed the current standard by which light sources are rated for their color rendering properties. Simply defined, the CRI is a numerical value given to the color comparison of one light source to that of a reference light source. The method by which the color rendering properties of a light source are determined is not a simple procedure. By having a better understanding of how the CRI value is obtained, illuminating engineers will be better able to use the CRI more correctly in their design process.

Color Preference Index (CPI). The CPI is determined by a similar procedure to that used for the CRI. The difference is that CPI recognizes the very real human ingredient of preference. This index is based on peoples' preference for the coloration of certain identifiable objects, such as complexions, meat, vegetables, fruit, and foliage, to be slightly different than their colors are in daylight. CPI indicates how a source will render colors with respect to how we best appreciate and remember that color.

Coefficient of Utilization (CU). CU is the ratio of the lumens reaching the working plane (assumed to be a horizontal plane 30 in.

above the floor) to the total lumens generated by the light source. This is a factor that takes into account the efficiency and distribution of the luminaire, its mounting height, the room proportions, and the reflectances of the walls, ceiling, and floor.

Light Loss Factor (LLF). The final LLF is the product of all the contributing loss factors. It is the ratio of the illumination when it reaches its lowest level at the task just before corrective action is taken, to the initial level if none of the contributing loss factors were considered. There are eight contributing loss factors that require consideration:

1. Ballast performance
2. Voltage to luminaires
3. Luminaire reflectance and transmittance changes
4. Lamp outages
5. Luminaire ambient temperature
6. Heat-exchange luminaires
7. Lamp lumen depreciation (LLD)
8. Luminaire dirt depreciation (LDD)

Equivalent Sphere Illumination (ESI). ESI is a means of determining how well a lighting system will provide task visibility in a given situation. ESI may be predicted for many points in a lighting system through the use of any of several available computer programs; or measured in an installation with any of several different types of meters.

Visual Comfort Probability (VCP). Discomfort glare is most often produced by direct glare from luminances that are excessively bright. Discomfort glare can also be caused by reflected glare (i.e., by annoying bright reflections in specular surfaces). Reflected glare should not be confused with veiling reflections, which cause a reduction in visual performance rather than discomfort. VCP rating is based in terms of the percentage of people who will be expected to find the given lighting system acceptable when they are seated in the most undesirable location.

12.2 NEW CONCEPTS IN LIGHTING DESIGN

During the past several decades, most illuminating engineers have been using the so-called "lumen method" to design lighting systems for interior space. For a design that yields average illumination, the coefficient of utilization (CU) is the one single most useful factor. However, this design method will only provide a mediocre lighting environment and would probably result in wasted energy. During the last decade, rapid changes were taking place in lighting design techniques. As people become more energy conscious and better understand human visual performance, illuminating engineers find that the conventional techniques are becoming less and less satisfactory, essentially because the lumen method procedure lacks two desirable characteristics:

1. Light should be placed specifically on the task where it is most needed, while reducing the illumination level in the surrounding areas. Failure to do so often results in a waste of energy.
2. The lumen method does not allow the engineers to increase the visual effectiveness of their lighting by optimizing available candlepower.

12.2.1 New Method of Determining Illuminance Levels

Among the many new concepts for lighting design, the first to be discussed is the new method of determining illuminance levels introduced in the 1981 IES Lighting Handbook, Application Volume. In the past, when illuminating engineers wanted to find the recommended illuminance for a given task, they would look in the Lighting Handbook to find a recommended level and then design a lighting system for the task using the value as a minimum. This procedure provides very little latitude for fine-tuning a lighting design. The new method for determining required illuminance provides three categories based on the room function and visual task available to the design engineers:

1. Rooms involving visual tasks for which there are tables of prescribed illuminance
2. Rooms involving visual tasks for which there are tables of measured visibility or for which visibility can be measured
3. Rooms not involving visual tasks

In each case, a more comprehensive investigation of required illuminance is performed according to the following steps:

1. Instead of a single recommended illuminance value, a category letter is assigned. Table 12.1 shows different category letters for a selected group of industries.
2. The category letters are used to define a range of illuminance. Table 12.2 details illuminance categories and illuminance values for general types of activities in interiors.
3. From within the recommended range of illuminance, a specific value of illuminance is selected after consideration is given to the average age of workers, the importance of speed and accuracy, and the reflectance of task background.

The importance of acknowledging the speed and accuracy with which a task must be performed is readily recognized. Less obvious is the need to consider the age of workers and the reflectance of task background. To compensate for reduced visual acuity, more illuminance is needed. Using the average age of workers as the age criterion is a compromise between the need of the young and the older workers and, therefore, a valid criterion.

Task background affects the ability to see because it affects contrast, an important aspect of visibility. More illuminance is required to enhance the visibility of tasks with poor contrast. Reflectance is calculated by dividing the reflected value by the incident value. The data given in Tables 12.3 and 12.4 are taken from the IES Lighting Handbook and are applied to provide a single value of illuminance from within the range recommended.

12.2.3 Examples Using the New Method to Determine the Required Illuminance

Rooms Involving Visual Tasks (Tabulated Data)

The required illuminance is to be determined for a room for which these data are given:

Task: Rough grinding
Average age of workers: 42
Speed and accuracy: important
Reflectance of task background: 25%

Design Steps

1. Consult Table 12.1 and find rough grinding to fall into category E.
2. Consult Table 12.2 and find that category E stipulates an illuminance range of 500–750–1000 lux (50–75–100 fc).
3. Consult Table 12.4 and obtain the following weighting factor:

   a. Worker's average age (42) 0
   b. Speed/accuracy (important) 0
   c. Reflectance of task background (25%) +1

4. Add the three weighting factors: 0 + 0 + 1 = +1
5. Determine the illuminance value within the prescribed range. The footnote to Table 12.4 states that when the algebraic sum of the weighting factors is −1 or +1, the midrange value should be used. Therefore, the recommended illuminance for this task is 750 lux (75 fc).

Table 12.1 Illuminance Categories for Selected Group of Industries

Area/Activity	Illuminance Category
Aircraft maintenance	[a]
Aircraft manufacturing	[a]
Assembly	
Simple	D
Moderately difficult	E
Difficult	F
Very difficult	G
Exacting	H
Automobile manufacturing	
Bakeries	
Mixing room	D
Face of shelves	D
Inside of mixing bowl	D
Fermentation room	D
Make-up room	
Bread	D
Sweet yeast-raised products	D
Proofing room	D
Oven room	D
Fillings and other ingredients	D
Decorating and icing	
Mechanical	D
Hand	E
Scales and thermometers	D
Wrapping	D
Book binding	
Folding, assembling, pasting	D
Cutting, punching, stitching	E
Embossing and inspection	F
Breweries	
Brew house	D
Boiling and keg washing	D
Filling (bottles, cans, kegs)	D
Candy making	
Box department	D
Chocolate department	
Husking, winnowing, fat extraction, crushing and refining, feeding	D
Bean cleaning, sorting, dipping, packing, wrapping	D
Milling	E
Cream making	
Mixing, cooking, molding	D
Gum drops and jellied forms	D
Hand decorating	D
Hard candy	
Mixing, cooking, molding	D
Die cutting and sorting	E
Kiss making and wrapping	E
Canning and preserving	
Initial grading raw material samples	D
Tomatoes	E
Color grading and cutting rooms	F
Preparation	
Preliminary sorting	
Apricots and peaches	D
Tomatoes	E
Olives	F
Cutting and pitting	E
Final sorting	E
Canning	
Continuous-belt canning	E
Sink canning	E
Hand packing	D
Olives	E
Examination of canned samples	F
Container handling	
Inspection	F
Can unscramblers	E
Labeling and cartoning	D
Casting (see **Foundries**)	
Central stations (see **Electric generating stations**)	
Chemical plants (see **Petroleum and chemical plants**)	
Clay and concrete products	
Grinding, filter presses, kiln rooms	C
Molding, pressing, cleaning, trimming	D
Enameling	E
Color and glazing—rough work	E
Color and glazing—fine work	F
Cleaning and pressing industry	
Checking and sorting	E
Dry and wet cleaning and steaming	E
Inspection and spotting	G
Pressing	F
Repair and alteration	F
Cloth products	
Cloth inspection	I
Cutting	G
Sewing	G
Pressing	F
Clothing manufacture (see **Sewn Products**)	
Receiving, opening, storing, shipping	D
Examining (perching)	I
Sponging, decating, winding, measuring	D
Piling up and marking	E
Cutting	G
Pattern making, preparation of trimming, piping, canvas and shoulder pads	E
Fitting, bundling, shading, stitching	D
Shops	F
Inspection	G
Pressing	F
Sewing	G
Control rooms (see **Electric generating stations—interior**)	
Corridors (see **Service spaces**)	
Cotton gin industry	
Overhead equipment—separators, driers, grid cleaners, stick machines, conveyers, feeders and catwalks	D
Gin stand	D
Control console	D
Lint cleaner	D
Bale press	D
Dairy farms (see **Farms**)	
Dairy products	
Fluid milk industry	
Boiler room	D
Bottle storage	D
Bottle sorting	E

Table 12.1 (Continued)

Area/Activity	Illuminance Category
Bottle washers	b
Can washers	D
Cooling equipment	D
Filling: inspection	E
Gauges (on face)	E
Laboratories	E
Meter panels (on face)	E
Pasteurizers	D
Separators	D
Storage refrigerator	D
Tanks, vats	
Light interiors	C
Dark interiors	E
Thermometer (on face)	E
Weighing room	D
Scales	E
Dispatch boards (see **Electric generating stations—interior**)	
Electrical equipment manufacturing	
Impregnating	D
Insulating: coil winding	E
Electric generating stations—interior (see also **Nuclear power plants**)	
Air-conditioning equipment, air preheater and fan floor, ash sluicing	B
Auxiliaries, pumps, tanks, compressors, gauge area	C
Battery rooms	D
Boiler platforms	B
Burner platforms	C
Cable room	B
Coal handling systems	B
Coal pulverizer	C
Condensers, deaerator floor, evaporator floor, heater floors	B
Control rooms	
Main control boards	D[c]
Auxiliary control panels	D[c]
Operator's station	E[c]
Maintenance and wiring areas	D
Emergency operating lighting	C
Gauge reading	D
Hydrogen and carbon dioxide manifold area	C
Laboratory	E
Precipitators	B
Screen house	C
Soot or slag blower platform	C
Steam headers and throttles	B
Switchgear and motor control centers	D
Telephone and communication eqűipment rooms	D
Tunnels or galleries, piping and electrical	B
Turbine building	
Operating floor	D
Below operating floor	C
Visitor's gallery	C
Water treating area	D
Elevators (see **Service spaces**)	
Explosives manufacturing	
Hand furnaces, boiling tanks, stationary driers, stationary and gravity crystallizers	D
Mechanical furnace, generators and stills, mechanical driers, evaporators, filtration, mechanical crystallizers	D
Tanks for cooking, extractors, percolators, nitrators	D
Farms—dairy	
Milking operation area (milking parlor and stall barn)	
General	C
Cow's udder	D
Milk handling equipment and storage area (milk house or milk room)	
General	C
Washing area	E
Bulk tank interior	E
Loading platform	C
Feeding area (stall barn feed alley, pens, loose housing feed area)	C
Feed storage area—forage	
Haymow	A
Hay inspection area	C
Ladders and stairs	C
Silo	A
Silo room	C
Feed storage area—grain and concentrate	
Grain bin	A
Concentrate storage area	B
Feed processing area	B
Livestock housing area (community, maternity, individual calf pens, and loose housing holding and resting areas)	B
Machine storage area (garage and machine shed)	B
Farm shop area	
Active storage area	B
General shop area (machinery repair, rough sawing)	D
Rough bench and machine work (painting, fine storage, ordinary sheet metal work, welding, medium benchwork)	D
Medium bench and machine work (fine woodworking, drill press, metal lathe, grinder)	E
Miscellaneous areas	
Farm office (see reference 11 in main text)	
Restrooms (see **Service spaces**)	
Pumphouse	C
Farms—poultry (see **Poultry industry**)	
Flour mills	
Rolling, sifting, purifying	E
Packing	D
Product control	F
Cleaning, screens, man lifts, aisleways and walkways, bin checking	D
Forge shops	E
Foundries	
Annealing (furnaces)	D
Cleaning	D
Core making	
Fine	F
Medium	E
Grinding and chipping	F
Inspection	
Fine	G
Medium	F
Molding	
Medium	F
Large	E
Pouring	E
Sorting	E
Cupola	C
Shakeout	D
Garages—parking (see reference 5)	
Garages—service	
Repairs	E
Active traffic areas	C
Write-up	D
Glass works	
Mix and furnace rooms, pressing and lehr, glassblowing machines	C
Grinding, cutting, silvering	D
Fine grinding, beveling, polishing	E
Inspection, etching and decorating	F
Glove manufacturing (see **Sewn Products**)	
Hangars (see **Aircraft manufacturing**)	
Hat manufacturing	
Dyeing, stiffening, braiding, cleaning, refining	E
Forming, sizing, pouncing, flanging, finishing, ironing	F
Sewing	G

Table 12.1 (Continued)

Area/Activity	Illuminance Category
Inspection	
Simple	D
Moderately difficult	E
Difficult	F
Very difficult	G
Exacting	H
Iron and steel manufacturing	[a]
Jewelry and watch manufacturing	G
Laundries	
Washing	D
Flat work ironing, weighing, listing, marking	D
Machine and press finishing, sorting	E
Fine hand ironing	E
Leather manufacturing	
Cleaning, tanning and stretching, vats	D
Cutting, fleshing and stuffing	D
Finishing and scarfing	E
Leather working	
Pressing, winding, glazing	F
Grading, matching, cutting, scarfing, sewing	G
Locker rooms	C
Machine shops	
Rough bench or machine work	D
Medium bench or machine work, ordinary automatic machines, rough grinding, medium buffing and polishing	E
Fine bench or machine work, fine automatic machines, medium grinding, fine buffing and polishing	G
Extra-fine bench or machine work, grinding, fine work	H
Materials handling	
Wrapping, packing, labeling	D
Picking stock, classifying	D
Loading, inside truck bodies and freight cars	C
Meat packing	
Slaughtering	D
Cleaning, cutting, cooking, grinding, canning, packing	D
Nuclear power plants (see also **Electric generating stations**)	
Auxiliary building, uncontrolled access areas	C
Controlled access areas	
Count room	E[c]
Laboratory	E
Health physics office	F
Medical aid room	F
Hot laundry	D
Storage room	C
Engineered safety features equipment	D
Diesel generator building	D
Fuel handling building	
Operating floor	D
Below operating floor	C
Off gas building	C
Radwaste building	D
Reactor building	
Operating floor	D
Below operating floor	C
Offices (see reference 11 in main text)	
Packing and boxing (see **Materials handling**)	
Paint manufacturing	
Processing	D
Mix comparison	F
Paint shops	
Dipping, simple spraying, firing	D
Rubbing, ordinary hand painting and finishing art, stencil and special spraying	D
Fine hand painting and finishing	E
Extra-fine hand painting and finishing	G
Paper-box manufacturing	E
Paper manufacturing	
Beaters, grinding, calendering	D
Finishing, cutting, trimming, papermaking machines	E
Hand counting, wet end of paper machine	E
Paper machine reel, paper inspection, and laboratories	F
Rewinder	F
Petroleum and chemical plants	[a]
Plating	D
Polishing and burnishing (see **Machine shops**)	
Power plants (see **Electric generating stations**)	
Poultry industry (see also **Farm—dairy**)	
Brooding, production, and laying houses	
Feeding, inspection, cleaning	C
Charts and records	D
Thermometers, thermostats, time clocks	D
Hatcheries	
General area and loading platform	C
Inside incubators	D
Dubbing station	F
Sexing	H
Egg handling, packing, and shipping	
General cleanliness	E
Egg quality inspection	E
Loading platform, egg storage area, etc.	C
Egg processing	
General lighting	E
Fowl processing plant	
General (excluding killing and unloading area)	E
Government inspection station and grading stations	E
Unloading and killing area	C
Feed storage	
Grain, feed rations	C
Processing	C
Charts and records	D
Machine storage area (garage and machine shed)	B
Printing industries	
Type foundries	
Matrix making, dressing type	E
Font assembly—sorting	D
Casting	E
Printing plants	
Color inspection and appraisal	F
Machine composition	E
Composing room	E
Presses	E
Imposing stones	F
Proofreading	F
Electrotyping	
Molding, routing, finishing, leveling molds, trimming	E
Blocking, tinning	D
Electroplating, washing, backing	D
Photoengraving	
Etching, staging, blocking	D
Routing, finishing, proofing	E
Tint laying, masking	E
Quality Control (see **Inspection**)	
Receiving and shipping (see **Materials handling**)	
Rubber goods—mechanical	(see Table B3)[a]
Rubber tire manufacturing	(see Table B3)[a]
Safety	(see Section 6 and Table 8)
Sawmills	
Secondary log deck	B
Head saw (cutting area viewed by sawyer)	E
Head saw outfeed	B
Machine in-feeds (bull edger, resaws, edgers, trim, hula saws, planers)	B
Main mill floor (base lighting)	A

Table 12.1 (Continued)

Area/Activity	Illuminance Category
Sorting tables	D
Rough lumber grading	D
Finished lumber grading	F
Dry lumber warehouse (planer)	C
Dry kiln colling shed	B
Chipper infeed	B
Basement areas	
Active	A
Inactive	A
Filing room (work areas)	E
Service spaces (see also **Storage rooms**)	
Stairways, corridors	D
Elevators, freight and passenger	B
Toilets and wash rooms	C
Sewn products	
Receiving, packing, shipping	E
Opening, raw goods storage	E
Designing, pattern-drafting, pattern grading and markermaking	F
Computerized designing, pattern-making and grading, digitizing, marker-making, and plotting	B
Cloth inspection and perching	I
Spreading and cutting (includes computerized cutting)	F[g]
Fitting, sorting and blunding, shading, stitch marking	G
Sewing	G
Pressing	F
In-process and final inspection	G
Finished goods storage and picking orders	F[h]
Trim preparation, piping, canvas and should pads	F
Machine repair shops	G
Knitting	F
Sponging, decating, rewinding, measuring	E
Hat manufacture (see **Hat Manufacture**)	
Leather working (see **Leather Working**)	
Shoe manufacturing (see **Shoe Manufacturing**)	
Sheet metal works	
Miscellaneous machines, ordinary bench work	E
Presses, shears, stamps, spinning, medium bench work	E
Punches	E
Tin plate inspection, galvanized	F
Scribing	F
Shoe manufacturing—leather	
Cutting and stitching	
Cutting tables	G
Marking, buttonholing, skiving, sorting, vamping, counting	G
Stitching, dark materials	G
Making and finishing, nailers, sole layers, welt beaters and scarfers, trimmers, welters, lasters, edge setters, sluggers, randers, wheelers, treers, cleaning, spraying, buffing, polishing, embossing	F
Shoe manufacturing—rubber	
Washing, coating, mill run compounding	D
Varnishing, vulcanizing, calendering, upper and sole cutting	D
Sole rolling, lining, making and finishing processes	E
Soap manufacturing	
Kettle houses, cutting, soap chip and powder	D
Stamping, wrapping and packing, filling and packing soap powder	D
Stairways (see **Service spaces**)	
Steel (see **Iron and steel**)	
Storage battery manufacturing	D
Storage rooms or warehouses	
Inactive	B
Active	
Rough, bulky items	C
Small items	D
Structural steel fabrication	E
Sugar refining	
Grading	E
Color inspection	F
Testing	
General	D
Exacting tests, extra-fine instruments, scales, etc.	F
Textile mills	
Staple fiber preparation	
Stock dyeing, tinting	D
Sorting and grading (wool and cotton)	E[d]
Yarn manufacturing	
Opening and picking (chute feed)	D
Carding (nonwoven web formation)	D[e]
Drawing (gilling, pin drafting)	D
Combing	D[e]
Roving (slubbing, fly frame)	E
Spinning (cap spinning, twisting, texturing)	E
Yarn preparation	
Winding, quilling, twisting	E
Warping (beaming, sizing)	F[d]
Warp tie-in or drawing-in (automatic)	E
Fabric production	
Weaving, knitting, tufting	F
Inspection	G[d]
Finishing	
Fabric preparation (desizing, scouring, bleaching, singeing, and mercerization)	D
Fabric dyeing (printing)	D
Fabric finishing (calendaring, sanforizing, sueding, chemical treatment)	E[d]
Inspection	G[d,f]
Tobacco products	
Drying, stripping	D
Grading and sorting	F
Toilets and wash rooms (see **Service spaces**)	
Upholstering	F
Warehouse (see **Storage rooms**)	
Welding	
Orientation	D
Precision manual arc-welding	H
Woodworking	
Rough sawing and bench work	D
Sizing, planing, rough sanding, medium quality machine and bench work, gluing, veneering, cooperage	D
Fine bench and machine work, fine sanding and finishing	E

[a] Industry representatives have established a table of single illuminance values which, in their opinion, can be used in preference to employing reference 6. Illuminance values for specific operations can also be determined using illuminance categories of similar tasks and activities found in this table and the application of the appropriate weighting factors in Table 3.

[b] Special lighting such that (1) the luminous area is large enough to cover the surface which is being inspected and (2) the luminance is within the limits necessary to obtain comfortable contrast conditions. This involves the use of sources of large area and relatively low luminance in which the source luminance is the principal factor rather than the illuminance produced at a given point.

[c] Maximum levels—controlled system.

[d] Supplementary lighting should be provided in this space to produce the higher levels required for specific seeing tasks involved.

[e] Additional lighting needs to be provided for maintenance only.

[f] Color temperature of the light source is important for color matching.

[g] Higher levels from local lighting may be required for manually operated cutting machines.

[h] If color matching is critical, use illuminance category G.

Table 12.2 Illuminance Categories and Illuminance Values for Generic Types of Activities in Interiors

Type of Activity	Illuminance Category	Ranges of Illuminances		Reference Work-Plane
		Lux	Footcandles	
Public spaces with dark surroundings	A	20–30–50	2–3–5	General lighting throughout spaces
Simple orientation for short temporary visits	B	50–75–100	5–7.5–10	
Working spaces where visual tasks are only occasionally performed	C	100–150–200	10–15–20	
Performance of visual tasks of high contrast or large size	D	200–300–500	20–30–50	Illuminance on task
Performance of visual tasks of medium contrast of small size	E	500–750–1000	50–75–100	
Performance of visual tasks of low contrast or very small size	F	1000–1500–2000	100–150–200	
Performance of visual tasks of low contrast and very small size over a prolonged period	G	2000–3000–5000	200–300–500	Illuminance on task, obtained by a combination of general and local (supplementary lighting)
Performance of very prolonged and exacting visual tasks	H	5000–7500–10000	500–750–1000	
Performance of very special visual tasks of extremely low contrast and small size	I	10000–15000–20000	1000–1500–2000	

Table 12.3 Weighting Factors for Selecting Specific Illuminance Within Ranges A, B, and C

Occupant and room characteristics	Weighting factor −1	0	+1
Workers' age (average)	Under 40	40 to 55	Over 55
Average room reflectance*	>70 percent	30 to 70 percent	<30 percent

SOURCE: *IES Lighting Handbook,* sixth edition.

NOTE: This table is used for assessing weighting factors in rooms where a task is not involved.

1. Assign the appropriate weighting factor for each characteristic.
2. Add the two weights; refer to Table 12.2, Categories A through C:
   a. If the algebraic sum is −1 or −2, use the lowest range value.
   b. If the algebraic sum is 0, use the middle range value.
   c. If the algebraic sum is +1 or +2, use the highest range value.

*To obtain average room reflectance: determine the areas of ceiling, walls and floor; add the three to establish room surface area; determine the proportion of each surface area to the total; multiply each proportion by the pertinent surface reflectance; and add the three numbers obtained.

Rooms Involving Visual Task (Visibility Measurements)

Over the years, IES has accumulated visibility data for a large number of disparate tasks. These data were accumulated by field and laboratory use of a complex instrument known as a visibility meter. The data obtained by visibility meter form the basis for a factor known as equivalent contrast, designated by the symbol $\tilde{C}$, usually referred to as C-wave or C-wiggle (Table 12.5). Using this table, illuminating engineers can apply judgement and determine the value of $\tilde{C}$ for comparable seeing tasks. In this example, the following data are given for an area whose illuminance requirements are to be determined:

Task: detecting hairline cracks in gray metal castings
Age of inspector: 57
Speed and accuracy: critical
Reflectance of task background: 25%

Design Steps

1. From Table 12.5, determine the value of $\tilde{C}$ for the task to be 0.79.

Table 12.4 Weighting Factors for Selecting Specific Illuminance Within Ranges D through I.

Task or worker characteristics	Weighting factor −1	0	+1
Workers' age (average)	Under 40	40 to 55	Over 55
Speed or accuracy*	Not important	Important	Critical
Reflectance of task background, percent	⟩70 percent	30 to 70 percent	⟨30 percent

SOURCE: *IES Lighting Handbook,* sixth edition.

NOTE: Weighting factors are based upon worker and task information.

1. Assign the appropriate factor for each characteristic.
2. Add the three weighting factors and refer to Table 12.2, Categories D through I:
   a. If the algebraic sum is −2 or −3, use the lowest range value.
   b. If the algebraic sum is −1, 0, or +1, use the middle range value.
   c. If the algebraic sum is +2 or +3, use the highest range value.

*Evaluation of speed and accuracy requires that time limitations, the effect of error on safety, quality, and cost, etc., be considered. For example, leisure reading imposes no restrictions on time, and errors are seldom costly or unsafe. Reading engineering drawings or a micrometer requires accuracy and, sometimes, speed. Properly positioning material in a press or mill can impose demands on safety, accuracy, and time.

2. Refer to Table 12.6. Because the C̆ value of 0.79 lies between 0.75 and 1.0, illuminance category D applies.
3. From Table 12.2, determine that category D stipulates 200−300−500 lux (20−30−50 fc).
4. Consult Table 12.4 to determine the weighting factors:

   a. Worker;s age (57) +1
   b. Speed/accuracy (critical) +1
   c. Reflectance of task background (25%) +1

5. Add the three weighting factors: 1 + 1 + 1 = +3
6. Determine the illuminance value within the prescribed range. The footnote to Table 12.4 states that when the algebraic sum of the weighting factors is +2 or +3, the highest value in the given range should be used. Therefore, the recommended value of maintained illuminance is 500 lux (50 fc).

Table 12.5 Typical Equivalent Contrast ($\tilde{C}$) Values

Task Description*	$\tilde{C}$
Detecting hairline crack on polished stainless steel vane	0.22
Reading white line on blueprint	0.28
Reading shelf label (black on pink)	0.44
Reading old micrometer	0.58
Reading metal nameplate	0.62
Reading shorthand notes (No. 3 pencil)	0.66
Detecting hairline crack in gray metal casting	0.79
Detecting bead defect in new tire	0.90
Seeing broken gray thread in bobbin	1.97
Seeing slot in head of Allen screw	2.63
Detecting defective solder joint	3.39
Detecting crack in rung of ladder	4.68

NOTE: Equivalent contrast ($\tilde{C}$) is the numerical description of a task's relative visibility. For an in-depth explanation, refer to the *IES Lighting Handbook.*

*These task descriptions and values are representative and may or may not be pertinent to similar tasks.

Table 12.6 Illuminance Categories for Measured Equivalent Contrast ($\tilde{C}$) Values

Equivalent Contrast $\tilde{C}$	**Illuminance Category**
over 1.0	*
.75–1.0	D
.62–.75	E
.50–.62	F
.40–.50	G
.30–.40	H
under .30	I

It should be noted that for tasks such as that described herein, the quality of light is just as important—possibly more important—than the quantity of light applied on the work plane.

Rooms Not Involving Visual Tasks

Rooms such as lobbies, entrance foyers, and hallways serve functions other than seeing tasks. The purpose of lighting these rooms is to provide comfort, safety, and well-being. Here there is no need to consider speed and accuracy. Consideration of reflectance applies to the average of the room surfaces rather than to a task. In this example, the illuminance requirements are to be determined for an industrial plant visitor's lobby. Room sizes are 40 × 25 × 12. Reflectances for ceiling, wall, and floor are 80, 60, and 20, respectively. The average age of occupants is assumed to be 56 years.

Design Steps

1. From Table 12.1 it is found that entrance lobbies fall into category C.
2. From Table 12.2 it is found that category C suggests a range of 100–150–200 lux (10–15–20 fc).
3. From Table 12.3 it is determined that the weighting factor for age 56 is +1. Calculations are then performed to determine the value of average reflectance to be used to determine the weighting factor of reflectance; the result of these calculations are summarized in Table 12.7. For the average reflectance of 54.4 obtained from these calculations, Table 12.3 indicates a zero weighting factor. The total weighting factor to be applied, then is +1.

Table 12.7 Summary of Eaxmple Calculations to Determine Average Room Reflectance

Surface	Area, sq ft	Proportion of total area, percent	Reflectance of surface	Proportion of total reflectance, percent
Ceiling	1000	28	0.80	22.4
Walls	1560	44	0.60	26.4
Floor	1000	28	0.20	5.6
Total	**3650**	**100**	—	—
Average reflectance (sum of proportions of total reflectance)				**54.4**

NOTE: Room dimensions are 40 x 25 x 12 ft high.

4. Select the proper value of illuminance from within the prescribed range. The footnote to Table 12.3 states that when the algebraic sum of the weighting factor is +1 or +2, the highest value within the prescribed range should be chosen. Therefore, the recommended maintained illuminance for the lobby area is 200 lux (20 fc).

Lighting system design can begin after the desired value of illuminance for a given task has been determined. Based on the IES Handbook, the zonal cavity method of determining the number of luminaires and lamps to yield a specified maintained luminance remains unchanged.

12.3 ZONAL CAVITY METHOD OF LIGHTING COMPUTATIONS

Introduced in 1964, the zonal cavity method of performing lighting calculations has gained rapid acceptance as the preferred way to calculate number and placement of luminaires required to satisfy a specified light-level requirement. Zonal cavity provides a higher degree of accuracy than does the old lumen method, because it gives individual consideration to factors that are glossed over empirically in the lumen system.

12.3.1 Definition of Cavities

With zonal cavity method, the room is considered to contain three vertical zones, or cavities. Figure 12.1 defines the various cavities used in this method of computation. Height from luminaire to ceiling is designated as the ceiling cavity (h_{cc}). Distance from luminaire to the work plane is the room cavity (h_{rc}), and the floor cavity (h_{fc}) is measured from the work plane to the floor.

To apply the zonal cavity method, it is necessary to determine a parameter known as the "cavity ratio" for each of the three cavities. Following is the formula for determining the cavity ratio:

$$\text{cavity ratio} = \frac{5h(\text{room length} + \text{room width})}{\text{room length} \times \text{room width}}$$

where

$h = h_{cc}$ for ceiling cavity ratio (CCR)
$= h_{rc}$ for room cavity ratio (RCR)
$= h_{fc}$ for floor cavity ratio (FCR)

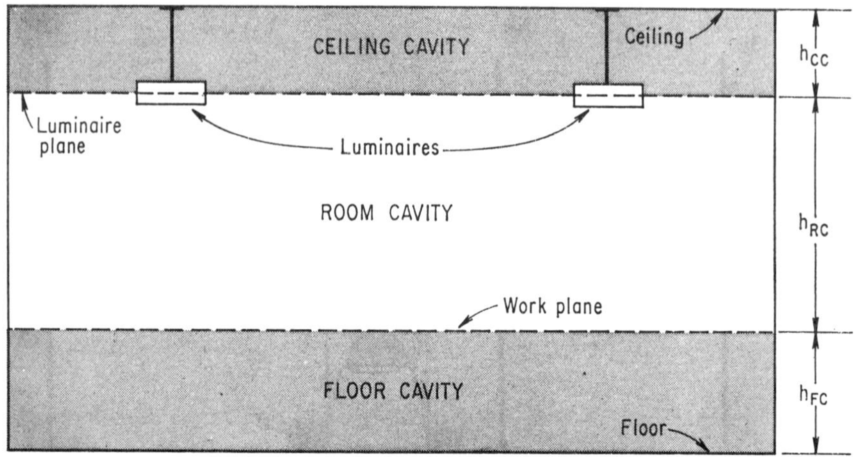

Figure 12.1 Basic cavity divisions of space.

12.3.2 Lumen Method Details

Because of the ease of application of the lumen method, which yields the average illumination in a room, it is usually employed for larger areas, where the illumination is substantially uniform. The lumen method is based on the definition of a footcandle equaling 1 lumen per square foot:

$$\text{footcandle} = \frac{\text{lumens striking an area}}{\text{square feet of area}}$$

In order to take into consideration such factors as dirt on the luminaire, general depreciation in lumen output of the lamp, and so on, the formula above is modified as follows:

$$\text{footcandle} = \frac{\text{lamps/luminaire} \times \text{lumens/lamp} \times \text{CU} \times \text{LLF}}{\text{area/luminaire}}$$

In using the lumen method, the following key steps should be taken:

1. Determine the required level of luminance. Use methods given in Section 12.2.1 and examples in Section 12.2.2 to determine the proper level of illuminance.
2. Determine the coefficient of utilization (CU). The coefficient of utilization is the ratio of the lumens reaching the working plane

to the total lumens generated by the lamps. It is a factor that takes into account the efficiency and the distribution of the luminaire, its mounting height, the room proportions, and the reflectances of the walls, ceiling, and floor. In general, the higher and narrower the room, the larger the percentage of light absorbed by the walls and the lower the coefficient of utilization. Rooms are classified according to shape by 10 room cavity numbers. The cavity ratio can be calculated using the formula given in Section 12.3.1. The coefficient of utilization should be selected from tables prepared for various luminaires by manufacturers. It can be determined for the proper room cavity ratio and the appropriate wall reflectance and ceiling cavity reflectance. For ceiling-mounted or recessed luminaires, the ceiling cavity reflectance is the same as the reflectance of the actual ceiling.

3. Determine the light loss factor (LLF). The final light loss factor is the product of all the contributing loss factors. Lamp manufacturers rate filement lamps in accordance with lumen output when the lamp is new; vapor discharge lamps (fluorescent, mercury, and other types) are rated in accordance with their output after 100 h of burning. The light loss factor is made up as the product of eight different contributing factors, as discussed in Section 12.1.6.
4. Calculate the number of lamps and luminaires required.

$$\text{number of lamps} = \frac{\text{footcandles} \times \text{area}}{\text{lumens/lamp} \times \text{CU} \times \text{LLF}}$$

$$\text{number of luminaires} = \frac{\text{number of lamps}}{\text{lamps/luminaire}}$$

5. Determine the location of the luminaires. Luminaire locations depend on the general architecture, size of bays, type of luminaire, position of previous outlets, and so on.

12.4 POINT-BY-POINT LIGHTING COMPUTATIONS

Although currently, lighting computations emphasize the zonal cavity method, there is still considerable merit in the point-by-point method. This method lends itself especially well to calculating the illumination level at a particular point where total illumination is the sum of general overhead lighting and supplementary lighting. In this method, information from luminaire candlepower distribution curves must be applied to the mathematical relationship. The total contribution from all luminaires to the illumination level on the task plane must be summed.

12.4.1 Computation of Direct Illumination Component

The angular coordinate system is most applicable to continuous rows of fluorescent luminaries. Two angles are involved; a longitudinal angle α and a lateral angle β. Angle α is the angle between a vertical line passing through the seeing task (point P) and a line from the seeing task to the end of the rows of luminaires. If the seeing task is not in the vertical plane of a row of luminaires, a parallel reference plane is created for the specification of angle α.

Angle α is easily determined graphically from a chart showing angles α and β for various combinations of V and H. Usually, all rows of luminaires have the same coordinates, one coordinate for each end of the row. Angle β is the angle between the vertical plane of the row of luminaires and a tilted plane containing both the seeing task and the luminaire or row of luminaires. In determining angle β, H is the horizontal distance from the seeing task to the row of luminaires, measured perpendicular to the luminaire. Each row has only one β coordinate. Figure 12.2 shows how angles α and β are defined. The direct illumination component for each luminaire or row of luminaires is determined by referring to the table of direct illumination components for the specific luminaire. The direct illumination components are based on the assumption that the luminaire is mounted 6 ft above the seeing task. If this mounting height is other than 6 ft, the direct illumination components shown in the Table 12.8 must

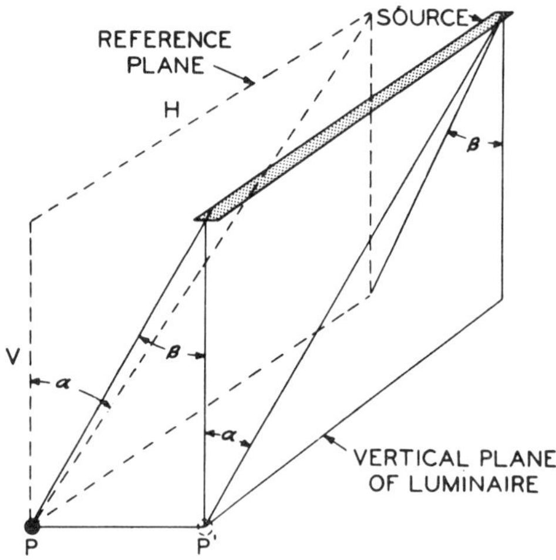

Figure 12.2 Definition of angular coordinate system for direct illumination component.

Table 12.8 Direct Illumination Components for Category III Luminaire (Based on F40 Lamps Producing 3100 Lumens)

Direct Illumination Components

β	5	15	25	35	45	55	65	75	5	15	25	35	45	55	65	75
∝	Vertical Surface Illumination Footcandles at a Point On a Plane Parallel to Luminaires								Vertical Surface Illumination Footcandles at a Point On a Plane Perpendicular to Luminaires							
0-10	.9	2.6	3.6	3.9	3.3	1.9	.7	.1	.9	.8	.7	.5	.3	.1	...	...
0-20	1.8	5.0	7.0	7.7	6.6	3.8	1.5	.2	3.6	3.2	2.7	1.9	1.2	.5	.1	...
0-30	2.6	7.2	10.1	11.3	9.8	5.7	2.3	.3	7.7	7.0	5.8	4.3	2.7	1.1	.3	...
0-40	3.2	9.0	12.8	14.5	12.9	7.7	3.2	.5	12.6	11.6	9.7	7.5	4.9	2.1	.6	...
0-50	3.7	10.3	14.9	17.1	15.7	9.6	4.3	.7	17.8	16.6	14.2	11.2	7.7	3.4	1.1	.1
0-60	4.0	11.2	16.3	18.8	17.6	11.3	5.5	1.0	22.6	21.2	18.4	14.7	10.4	5.1	1.9	.2
0-70	4.1	11.6	17.0	19.8	18.9	12.7	6.8	1.4	26.2	24.7	21.8	17.8	13.1	7.2	3.2	.3
0-80	4.1	11.7	17.3	20.2	19.4	13.3	7.4	1.9	28.2	26.7	23.8	19.7	14.9	8.7	4.3	.8
0-90	4.1	11.7	17.3	20.2	19.4	13.4	7.5	2.0	28.6	27.1	24.2	20.1	15.3	9.1	4.7	1.1

F.C. at a Point on Workplane

∝	5	15	25	35	45	55	65	75
0-10	10.6	9.5	7.6	5.5	3.3	1.3	.3	...
0-20	20.6	18.5	14.9	10.9	6.6	2.6	.7	...
0-30	29.4	26.5	21.6	16.0	9.8	4.0	1.1	...
0-40	36.5	33.1	27.4	20.6	12.9	5.4	1.5	...
0-50	41.8	38.1	31.9	24.3	15.7	6.7	2.0	.1
0-60	45.2	41.3	34.8	26.8	17.6	7.9	2.6	.2
0-70	46.9	43.0	36.4	28.3	18.9	8.9	3.2	.3
0-80	47.4	43.6	36.9	28.8	19.4	9.3	3.5	.4
0-90	47.5	43.7	37.0	28.8	19.4	9.3	3.5	.4

Category III

2 T-12 Lamps — Any Loading
For T-10 Lamps — C.U. × 1.02

Luminance Coefficients for 20% Effective Floor Cavity Reflectance

Ceiling Cavity		Reflectances 80		50		10		80		50		10	
Walls		50	30	50	30	50	30	50	30	50	30	50	30
WDRC	RCR	Wall Luminance Coefficients						Ceiling Cavity Luminance Coefficients					
.281	1	.246	.140	.220	.126	.190	.109	.230	.209	.135	.124	.025	.023
.266	2	.232	.127	.209	.115	.182	.102	.222	.190	.130	.113	.024	.021
.245	3	.216	115	.196	.105	.172	.095	.215	.176	.127	.105	.024	.020
.226	4	.202	.102	.183	.097	.161	.088	.209	.164	.124	.099	.023	.019
.212	5	.191	.097	.173	.090	.154	.082	.204	.156	.121	.094	.023	.018
.196	6	.178	.090	.163	.084	.145	.076	.200	.149	.118	.090	.022	.017
.182	7	.168	.083	.153	.078	.136	.071	.194	.144	.115	.087	.022	.017
.170	8	.158	.077	.145	.072	.130	.066	.190	.139	.113	.085	.021	.016
.159	9	.150	.072	.138	.068	.123	.062	.185	.135	.110	.082	.021	.016
.149	10	.141	.068	.130	.064	.116	.059	.180	.131	.107	.080	.020	.016

be multiplied by 6/V, where V is the mounting height above the task. Thus the total direct illumination component would be the product of 6/V and the sum of the individual direct illumination component of each row.

12.4.2 Computations of Reflected Illumination Component

On Horizontal Surfaces

The reflected illumination component on horizontal surfaces is calculated in exactly the same manner as the average illumination is computed using the lumen method, except that the RRC, the reflected radiation coefficient, is substituted for the coefficient of utilization.

$$FC_{RH} = \frac{\text{lamps/luminaire} \times \text{lumens/lamps} \times \text{RRC} \times \text{LLF}}{\text{area/luminaire (on work plane)}}$$

where

$$\text{reflected radiation coefficient} = LC_W + RPM(LC_{CC} - LC_W)$$

where

LC_W = wall luminance coefficient
LC_{CC} = ceiling cavity luminance coefficient
RPM = room position multiplier

The wall luminance coefficient and the ceiling cavity luminance coefficient are selected for the appropriate room cavity ratio and proper wall and ceiling cavity reflectances from the table of luminance coefficients in the same manner as the coefficient of utilization is selected from the coefficient of utilization table in using the lumen method. The room position multiplier is a function of the room cavity ratio and of the location in the room of the point where the illumination is desired. Table 12.9 lists the value of the RPM for each possible location of the part in the rooms of all room cavity ratios. Figure 12.3 shows a grid diagram that illustrates the method of designating the location in the room by a letter and a number.

On Vertical Surfaces

To determine illumination reflected to vertical surfaces, the approximate average value is determined using the same general formula, but substituting the wall reflected radiation coefficient (WRRC) for the coefficient of utilization.

Table 12.9 Room Position Multipliers

	A	B	C	D	E	F
Room Cavity Ratio = 1						
0	.24	.42	.47	.48	.44	.48
1	.42	.74	.81	.83	.84	.84
2	.47	.81	.90	.92	.93	.93
3	.48	.83	.92	.94	.95	.95
4	.48	.84	.93	.95	.96	.97
5	.48	.84	.93	.95	.97	.97
Room Cavity Ratio = 2						
0	.24	.36	.42	.44	.46	.46
1	.36	.51	.60	.63	.66	.68
2	.42	.60	.68	.72	.78	.83
3	.44	.63	.72	.77	.82	.85
4	.46	.66	.78	.82	.85	.86
5	.46	.68	.83	.85	.86	.87
Room Cavity Ratio = 3						
0	.23	.32	.37	.40	.42	.42
1	.32	.40	.48	.51	.53	.57
2	.37	.48	.58	.61	.64	.67
3	.40	.51	.61	.65	.69	.71
4	.42	.53	.64	.69	.73	.75
5	.42	.57	.67	.71	.75	.77
Room Cavity Ratio = 4						
0	.22	.28	.32	.35	.37	.37
1	.28	.33	.40	.42	.44	.48
2	.32	.40	.48	.50	.52	.57
3	.35	.42	.50	.54	.58	.61
4	.37	.44	.52	.58	.62	.64
5	.37	.48	.57	.61	.64	.66
Room Cavity Ratio = 5						
0	.21	.25	.28	.31	.33	.33
1	.25	.29	.33	.36	.38	.42
2	.28	.33	.40	.42	.44	.48
3	.31	.36	.42	.46	.49	.52
4	.33	.38	.44	.49	.52	.54
5	.33	.42	.48	.52	.54	.56

	A	B	C	D	E	F
Room Cavity Ratio = 6						
0	.20	.23	.26	.28	.29	.30
1	.23	.26	.29	.31	.33	.36
2	.26	.29	.35	.37	.38	.40
3	.28	.31	.37	.39	.41	.43
4	.29	.33	.38	.41	.43	.45
5	.30	.36	.40	.43	.45	.47
Room Cavity Ratio = 7						
0	.18	.21	.23	.25	.26	.27
1	.21	.23	.26	.28	.29	.30
2	.23	.26	.30	.32	.33	.34
3	.25	.28	.32	.34	.35	.36
4	.26	.29	.33	.35	.37	.37
5	.27	.30	.34	.36	.37	.38
Room Cavity Ratio = 8						
0	.17	.18	.21	.22	.22	.23
1	.18	.20	.23	.25	.26	.26
2	.21	.23	.26	.27	.28	.29
3	.22	.25	.27	.29	.30	.30
4	.22	.26	.28	.30	.31	.32
5	.23	.26	.29	.30	.31	.32
Room Cavity Ratio = 9						
0	.15	.17	.18	.19	.20	.20
1	.17	.18	.20	.21	.22	.23
2	.18	.20	.23	.24	.25	.25
3	.19	.21	.24	.25	.26	.26
4	.20	.22	.25	.26	.26	.27
5	.20	.23	.25	.26	.27	.27
Room Cavity Ratio = 10						
0	.14	.16	.16	.17	.18	.18
1	.16	.17	.18	.19	.19	.20
2	.16	.18	.19	.21	.22	.22
3	.17	.19	.21	.22	.23	.23
4	.18	.19	.22	.23	.23	.24
5	.18	.20	.22	.23	.24	.25

$$FC_{RV} = \frac{\text{lamps/luminaire} \times \text{lumens/lamp} \times \text{WRRC} \times \text{LLF}}{\text{area/luminaire (on work plane)}}$$

where the wall reflected radiation coefficient (WRRC) is found as follows:

$$\text{WRRC} = \frac{\text{wall luminance coefficient}}{\text{average wall reflectance}} - \text{WDRC}$$

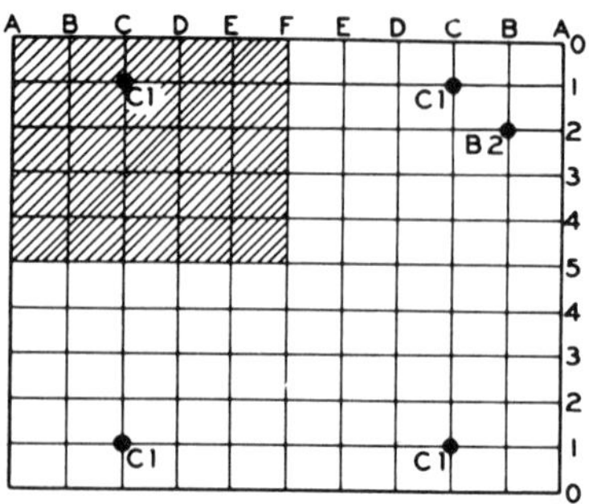

Figure 12.3 Grid diagram for locating points on the work plane.

where WDRC, the wall direct radiation coefficient, is published for each room cavity ratio together with a table of wall luminance coefficients. (See Table 12.8 for a specific type of luminaire.)

Typical Example

As an example of the calculation of the illumination at a point, assume that four rows of six 4-ft luminaires (for which data are shown in Table 12.8) are surface mounted on 8-ft centers in a room 28 by 30 ft. Assume that the ceiling reflectance (and also that of the ceiling cavity since the luminaires are ceiling mounted) is 80% and that of the walls is 50%. Floor cavity reflectance is 20%. The mounting height of the luminaire is 8½ ft above the work plane. The initial illumination on the horizontal work plane at point P is desired. (See Figure 12.4, a typical luminaire layout plan for this example.)

Calculation of Direct Component. First let us determine angle α for both ends of the rows of luminaires, and angle β. For angle α, H is 10 ft for α_1 and 14 ft for α_2. The vertical distance V is 8½ ft. For angle β, H is 12 ft for rows A and D and is 4 ft for rows B and C. The vertical distance is still 8½ ft. Refer to Table 12.8 for data on the direct illumination component. Table 12.10 summarizes the results of various components as found from the data in Table 12.8. Since the direct illumination component table is all based on a mounting height of 6 ft above the point, and in this case the luminaires are actually 8½ ft above point P, it is necessary to multiply the total fc by 6/8.5. The resultant direct component is 114.8 fc.

Calculation of Reflected Component. The RCR for this room is 3.0 and the area per luminaire is 35 ft^2. Using the formula for computing the initial value of the reflected illumination component on the horizontal, FC_{RH},

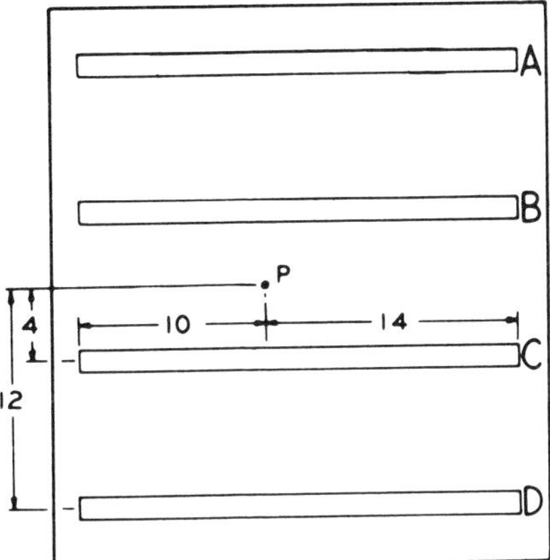

Figure 12.4 A typical luminaire layout plan.

Table 12.10 Summary of Direct Illumination Components

Row	α_1	α_2	β	Direct illumination component: Front left end	Front right end	Total
A	50	60[a]	55	6.7	7.9	14.6
B	50	60	25	31.9	34.8	66.7
C	50	60	25	31.9	34.8	66.7
D	50	60	55	6.7	7.9	14.6
						162.6

[a]Actually α_2 is 59, but is rounded off to 60.

$$FC_{RH} = \frac{2 \times 3200 \times \text{reflected radiation coefficient}}{35}$$

The reflected radiation coefficient

$$\begin{aligned} RRC &= LC_W + RPM(LC_{CC} - LC_W) \\ &= 0.155 + RPM(0.157 - 0.155) \\ &= 0.155 + 0.75 \times 0.002 \\ &= 0.157 \end{aligned}$$

RPM is taken from Table 12.9 at point E5, and $FC_{RH} = 28.7$. The total illumination at point P is 114.8 + 28.7 = 143.5 fc.

12.5 COMPUTER PROGRAMS FOR LIGHTING DESIGN

Several methods of point-by-point computation are available. Such computations can best be handled by computer. Assuming that the necessary programing has been worked out, the computer operator needs only to supply the details of the room, the geometry of the lighting installation, room surface reflectances, and the candlepower data for the luminaire in order to provide a printout of footcandles at the desired points.

12.5.1 Computation of ESI Values

Despite the fact that point-by-point computation allows for a particular purpose, this lighting design approach still restricts the illuminating engineers to determining "raw" footcandles and does not provide a measure of whether satisfactory visibility will be produced. With typical reading and writing tasks, visual performance loss caused by veiling reflections should be taken into account. This can only be achieved by equivalent sphere illumination (ESI) computations. The IES report provided the mathematical details for computing ESI from candlepower data which are the same as those for the point-by-point computations. However, the viewing direction of the observer must be specified, as ESI can change substantially with the viewer's orientation. ESI computation is similar to point-by-point computation except that the direction of all light rays is analyzed and associated with the reflecting characteristics of the task.

The fundamental qualities involved in the computation of ESI due to a single luminaire are expressed in a form that permits separation of the two variables that give the luminaire's position with respect to an observer. The total effect of all the luminaires in the layout is expressed in a single easily evaluated equation, which is a function of the two variables that give the observer's position. Specification of luminaire placement and orientation is achieved by a simple data input technique. For each of the four viewing directions, the maximum, minimum, average, and mean deviation of ESI values are calculated. These quantities are also calculated for all four directions taken together. The grid point may be as large as 20 by 20, which will give 400 points. This will yield a total 1600 values of ESI. All ESI are printed in an array that corresponds to their positions in the plan view of the room. The value of background luminance (L_B), contrast rendition factor (CRF), lighting effectiveness factor (LEF), and effective visibility level are printed in the same format. Figure 12.5 shows a typical room with six points marked, and Table 12.11 gives the resultant sample computer printout.

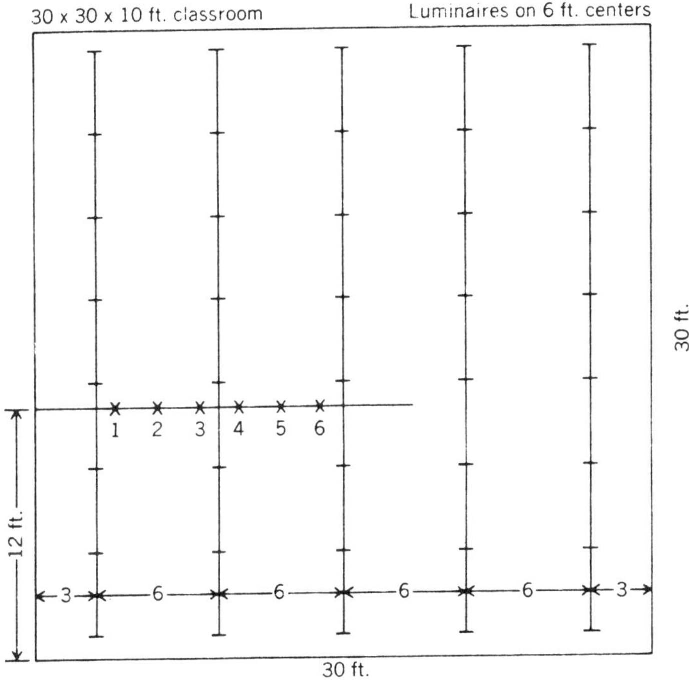

Figure 12.5 Typical 30 × 30 × 10 room with six points as marked.

Table 12.11 Computer Printout on ESI and Related Computations

	LOCATION X	Y	ORIENTATION ANGLE	FC*	FC	LB	LT	CRF	ESI	LEF
1	4.00	12.00	0.0	128.87	116.26	101.28	86.54	.869	40.02	.344
			90.0		114.95	96.17	81.35	.920	61.15	.532
2	6.00	12.00	0.0	139.02	135.63	116.16	97.89	.939	82.02	.605
			90.0		122.48	105.09	89.56	.882	46.61	.381
3	8.00	12.00	0.0	145.06	132.43	114.75	97.63	.890	53.49	.404
			90.0		134.84	115.13	97.37	.921	70.83	.525
4	10.00	12.00	0.0	147.29	134.71	116.61	99.17	.893	56.05	.414
			90.0		133.86	114.68	97.19	.911	64.88	.485
5	12.00	12.00	0.0	147.44	144.18	123.15	103.61	.947	92.16	.639
			90.0		131.47	113.89	96.95	.888	52.03	.396
6	14.00	12.00	0.0	148.68	136.10	117.75	100.09	.895	57.84	.425
			90.0		139.29	119.49	100.99	.924	74.99	.538

FC* IS THE FOOTCANDLE LEVEL FOR NO BODY SHADOW. ALL OTHER VALUES INCLUDE BODY SHADOW. LB AND LT ARE THE LUMINANCES OF THE BACKGROUND AND TASK RESPECTIVELY.

Lighting design procedures that make full use of the computer programs are still evolving. However, the following effects have proven useful:

1. In many typical design problems, the locations of the workstations are not known at design time. The iso-ESI plots can be used to help with furniture placement recommendations.
2. Knowledge of specific workstation locations eliminates some uncertainty and allows the design to be based on the performance calculated at specific points. Calculations at several points in the neighborhood of each of the design locations will reveal the magnitude of change in ESI with change in the observer's position. Knowledge of these deviations is essential for task-oriented lighting design.
3. The effect of other design parameter uncertainties can be analyzed by recomputing with different parameter values. Examples of this are surface reflectances or lamp lumen output. It allows the design engineers to appraise the effect of possible variations in performance introduced after the design is completed.

12.5.2 Limitations of ESI

ESI is a way to combine contrast and background luminance to define quantity visibility, but is not a perfect way to do so. ESI is a measure of visibility. Increases or decreases in the visual performance are important, yet increases or decreases in visibility may or may not be important. Experimental work done since the ESI system was formulated has shown that the relationship between visibility and visual performance is very nonlinear. The relationship between them is different for difficult tasks; even a small amount of visibility is sufficient to provide satisfactory visual performance. Other tasks require large amounts of visibility for a satisfactory visual performance.

Using ESI to rank lighting systems' abilities to produce visibility can provide important information, which is certainly better than a raw illumination level. Much more has been known about visual performance and lighting-controlled task characteristics (contrast and luminance) since the formulation of the ESI system. At present it appears that the next step will be to move beyond ESI and deal with relative visual performance. However, in the meantime, ESI remains the engineering system of choice for investigating lighting system performance with respect to visibility.

12.5.3 Computer Programs for Lighting Design in General

Computation of ESI values by computer is only one of many computer programs available today. Some programs will display results graphically on a PC screen and one program can make a really creditable "picture" using a dot matrix printer. The computer merges the processes of design and drafting by requiring the data needed in these two processes to be entered only once, resulting in a two-way information flow between calculations and graphics. The software is able to interpret graphic information, which exists in its database as numeric data, and insert these numeric data into calculations. The results are then displayed alphanumeric information in tabular or report form. This type of program is commonly known as a computer-aided deisgn and drafting (CADD) system.

A standardized form of presenting luminaire data has been established by the IES and a number of lighting fixture manufacturers. Data needed by the computer to make the calculations are voluminous. As a minimum, lamp data and coefficient of utilization tables, and light loss factors for each luminaire, need to be entered. Availability of these data in an electronic medium is essential in making interactive CADD practical.

12.5.4 VCP Values

Many factors are involved in evaluation of the relative comfort of a lighting installation: shape and size of room; reflectances of room surfaces; illumination level; type, size, and light distribution of luminaire used; number and location of luminaires; luminances and their relationship in the entire field of view; location and line of sight

Table 12.12 Typical VCP Values

Wall reflectance: 50%

Ceiling cavity reflectance: 80%

Effective floor cavity reflectance: 20%

Work plane illumination: 100 fc

Luminaire no.: 00

Room size: 60 × 30 ft

Mounting height above the floor (ft)	Luminaires	
	Lengthwise	Crosswise
8.5	68	67
10	69	69
13	71	70
16	74	73

of observer; and differences in observer sensitivity to glare. A comprehensive standard evaluation procedure taking all of the foregoing factors into account has come about as the result of numerous extensive investigations. This procedure provides a visual comfort probability (VCP) rating of a given system of lighting. The rating is in terms of the precentage of people who will be expected to find the given lighting system acceptable when they are seated in the most undesirable location.

By means of several procedures outlined in the IES Lighting Handbook, it is possible and useful to study proposed lighting system designs from the standpoint of visual comfort probability (VCP) by preparing tables such as Table 12.12.

BIBLIOGRAPHY

Chen, Kao, New Concepts in Interior Lighting Design, *IEEE Transactions on Industry Applications*, Sept./Oct. 1984, pp. 1179–1184.

DiLaura, D. L., Whatever Happened to Equivalent Sphere Illumination, *Lighting Design and Application*, Nov. 1982, pp. 17–18.

IES Computer Committee Report, Available Lighting Computer Programs, *Lighting Design and Application*, Sept. 1986.

Lighting Handbook, Chap. 1 of Application Volume, Illuminating Engineering Society, New York, 1987.

Lighting Handbook, Westinghouse Electric Corporation, Bloomfield, N.J., Jan. 1976.

Rebane, Henn, Applying Computers for Lighting Design, *Electrical System Design*, Sept./Oct. 1986, pp. 38–42.

Rowe, G. D., Determining Illumination Requirements, *Plant Engineering*, Feb. 4, 1982, pp. 69–72.

Sisson, William, Determining Cavity Ratios for Zonal Cavity Lighting Calculations, *Plant Engineering*, Nov. 27, 1970, pp. 68–69.

13
Factors Affecting Industrial Illumination

13.1 INTRODUCTION

The purposes of industrial illumination are to help provide a safe working environment, to provide efficient and comfortable seeing as an aid to all types of industrial operation, and to reduce losses in visual performance. It must be realized that several of the factors that contribute to seeing are the task, the environment, and the lighting.

13.2 FACTORS AND REMEDIES

In general, one sees by reflection, transmission, and silhouette. Silhouette seeing involves detection of the presence of an object and its contour because its darker outline is revealed by a contrast against lighted surroundings. Transmission concerns the revealing of details through the variation of transmission of white light, or the changing of color through materials that are susceptible to penetration. By far the most common method of seeing is by reflected light, where light and dark areas or details are revealed by a difference in reflection. The principle of silhouette lighting is involved in locations where low-level safety lighting is used. This is involved in protective lighting, emergency lighting, and outdoor passageways and roadways. Seeing by transmission generally involves the inspection of translucent materials.

13.2.1 Specific Factors and Remedies

Quality of illumination pertains to the distribution of luminaires in the visual environment. The term is used in a positive sense and implies that all luminaires contribute favorably to visual performance. However, glare, diffusion, reflection, uniformity, color, luminance, and luminance ratio all have a significant effect on visibility and the ability to see easily, accurately, and quickly. Certain seeing tasks require much more careful analysis than others. Industrial installations of poor quality are easily recognized as uncomfortable and possibly hazardous. The cumulative effect of even slightly glaring conditions can result in material loss of seeing efficiency and undue fatigue. Some of the foregoing factors are discussed in more detail below.

Direct Glare

When glare is caused by the source of lighting within the field of view, whether daylight or electric, it is defined as direct glare. To reduce direct glare, the following steps may be useful:

1. Decrease the brightness of light sources or lighting equipment, or both.
2. Reduce the area of high luminance causing the glare condition.
3. Increase the angle between the glare source and the line of vision.
4. Increase the luminance of the area surrounding the glare source and against which it is seen.

Unshaded factory windows are frequent causes of direct glare. Luminaires that are too bright for their environment will often produce glare. This glare may be in the form of discomfort direct glare or disability glare, or both. The former produces visual discomfort without necessarily interfering with visual performance or visibility. Disability glare reduces visual performance and is often accompanied by visual discomfort.

To reduce direct glare, luminaires should be mounted as far above the normal line of sight as possible and should be designed to limit both the luminance and the quality of light emitted in the 45- to 85-degree zone because such light may interfere with vision. This precaution includes the use of supplementary lighting equipment. There is such a wide divergence of tasks and environmental conditions that it may not be possible to recommend a degree of quality satisfactory to all needs. In production areas, luminaires within the normal field of view should be shielded to at least 25 degrees from the horizontal, preferably to 45 degrees.

Reflected Glare

Reflected glare is caused by the reflection of high-luminance light sources from shiny surfaces. In a manufacturing area this may be a particularly serious problem where critical seeing is involved with highly polished surfaces, such as polished sheet metal, vernier scales, and machined metal surfaces. There are several ways in which reflected glare can be minimized or eliminated:

1. Use a light source of low luminance, consistent with the type of work in process and the surroundings.
2. If the luminance of the light source cannot be reduced to a desirable level, it may be possible to orient the work so that reflections are not directed in the normal line of vision.
3. Increasing the level of illumination by increasing the number of sources will reduce the effect of reflected glare by reducing the proportion of illumination provided on the task by sources located in positions causing reflections.
4. In special cases it may be practical to reduce the specular reflection (and the resultant reflected glare) by changing the specular character of the offending surface.

Distribution, Reflection, and Shadows

Uniform horizontal illuminance (maximum and minimum not more than one-sixth above or below the average level) is usually desirable for industrial interiors to permit flexible arrangements of operations and equipment, and to assure more uniform luminance in the entire area. Alternate areas of extreme luminance differences are undesirable because it tires the eyes to adjust to them. Reflections of light sources in the task can be useful provided that the reflection does not create reflected glare. In the machining and inspection of small metal parts, reflections can indicate faults in contours, make scribe marks more visible, and so on.

Shadows from the general illumination systems can be desirable for accenting the depth and form of various objects, but harsh shadows should be avoided. Shadows are softer and less pronounced when large diffusing luminaires are used or the object is illuminated from many sources. Clearly defined shadows are distinct aids in some specialized operations, such as engraving on polished surfaces, some type of bench layout work, or certain textile inspections. This type of shadow effect can best be obtained by supplementary directional lighting combined with ample diffused general illumination.

Luminance and Luminance Ratios

The ability to see details depends on the contrast between the detail and its background. The greater the contrast difference in luminance, the more readily the seeing task is performed. The eye functions most comfortably and efficiently when the luminances within the remainder of the environment are relatively uniform. In manufacturing, there are many areas where it is not practical to achieve the same luminance relationships as easily as in offices. Table 13.1 is shown as a practical guide to recommended maximum luminance ratios for industrial areas. To achieve the recommended luminance relationships, it is necessary to select the reflectances of all the finishes of the room surfaces and equipment as well as control of the luminance distribution of the lighting equipment. Table 13.2 lists the recommended reflectance values for industrial interiors and equipment. High-reflectance surfaces are desirable to provide the recommended luminance relationships and high utilization of light.

Color Quality of Light

In general, for seeing tasks in industrial areas, there appears to be no effect upon visual acuity by variations in color of light. However, where color discrimination or color matching are a part of the work process, such as in the printing and textile industries, the color of light should be carefully selected. Color always has an effect on the appearance of the work space and on the complexions of personnel. Therefore, the illuminating system and the decorative scheme should be carefully coordinated.

Table 13.1 Recommended Maximum Luminance Ratios for Industrial Areas

	Environmental Classification		
	A	B	C
(1) Between tasks and adjacent darker surroundings	3 to 1	3 to 1	5 to 1
(2) Between tasks and adjacent lighter surroundings	1 to 3	1 to 3	1 to 5
(3) Between tasks and more remote darker surfaces	10 to 1	20 to 1	*
(4) Between tasks and more remote lighter surfaces	1 to 10	1 to 20	*
(5) Between luminaires (or windows, skylights, etc.) and surfaces adjacent to them	20 to 1	*	*
(6) Anywhere within normal field of view	40 to 1	*	*

* Luminance ratio control not practical.

A—Interior areas where reflectances of entire space can be controlled in line with recommendations for optimum seeing conditions.

B—Areas where reflectances of immediate work area can be controlled, but control of remote surround is limited.

C—Areas (indoor and outdoor) where it is completely impractical to control reflectances and difficult to alter environmental conditions.

Table 13.2 Recommended Reflectance Values for Industrial Interiors and Equipment

Surfaces	Reflectance* (percent)
Ceiling	80 to 90
Walls	40 to 60
Desk and bench tops, machines and equipment	25 to 45
Floors	not less than 20

* Reflectance should be maintained as near as practical to recommended values.

Veiling Reflections

Since contrast is one of the chief factors affecting the visibility of a seeing task, and since veiling reflections are directly involved with loss of contrast, there is a subsequent loss of visibility. Figure 13.1 shows that light would reflect into eyes of viewer from the "offending zone" and defines the zone of veiling reflection. Veiling reflection would diminish visibility, but the viewer would be unaware of it. In its RQQ Report 4, IES summed up that the contrast rendition factor (CRF) can be applied as a measure of the amount of veiling reflection. The CRF is the task contrast under the lighting system being studied, divided by the contrast the task would have under equivalent sphere illumination. Thus a CRF value of 0.62 means that visibility contrast is roughly 62% of what the contrast would be under a bright, but uniformly cloudy sky. It should be noted that the equivalent sphere illumination does not necessarily minimize veiling reflections.

Another important factor developed by the IES-RQQ Committee is the lighting effectiveness factor (LEF). An overall lighting system efficiency factor considers both the quality of light as reference to equivalent sphere illumination, and the effects of veiling reflections. Increased awareness of the need to design for elimination of veiling reflections has prompted manufacturers of light equipment to develop luminaires with light distribution patterns that will help solve veiling reflection problems. Light patterns cast by such luminaires are called "batwing" patterns. Figure 13.2 shows the light distribution curves of a typical batwing luminaire.

13.2.2 Daylighting

The daylight contribution should be carefully evaluated and should always be coordinated with a planned electric lighting system.

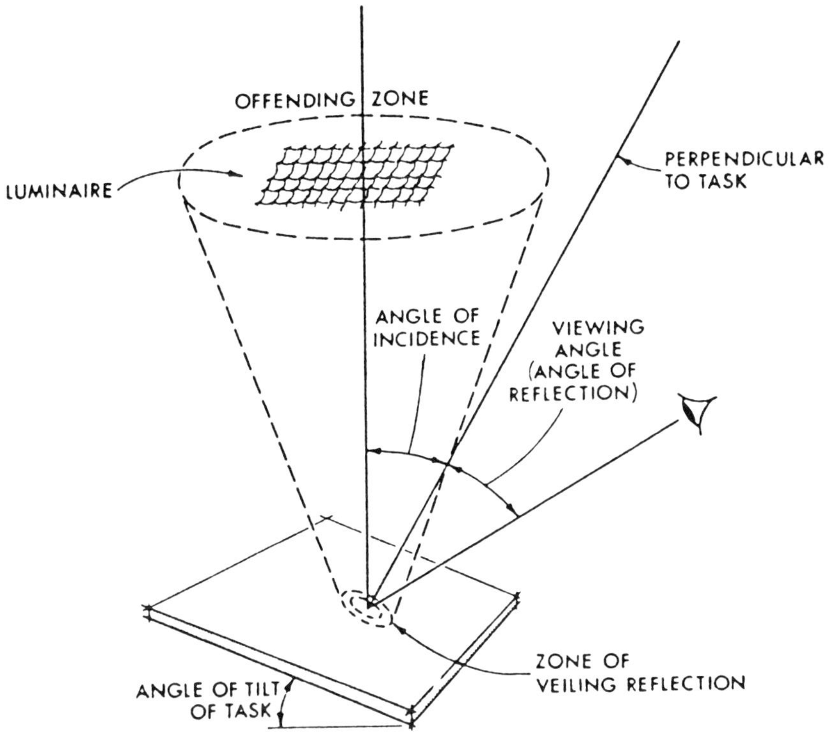

Figure 13.1 Diagram showing "offending zone" and zone of veiling reflection.

Advantageous Factors

Several factors affect the advantageous use of daylight. These include variations in the amount and direction of the incident sunlight; luminance distribution of clear, partly cloudy, or overcast skies; effect of local terrain, landscaping, and nearby buildings.

Fenestration

Fenestration has at least three useful purposes in industrial buildings:

1. For the admission, control, and distribution of daylight
2. For a distant focus for the eyes, which relaxes the eye muscles
3. To eliminate the dissatisfaction many people experience in completely closed-in areas

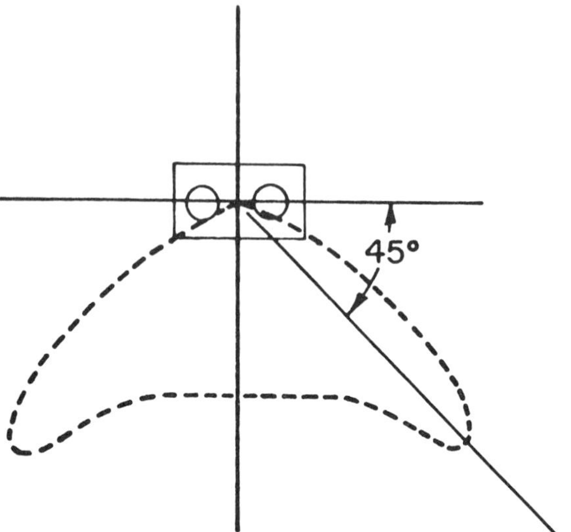

Figure 13.2 A typical "batwing" light distribution.

An adequate electric lighting system should always be provided because of the wide variation in daylight. The basic requirements of the principal architectural and work surfaces are the same whether the lighting is daylight, electric, or a combination of the two.

Building Orientation

Fenestration should suit the orientation, variations in topography, and landscaping related to each exterior wall. All fenestration should be equipped with control device appropriate to any luminance problems. Special attention should be given to glare control for latitudes where fenestration frequently receives direct sunlight. Diffuse-glaring fixed or adjustable louvers are some of the control means that may be applied.

For an industrial building, windows in the sidewalls admit daylight and natural ventilation, and afford occupants a view out, all of which may be desirable. Nevertheless, their uncontrolled luminance may be troublesome. There are many forms of lighting sections used by architects to admit some daylight into a factory. Each will require some control means to make the daylight useful to worker's seeing tasks, thus resulting in energy savings as the ultimate goal. More discussions on Energy-saving aspects of daylighting are discussed further in Chapter 17.

BIBLIOGRAPHY

Allphin, Willard, Minimizing Veiling Reflections in Lighting Installations, *Plant Engineering,* June 1, 1972, pp. 62–63.

Lighting Handbook, Application Volume, Illuminating Engineering Society, New York, 1987.

14
System Components for Industrial Illumination

14.1 LIGHT SOURCES

Incandescent, fluorescent, and/or high-intensity-discharge (HID) lamps are used in industrial lighting. They differ considerably in physical dimensions, electrical characteristics, spectral power distribution, and operating performance. Some are better suited than others to certain applications; however, sometimes two or more sources may qualify to fulfill a specific lighting requirement.

14.1.1 Incandescant Filament Lamps

Initial efficacy of typical incandescent lamps (25 to 1000 W) ranges from approximately 10 to 23 lumens per watt. They are commonly designed for approximately 1000 h of life. It has been pointed out in Chapter 11 that incandescent lamps conform to the supply voltage. A change of only a few volts can seriously affect both life and light output. There are special types of incandescent lamps:

1. *Reflectorized (R, PAR, and ER) lamps.* These lamps have self-contained reflectors and are manufactured in a number of sizes, from 30 to 1500 W, and in various light distributions. These lamps in general have a better-maintained illuminance. ER lamps (50, 75, and 120 W) control their beams such that they focus about 2 in. in front of their faces. Especially useful in "baffled downlight" luminaires, they permit high light utilization with attendant savings in energy. Figure 14.1 shows an ER30 lamp and its beam focus.
2. *Rough service and vibration service lamps.* These are special types used in industry. Rough service lamps (from 25 to 500 W)

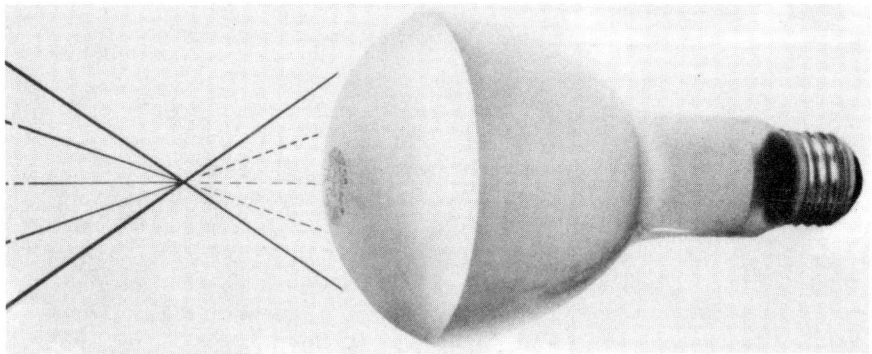

Figure 14.1 ER30 lamp.

are made with extra filament supports to withstand mechanical shock, and are used principally with extension cords. Vibration service lamps (25 to 150 W) are made with a flexible filament support which precludes early failure that vibration would cause to general service lamps. Lower-voltage lamps, generally operated from transformers, are much more resistant to both shock and vibration than are standard voltage types.

3. *Silicone-rubber coated lamps.* These lamps have special rubber-like coatings that serve to reduce breakage from both thermal and mechanical shock; or should breakage occur, the glass fragments nearly always remain intact. Available in sizes from 25 to 200 W, they are especially suited to food-packaging industries and to others where manufacturing functions may subject lamps to mechanical damage.

4. *Extended service lamps.* These operate for approximately two to three times the normal rated life. They are useful where cost of lamp replacement is high and cost of power is low.

5. *Thermal shock resistant or special service lamps.* These are available in various wattages and bulb shapes and are recommended for applications where moisture may fracture the hot bulb.

6. *Tungsten-halogen lamps.* These employ halogens to preclude blackening of the tubular envelope. They have extremely good lumen maintenance (approximately 97%) over a life of 2000 h or more. The shape of the lamps enables the luminaire to provide excellent beam control.

7. *High-voltage general lighting lamps.* These are available in 100 to 1500 W for 230- and 250 V circuits. They have less rugged filament, require more supports, and are less efficient than are 120 V lamps of equal wattage.

14.1.2 Fluorescent Lamps

The fluorescent lamp is an electric discharge source in which light is produced by the fluorescence of phosphors activated by ultraviolet energy from a low-pressure mercury arc. The lamp requires a ballast to limit the current and, in many instances, to transform the supply voltage. Lamp performance is influenced by the character of the ballast and luminaire, line voltage, ambient temperature, burning hours per start, and air movement. Fluorescent lamps are available in many variations of "white" and in a number of colors. Standard "cool white" is most popular for industrial lighting. The efficacy of cool white lamps varies between 30 and 100 lumens per watt (exclusive of 20% power loss in the ballast). Although most fluorescent lamps have tubular envelopes, there are special types, such as circular, U-shaped, reflectorized, and jacketed. Figure 14.2 shows major components that make up a compact fluorescent lamp.

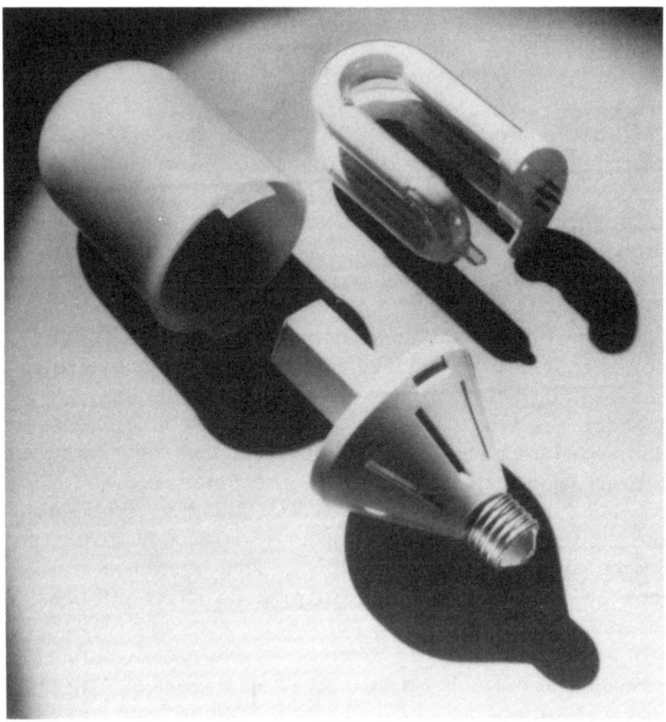

Figure 14.2 Major components of a compact fluorescent lamp.

Ballasts

Most fluorescent lamps operate on one of three types of ballast circuit: preheat, instant start, or rapid start (Figures 14.3 to 14.5). A few can be operated with either preheat or rapid-start ballasts. Preheat lamps up to 20 W can be operated on special rapid-start (trigger-start) ballasts. Preheat lamps operated on preheat ballasts require auxiliary "starters" to allow current to flow through the electrodes for a few moments before the arc is established across the length of the lamps. Instant start and slimline lamps require no starters. The ballasts provide enough voltage to light the lamps instantly. Rapid-start lamp operation is also starterless. Rapid-start lamps are most popular for new fluorescent lighting installations. They are available for ballasts that provide 430-, 800-, 1000-, and 1500-mA loadings. Fluorescent lamp ballasts are available for most secondary distribution voltages.

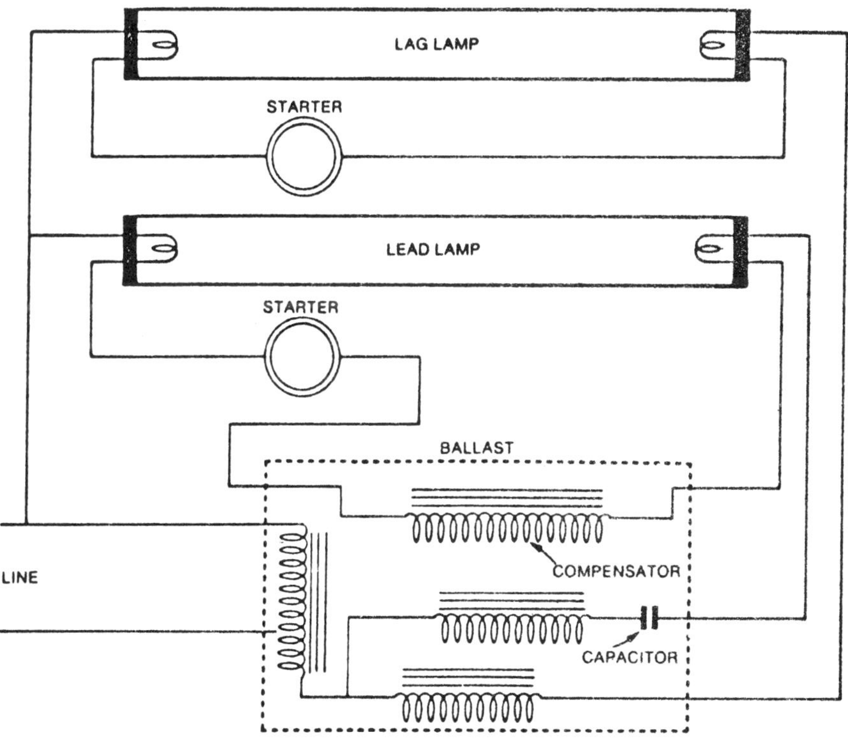

Figure 14.3 Preheat fluorescent ballast circuit.

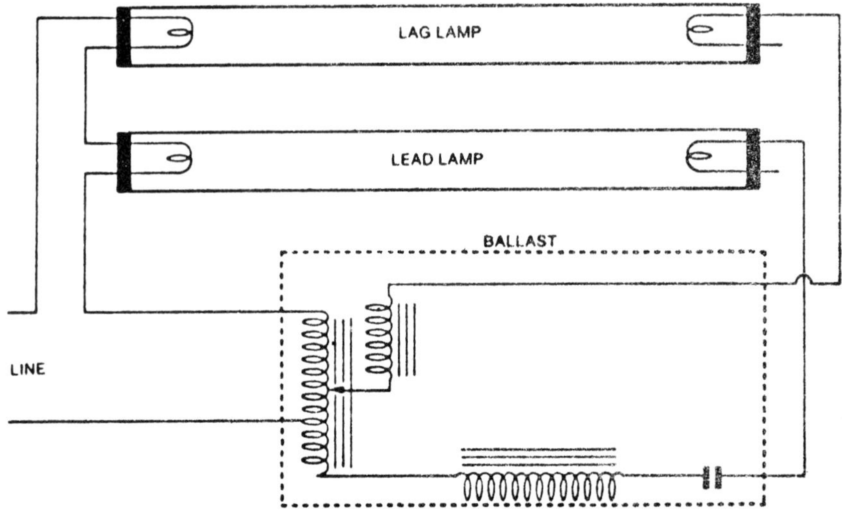

Figure 14.4 Instant-start fluorescent ballast circuit.

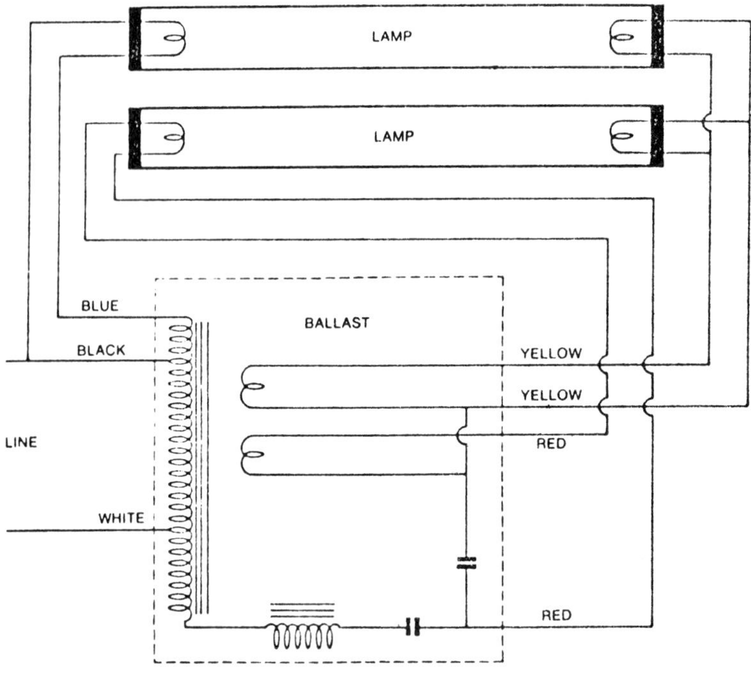

Figure 14.5 Rapid-start fluorescent ballast circuit.

Lamp Performance Factors

As discussed in Chapter 1, because voltage changes affect lamp performance, the specified voltage rating of the ballast should be maintained as close as possible. Low voltage as well as high voltage is undesirable. Both conditions may reduce lamp life. Extreme low voltage may even damage ballasts; excessive voltage may shorten ballast life. Cold temperature reduces lamp lumen output, as does high temperature. Lamp life is affected by the number of hours per start; a minimum number of starts favors lamp life.

14.1.3 High Intensity Discharge Lamps

High intensity discharge (HID) lamps are electric discharge sources. The basic difference from fluorescent lamps is that HID lamps operate at a much higher arc pressure. HID lamps include mercury, metal halide, and high-pressure sodium lamps. Spectral characteristics differ from those of fluorescent lamps because the higher pressure arc emits a large portion of its visible light. HID lamps produce full light output only at full operating pressure usually several minutes after starting. Most HID lamps contain both an inner and an outer bulb. The inner bulb is made of quartz or polycrystalline aluminum; the outer bulb is generally of thermal shock-resistant glass. HID lamps require current-limiting devices, which consume 10 to 20% additional watts.

Lamps

Mercury Lamps. These are low in efficacy compared to other HID sources, and are obsolescent for most industrial lighting applications. They are available with either "clear" or phosphor-coated bulbs of 40 to 1000 W, and in various sizes and shapes. Typical efficacy ranges from 30 to 63 lumens per watt, not including ballast loss. "Clear" mercury lamps produce light rich in yellow and green tones while lacking in red. Phosphor-coated lamps provide improved color. Special types include semireflector, reflectorized, and self-ballasted lamps.

Metal Halide (MH) Lamps. These are similar in construction to mercury lamps. The difference is in the arc tube, which contains various metal halides in addition to mercury. They are available in either clear or phosphor-coated bulbs from 175 to 1500 W. Present efficacies range from 70 to 125 lumens per watt, not including ballast power loss. Color improvement is achieved by the metal halide additives.

High Pressure Sodium (HPS) Lamps. Light is produced by electricity passing through sodium vapor. They are presently available in sizes of 35 to 1000 W. Typical initial efficacies are about twice that of mercury vapor: from 80 to 140 lumens per watt, not including ballast power loss. Normally with clear outer envelopes, they may also be obtained with coatings that improve diffusion. The color of light produced is golden white. Figure 14.6 shows several 250-W HPS lamps.

Low Pressure Sodium Lamps. These are presently available in 35 to 180 W. Typical initial efficacies are high: 137 to 183 lumens per watt, exclusive of ballast power loss. Applications are limited by virtue of their monochromatic yellow color. Figure 14.7 shows a typical low-pressure sodium lamp.

Ballasts

The proper ballast will operate a HID lamp from any supply voltage, usually 120 to 480 V. Certain lamp types are, however, particularly suited for operation on 480 V circuits. Ballast designs vary widely and include types for operating single lamps; others, for two lamps.

Mercury Lamp Ballasts. The choice of a ballast depends mainly on economic considerations versus performance. If the voltage can

Figure 14.6 250 W high pressure sodium lamps.

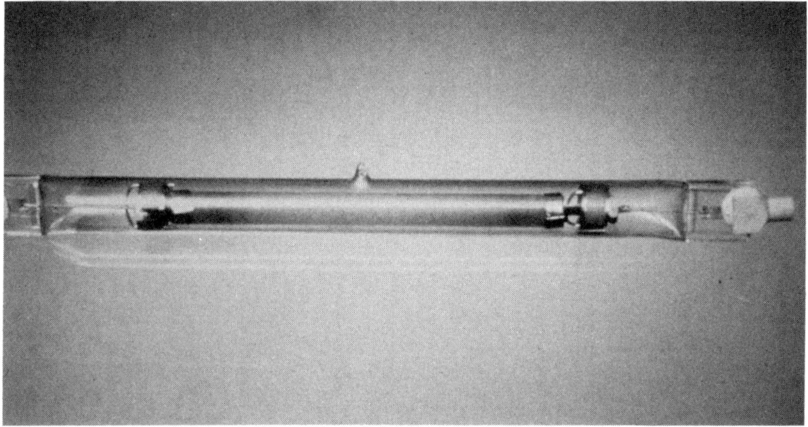

Figure 14.7 Typical low pressure sodium lamp.

be expected to remain within ±5%, reactor or high-reactance types are least costly. Regulated types hold average lamp watts closer to normal even with considerable variation of primary voltage. Mercury lamp will operate from metal halide ballasts, but the converse is not always true. Figure 14.8 shows electric circuit diagrams for several different types of ballasts for mercury lamps.

Metal Halide Lamp Ballasts. Most MH lamps operate on a special ballast designed for metal halide lamps. The 1000-W type may be operated on a mercury vapor lamp reactor ballast if the ambient is over 50°F. The metal halide ballast (Figure 14.9) is similar in circuitry to the mercury vapor lamp CWA ballast, but with modifications to provide the higher starting voltage required, and wave-shape characteristics to assure reignition of the arc each half-cycle.

High Pressure Sodium Lamp Ballasts. Since no starting electrode is incorporated in HPS lamps, the ballast must supply a high voltage pulse of 2500 to 4000 V at least once per cycle for wattages other than 1000 W. The 1000-W lamps require 4000 to 6000 V. The element that does this is called a starter or an ignitor. At present, four general types of ballasts are available to the HPS lamp user. Each has its advantages and disadvantages compared with the others in terms of lamp performance, cost, and energy consumption.

1. *Reactor or lag ballast.* This can be made to have good lamp voltage regulation for changes in lamp voltages, but has poor regulation for changes in line voltage. It has a relatively high starting

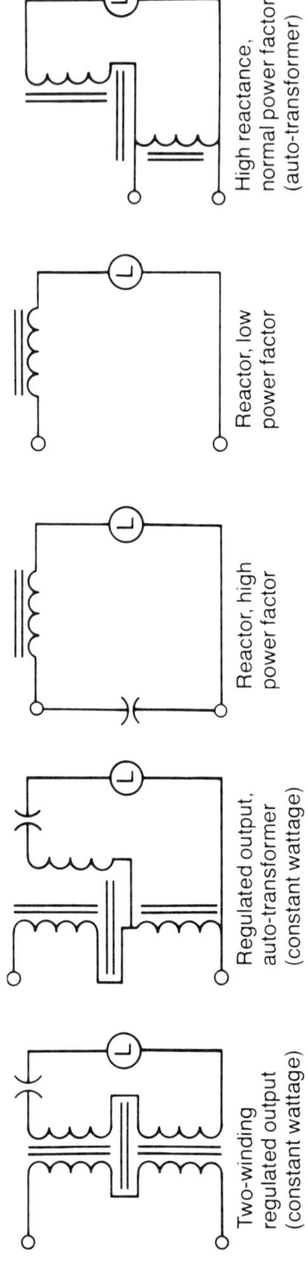

Figure 14.8 Electric circuit diagrams for mercury lamp ballasts.

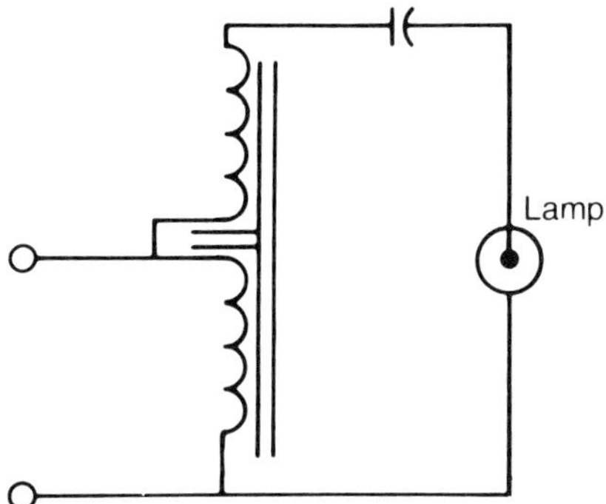

Figure 14.9 Electric circuit diagram for metal halide ballast.

current, which can produce a desirably faster lamp warm-up. It is relatively inexpensive, having low power losses and is small in size.

2. *Lead ballast.* This has fairly good regulation for line voltage variation and has fair regulation for lamp voltage variation.

3. *Magnetic-regulated ballast.* This is essentially a voltage-regulating isolation transformer with its primary and secondary windings mounted on the same core, and containing a third capacitive winding, which adjusts magnetic flux with changes in either primary or secondary voltage. It provides the best wattage regulation with change of either input voltage or lamp voltage. It has a low line starting current and a high power factor. It is, however, the most costly and has the greatest wattage loss. Figure 14.10 shows these three types of ballast circuits.

4. *Solid-state electronic ballasts.* The major problem with existing HPS lamp ballasts is their inability to operate the lamps at rated power. Electronically controlled ballasts have been designed and built with a solid-state control circuit and a reactor. Figure 14.11 shows an electronically controlled HPS ballas, or "energy-efficient" ballast. The use of a solid-state switching device permits the control winding to be shorted in a "phase-controlled" manner, thus providing a smooth and continuous variation in the average inductance of the ballast. The solid-state control circuit monitors lamp and line operating conditions and then establishes the proper value of ballast required to operate the lamp at its rated power.

Lamp Performance Factors

Ballast characteristics affect HID lamp performance. Ballast design determines the ability to start the lamp at low temperatures, controls the time required for the lamp to reach full output, and greatly determines the tolerance of a lamp to voltage dips. Serious voltage

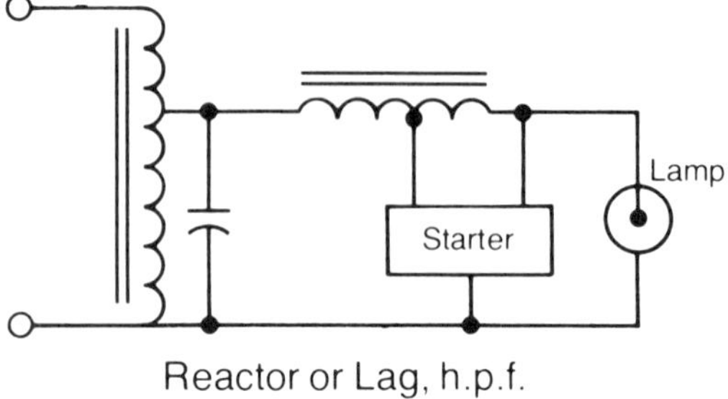

a

Starter

Lamp

Lead

b

Starter

Lamp

Regulated, h.p.f.

c

Figure 14.10 Electric circuit diagram for HPS lamp ballasts. (a) Reactor ballast circuit; (b) Lead-peaked ballast circuit; (c) Magnetic-regulated ballast circuit.

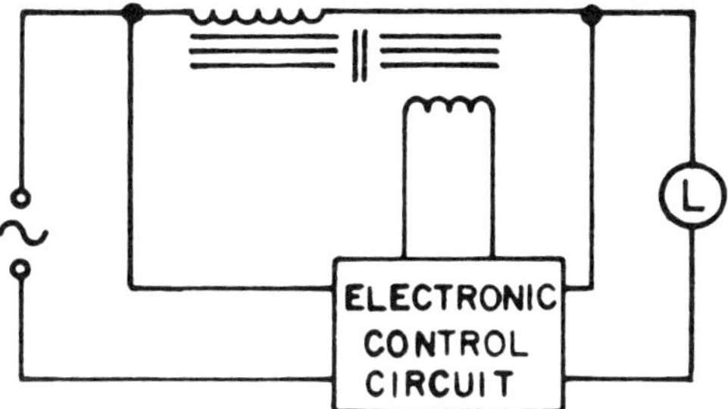

Figure 14.11 Electronically controlled HPS lamp ballast circuit.

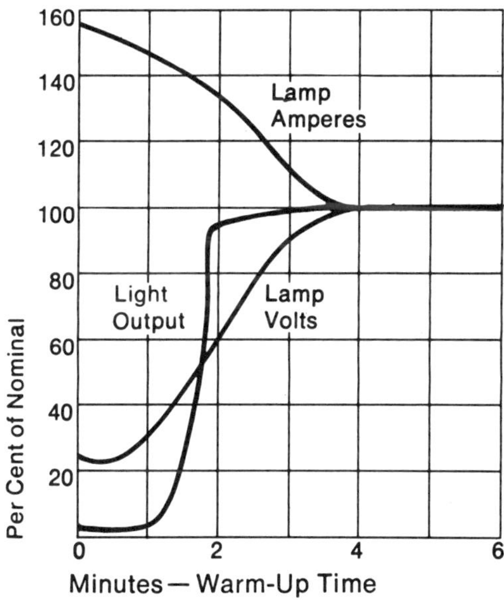

a

Figure 14.12 Warm-up characteristics of three types of HID lamps. (a) Mercury lamps; (b) Metal halide lamps; (c) High pressure sodium lamps.

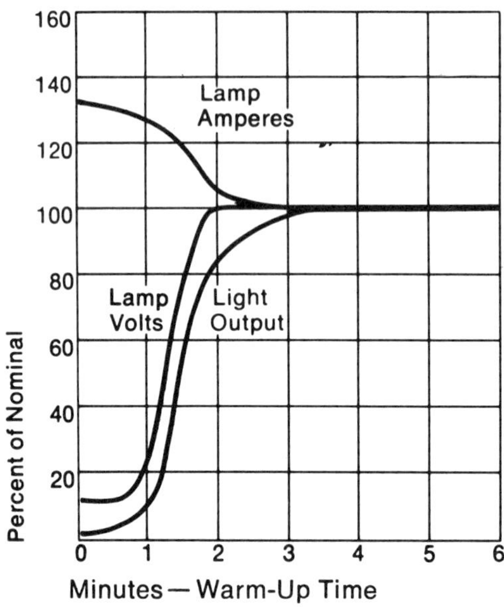

b

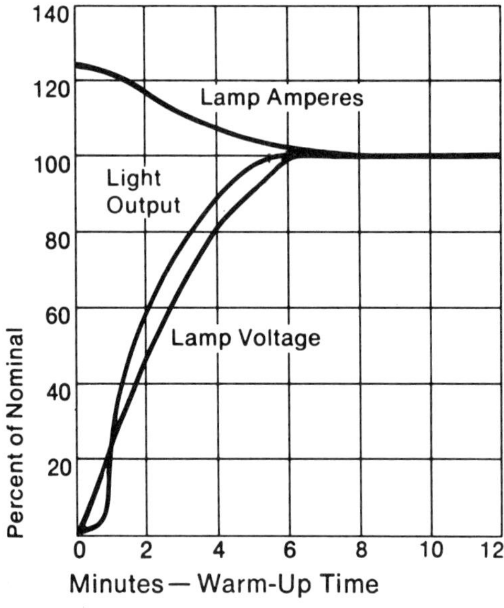

c

Figure 14.12 (continued)

dips or any power interruption will extinguish the lamp, after which the lamp must cool for several minutes before it can restart. Figure 14.12 shows the warm-up characteristics of three types of HID lamps.

Because the voltage of HPS lamps is dependent on lamp wattage, unlike the other HID lamps, which have relatively constant voltage regardless of wattage, and since lamp voltage rises during life, lamp manufacturers have established trapezoid diagrams which define the lamp voltage and wattage limits. Figure 14.13 is a trapezoid diagram for a 400-W HPS lamp. It is noted that the upper and lower lines define the maximum and minimum wattage at which the lamp may be operated under any conditions. The left line represents the minimum anticipated lamp voltages, while the right line represents the voltages at which the lamp may begin to cycle.

14.2 LUMINAIRES

There are many types of industrial luminaires. Selection of specific types for an installation requires consideration of many factors: candlepower distribution, efficiency, shielding and brightness control, mounting height, lumen maintenance characteristics, mechanical construction, environmental suitability for use in normal, hazardous, or special areas. In general, there are five types of luminaires, in accordance with CIE classification for interior applications:

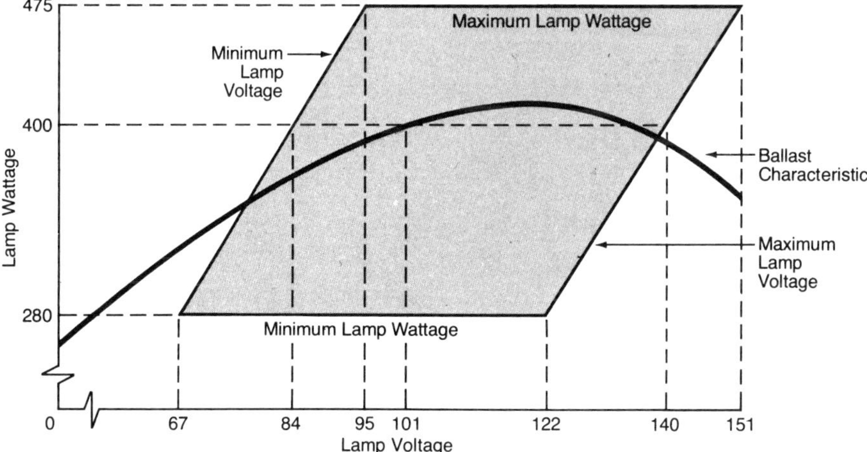

Figure 14.13 Trapezoid diagram for a 400 W high pressure sodium lamp.

14.2.1 Direct Type

Direct-type units emit practically all (90 to 100%) of the light downward to the working area. Although such luminaires usually provide the most efficient illumination on working surfaces, it is usually at the expense of other factors. For example, shadows may be excessive unless the units have relatively large luminous areas or are mounted closer together than suggested maximum spacing-to-mounting height ratios. But direct and reflected glare may be disturbing because of the higher luminance difference between the bright source and the darker surround. Direct glare can be made low from a well-designed luminaire.

Direct industrial lighting equipment is usually classified according to the distribution of the downward component from "highly concentrating" to "widespread." This classification of luminaires is expressed in terms of suggested spacing-to-mounting height ratios, which are shown in Table 14.1. The widespread category includes high-intensity discharge (HID) luminaires that have optical assemblies consisting of a refractor/reflector design that can provide lamp concealment and reduce luminaire sufficiently to permit a lower mounting height than would be acceptable for conventional HID luminaires. The distribution of low-bay units tends to improve vertical illumination (because of their wide-angle component) and to permit spacing as much as two or more times their mounting height above the work plane.

Prismatic or mirrored glass or specular aluminum reflectors produce the more concentrating distributions. These are useful when luminaires for general lighting must be mounted at a height equal to or greater than the width of the room, or where high machinery or processing equipment necessitates directional control for efficient illumination between the equipment. They are also useful for supple-

Table 14.1 Classification of Luminaire Direct Component Expressed in Terms of Permissible Spacing Criteria

Spacing to Mounting Height Ratio (Above Work-Plane)	Luminaire Classification
Up to 0.5	Highly Concentrating
0.5 to 0.7	Concentrating
0.7 to 1.0	Medium Spread
1.0 to 1.5	Spread
Over 1.5	Wide Spread

mentary illumination. Spread types are comprised of procelain-enameled reflectors, other white reflecting surfaces, diffuse aluminum, mirrored or prismatic glass or plastic, and similar materials. Spread distributions are advantageous in low-bay areas or where there are many vertical or near-vertical seeing tasks.

Generally speaking, concentrating and medium spread distributions are best suited to high-bay areas. Wherever there is a need for higher-than-average general illumination for an inspection or special work area, highly concentrating luminaires should be installed above cranes at mounting heights where the basic high-bay lighting system is located. For large areas, low-luminance luminaires are preferred to provide low-reflected luminance. Such luminaires may consist of a diffusing panel on a standard type of fluorescent reflector, an indirect lighting hood, or a large luminous area. In a very dusty or corrosive area, luminaires with gasketed glass or plastic covers are recommended.

Area lighting extending from wall to wall is another form of direct lighting in which light from sources in a large cavity of high reflectanse is directed downward through cellular louvers or translucent or refracting glass or plastic. When these materials canceal the lamps completely, the illumination characteristics are similar to those of indirect lighting system. Cellular louvers used as the shielding medium may present a reflected glare problem. This should be minimized in the design. Figure 14.14 shows a typical 400-W high-pressure-sodium luminaire that is popular in general factory illumination.

14.2.2 Semi-direct Type

Semi-direct units are those that emit 60 to 90% of their light downward. Utilization of light from these luminaires depends greatly on ceiling reflectance. Light-colored ceilings usually result in improved utilization and visual comfort. The increased ceiling illumination from the semi-direct distribution reduces the luminance difference between ceiling and luminaire, increases diffusion, and softens shadows. Appropriately designed reflectors or refractors will reduce luminaire luminance and provide additional comfort. Most fluorescent and some HID and incandescent luminaires may be equipped with louvers to further increase shielding and reduce direct glare.

14.2.3 General Diffuse or Direct-Indirect Type

In these luminaires, the downward and upward components are approzimately equal: 40 to 60% of the total luminaire output. General diffuse-type luminaires emit light about equally in all directions; direct-indirect luminaires emit very little light at angles near the horizontal, which is preferred because of their lower luminance in the direct glare zone. Luminaires with such a distribution are widely used in offices and laboratories, and their use in clean manufacturing areas is increasing.

14.2.4 Semi-indirect Type

This type of luminaire emits most of the light (60 to 90%) upward. The major portion of the light reaching the horizontal work plane must be reflected from the ceiling and upper walls; therefore, it is necessary that these surfaces have high reflectance. The need for high reflectances and good maintenance limits the use of industrial semi-indirect systems to areas where it is necessary to minimize reflected glare from specular work surfaces.

Figure 14.14 A typical 400 W HPS luminaire for industrial application.

14.2.5 Indirect Type

Indirect luminaires emitting from 90 to 100% of their light upward are seldom used in industry. These units have the lowest utilization and are more difficult to maintain. Figure 14.15 shows luminaires for general lighting as classified by the CIE in accordance with the percentage of total luminaire output emitted above and below horizontal.

14.2.6 Supplementary Luminaire Types

Supplementary lighting units can be divided into five major types accoridng to candlepower distribution and luminance:

1. *Type S-I—directional:* includes all concentrating units, such as a reflector spot lamp or units employing concentrating reflectors or lenses.
2. *Type S-II—spread, high–luminance:* includes small area sources such as incandescent or high intensity discharge. An open-bottom, deep-bowl diffusing reflector with a high intensity discharge lamp is an example.
3. *Type S-III—spread, moderate–luminance:* includes all fluorescent units having a variation in luminance greater than 2:1.
4. *Type S-IV—uniform–luminance:* includes all units having less than a 2:1 variation of luminance. Usually, this luminance is

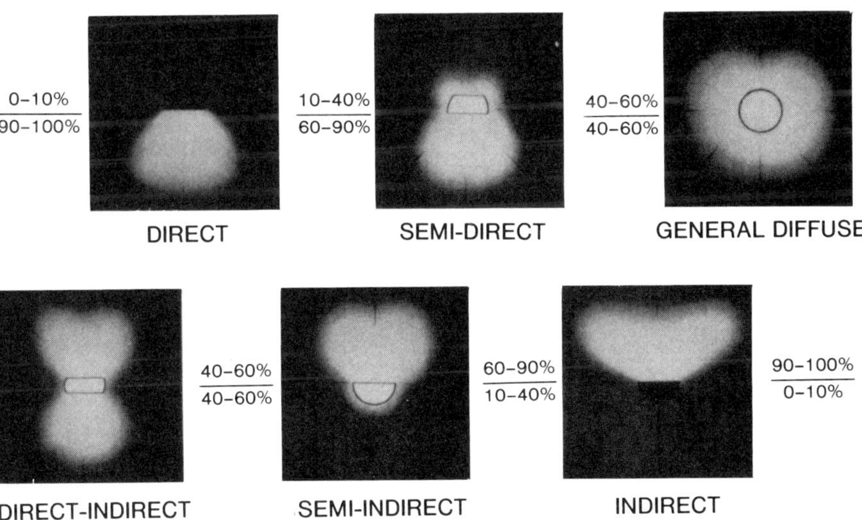

Figure 14.15 General lighting luminance classifications.

less than 6800 candelas per square meter (2000 footlamberts). An example of this type is an arrangement of lamps behind a diffusing panel.

5. *Type S-V—uniform–luminance with pattern:* a luminaire similar to type S-IV, except that a pattern of stripes or lines is superimposed.

14.3 TYPES OF INDUSTRIAL LIGHTING SYSTEMS

14.3.1 Factory Lighting for Visual Tasks

The prime requirement for industrial lighting is to facilitate the performance of visual tasks through high-quality illumination. With such lighting, personnel will be able to observe and effectively control the operation and maintenance of various types of machines and processes. There are three types of lighting used in industrial areas: (1) general, (2) localized general, and (3) supplementary.

General Lighting

General lighting should be designed to provide the desired level of illumination uniformly over the entire area. The variation of light level from point to point within the area should be within 17% of the selected level. A good general lighting system makes it possible to change the location of machinery without rearranging the lighting, and also permits full utilization of floor space. In general, uniform spacing and continuous row installation of fixtures is usually the most economical method where fluorescent is considered. Exceptionally large machines with one or more areas of operation or control should be studied to make sure that proper lighting levels are delivered to these critical seeing areas. Local lighting for bench work, assembly, inspection, and so on, may sometimes be needed. But it will lose a great deal of its effectiveness in the absence of good general lighting.

Localized General Lighting

Within a general area there may be a few areas where the tasks performed require a greater quantity of light and a different quality of light. An example would be a packing table within a warehouse, where it is necessary to identify and count small parts quickly, and read or prepare orders to packing slips. In these cases, additional luminaires for local lighting will be required. The specifications of local lighting equipment should be given careful consideration since improper shielding or design could adversely affect the visual comfort within the area. Care must be exercised to eliminate direct or reflected glare from the task and from other workers. In cutting rooms in the garment industry, the tables may be 100 ft long and 4 ft wide, separated by narrow aisles possibly 3 ft wide. By lighting each table separately, the overall result is a general lighting system, even though the space has been lighted using a local or supplementary approach.

Supplementary Lighting

Supplementary lighting is specified for difficult seeing tasks that require a specific amount or quality of light not readily obtained by standard general lighting methods. Supplementary lighting is often used to provide higher illumination levels for small or restricted areas.

Supplementary lighting is a valuable industrial lighting tool. Many machine, assembly, and inspection operations involving difficult seeing tasks may require more illumination or a different type of illumination than that provided by general lighting alone. Typical problems arise where work is shielded from the general lighting system by an obstruction, or its brightness is otherwise lowered; where low contrast, such as scribe marks on steel, may lead to visual errors; and where the product moves too rapidly to be seen clearly by the unaided eye.

Supplementary equipment must be carefully shielded to prevent glare from users and their co-workers. The luminance ratio should be carefully controlled. To attain a good balance, it is important to coordinate the design of supplementary and general lighting with great care.

14.3.2 Security Lighting

Security lighting pertains to the lighting of building exterior and surrounding areas—out to and including the boundaries of the property. In some areas it may be an integral part of the industrial lighting design. Security lighting contributes to a sense of personal security and to the protection of property. This may be accomplished through:

1. *Surveillance lighting:* lighting to detect and observe intruders.
2. *Protective lighting:* lighting to discourage or deter attempts at entrance, vandalism, and so on. It may lead a potential intruder to believe detection highly possible and so not attempt entry.
3. *Lighting for safety:* lighting to permit safe movement of guards and other authorized persons.

14.3.3 Emergency Lighting

Emergency lighting is the lighting provided for use when the power supply for the normal lighting fails to ensure that escape routes can be effectively identified and used. Standby lighting is that part of emergency lighting that is sometimes provided to enable normal activities to continue. When the normal lighting of an occupied building fails, regardless of the cause, emergency lighting is required to fulfill the following functions:

1. Indicating clearly the escape routes
2. Providing illumination and a comforting visual environment along the escape routes sufficient to facilitate safe movement along them forward and through the exits and emergency exits provided
3. Permitting reading identification of all fire alarm call points and firefighting equipment provided along the escape routes

The following are recommended minimum illumination requirements for exit signs and egress route:

1. *Internally illuminated signs.* Where codes exist, an illuminance of 54 lux (5 fc) on the face of the sign is usually specified.
2. *Externally illuminated exit signs.* These vary so greatly in design, material, color, and printing that standards are difficult to establish. NFPA 101 requires 54 lux (5 fc) on the face of the sign. However, consideration must be given to contrast, glare, veiling reflectance, and reliability of the emergency power source.
3. *Egress route.* The horizontal illuminance of any escape route should not be less than 1% of the average provided by the normal lighting, with a minimum average of 5 lux (0.5 fc) at floor level. A uniformity ratio (E_{max}/E_{min}) of up to 20:1 along the centerline of an escape route is desirable for safe movement. A value of 40:1 should not be exceeded.
4. *Location of egress luminaires.* A luminaire should be provided for each exit door and emergency exit door and at points where it is necessary to emphasize the position of potential hazards, sufficient to light that area to a level of 30 lux (3 fc).
5. *Power supply systems.* Emergency lighting systems should provide the required illuminance within 10 s of interruption of the normal lighting. If such a system is to be powered by a generator, it is essential that the generator can be run up to its required output within the specified period, and that startup be automatic on failure of the normal lighting. A battery-powered emergency lighting system utilizing suitable rechargeable secondary batteries may be designed for operation from a centrally located battery and charger combination (central system) or from batteries located at lamps themselves. The battery/charger combination should be capable of supporting the system for 1-hour following a 24-h recharge period.

All emergency lighting systems should be tested and inspected at least every 30 days, no matter what type of emergency power is used.

14.3.4 General Summaries

In large industrial areas, all three lighting systems (see Section 14.3.1) may be used. In smaller areas, however, localized general lighting may also serve as a substitute for general lighting. In this case additional supplementary lighting may be required to increase the quantity or improve the quality of the illumination. Many factors must be considered in selecting a lighting system, including cost, comfort, ease of maintenance, appearance, amount of light, color rendition, power distribution, and seeing task. Since all of these factors are interrelated, it is not feasible to recommend one or two systems for all conditions. Because of the relationship of ceiling heights to light utilization, most industrial applications call for either direct or semi-direct lighting systems. For characteristics of these two types of luminaires, a review of Sections 14.2.1 and 14.2.2 is highly recommended.

14.3.5 Selection of the Equipment

In the selection of equipment—light sources and luminaires—many variables must be considered. Some of the basic factors are lamp efficacy and life, lumen maintenance, burning hours per start, luminaire efficiency, finish and distribution, brightness and glare control, the accessibility of equipment for maintenance, and the amount of dust and dirt in the air. As with any list of variables, it is necessary for purposes of comparison to hold some factors constant. In industrial lighting that factor is usually mounting height and location.

High-Bay Areas

In high-bay areas the work generally presents visual tasks that are not difficult because of large machinery and other objects. There is little problem with reflected glare, and color discrimination is not important. Usually, overhead cranes are needed to handle large pieces of equipment.

Illuminance levels for high-bay areas generally range from 50 to 150 fc, although more and more areas are being lighted with 200 and 300 fc. At high mounting height, it is possible to obtain uniform illumination by using a few high-wattage sources rather than a larger number of low-wattage sources. For luminaires with medium or narrow distribution, greater mounting height or closer spacing is ordinarily required for uniform general illumination.

Regardless of mounting height, wide-distribution luminaires are well suited for use in areas that are wide in respect to mounting height. Unless the seeing task is largely on vertical surfaces, narrow-distribution luminaires are advisable for high and narrow areas. Illumination is then produced on the horizontal working area rather than projected high on the walls where it is less useful. Large machinery and objects do tend to cut off light and cast shadows. Since this makes it difficult to see important vertical and angular surfaces, broad light distribution is essential.

High intensity discharge or fluorescent luminaires for high-bay lighting may be enclosed, ventilated open, or nonventilated open. Enclosed luminaires are usually of a heavy-duty type with a gasketed glass cover to protect the reflector and light source from collection of dirt. The initial luminaire efficiency is lower and the equipment is more costly. Ventilated-open luminaires have largely replaced the nonventilated type. Ventilated units are recommended for all high-bay applications except where the air has extremely high dirt content or where fume might attack the reflector.

As far as choices of lamps are concerned, metal halide and HPS are preferred over the mercury type. The use of fluorescent lamps in high-bay areas is limited. Only where the area proportions are such that the room cavity ratios are in the range of 1 to 3, fluorescent lamps may be acceptable. But only high or extra high output fluorescent in 8-ft sizes are recommended.

Medium- and Low-Bay Areas

Seeing tasks in medium- and low-bay areas are usually more difficult than those encountered in high-bay areas. The tasks include precision and inspection work with smaller objects and machinery. The provisions of good visibility and visual comfort are not always satisfied by the same characteristics of the system. Increasing the size and reducing the brightness of the luminaires will improve visual comfort and will usually improve the visibility of specular objects. It may not improve the visibility of diffuse three-dimensional objects.

Luminaires used for general lighting in medium-bay areas are nearly always of the direct or semi-direct type, either fluorescent or wide-distribution HID. They may be of the ventilated or nonventilated type and may have the lamps shielded by louvers, baffles, or other devices. All these accessories improve visual comfort, louvers usually being the most effective. For lower mounting, the trend is toward the semi-direct type.

In some areas, the seeing task consists of diffuse three-dimensional objects that can be well lighted with directional sources. Some of the visual tasks involve specular or semispecular objects, for which optimum lighting might be an indirect system. The quality of fluorescent sources, with their broad distribution of light, makes them a prime selection for medium- and low-bay lighting. When the proper quality control can be attained, low-wattage HID sources are finding an increasing number of low-bay applications.

14.4 SPECIAL CONSIDERATIONS

14.4.1 Lighting and Space Conditioning

With the use of higher illuminances, it is often practical to combine the lighting, heating, cooling, and atmospheric control requirements in an integrated system. The lighting system can often provide most of the energy during the heating season.

14.4.2 High-Humidity or Corrosive Atmosphere and Hazardous Location Lighting

Enclosed gasketed luminaries are used in non-hazardous areas where atmospheres contain non-inflammable dusts and vapors. Steam processing, plating areas, wash and shower rooms, and other areas of high humidity are typical areas that require enclosed luminaires. Hazardous locations are areas where atmospheres contain inflammable dusts, vapors, or gases in explosive concentrations. They are grouped by the National Electrical Code on the basis of their hazardous characteristics. Luminaires are available specifically designed to operate in these areas, which are noted in Article 500 of the NEC as class I, class II, and class III locations. Luminaires used in these areas are defined as explosion-proof, dust-tight, dust-proof, and enclosed and gasketed.

14.4.3 Abnormal Temperature Conditions

Low ambient temperature must be recognized as existing in such areas as unheated heavy industrial plants, frozen-food plants, and cold storage warehouses. Equipment should be selected to operate under such conditions, and particular attention should be paid to lamp starting and lumen output characteristics if fluorescent equipment is considered. With HID equipment, temperature variation has little effect on lumen output, but the proper starting characteristics must be provided. There are starting or operational problems with incandescent lamp equipment.

Abnormally high temperatures may be common at truss height in foundries, steel mills, forge shops, and the like. Caution should be exercised in selecting lighting equipment for mounting in such locations. The temperature limitations of fluorescent and HID ballasts under such conditions should be observed. They may be mounted remotely to avoid excessive high temperature.

14.4.4 Maintenance

Regular cleaning and prompt replacement of lamp outages are essential in any well-operated industrial illuminating system. It is important for the illuminating engineers to analyze luminaire construction and reflector finish and to make provisions for maintenance access so that the system can be serviced properly.

BIBLIOGRAPHY

ANSI/IES Standard Practice for Industrial Lighting, Illuminating Engineering Society Publication 0-87995-014-5, 1983.

Chen, Kao, Fundamentals of Circuit Design and Characteristics of Ballasts, lecture in advanced lighting course, 1970, sponsored by New Jersey Section of Illuminating Engineering Society.

Chen, Kao, and Kane, R. M., Achieving Optimum Performance in a High Pressure Sodium Lighting System, *IEEE Transactions on Industry Applications*, July/Aug. 1982, pp. 416–423.

Lighting Handbook, Westinghouse Electric Corporation, Bloomfield, N.J., Jan. 1976.

Various lamp catalogs, Westinghouse Electric Corporation, Bloomfield, N.J. and North American Philips Lighting Corp., New York.

15

Applications and Installations of Industrial Illuminating Systems

15.1 INTRODUCTION

Chapter 14 has discussed the most important components that make up an industrial illuminating system. This chapter will cover examples of typical installations which have achieved their objectives of various seeing tasks.

15.2 MACHINE SHOPS

Machining of metal parts consists of setting up and operating machines such as lathes, grinders, millers, shapers, and drill presses, bench work, and inspection of metal surfaces. The precision of such machine operation usually depends on the accuracy of the setup and careful use of the graduated feed-indicating dials rather than observation of the cutting tool or its path. The fundamental seeing problem is the discrimination of detail on plane or curved metallic surfaces.

The visibility of scribed marks depends on the characteristics of the surface, the orientation of the scribed mark, and the nature of the light source. Directional light produces good visibility of scribed marks on untreated cold-rolled steel if the marks are oriented for maximum visibility, such that the brightness of the source is reflected from the side of the scribed mark to the observer's eye. Unfortunately, this technique reduces the visibility of other scribed marks. Better average results are obtained with a large-area low-luminance source.

There is an evident advantage in the use of large-area low-luminance sources for most visual tasks in the machining of metal parts.

The ideal general lighting system is one having a large indirect component. Both fluorescent and HID sources are used for general lighting; fluorescent luminaires in a grid pattern are usually preferred. High-reflectance room surfaces improve visual performance. Figure 15.1 shows the result of such a lighting system, which provides a pleasant environment for die-making in a machine shop. Supplementary lighting is used to maintain close tolerances of die production. Green plants and modern wall treatment contribute to making a stimulating area.

Special types of luminaires are designed to illuminate three-dimensional objects such as machinery, dials, spindles, presses, panels, and stacked materials. Light is contributed from many locations and distances to minimize shadow, reduce glare, and optimize visual comfort. This type of luminaire has proven to be satisfactory for lighting machine shops. Figure 15.2 shows a typical industrial plant machine shop lighted with 250 W metal halide luminaires, with emphasis on vertical surface illumination.

Figure 15.1 Fluorescent lighting for a machine shop with supplementary lighting for die production.

Figure 15.2 Metal halide lighting for an industrial plant machine shop.

15.3 GENERATING STATIONS

Turbine-generator buildings usually have medium-high to high ceilings. Seeing tasks include general inspection, meter and gauge reading, and pedestrian movement. In low- and medium-bay areas, low-wattage high intensity discharge or fluorescent industrial luminaires suspended below major obstructions are appropriate for general illumination. When HID luminaires are utilized, reflectors with wide distribution should be selected to assure eveness and a good vertical component, and to minimize shadowing. Supplementary lighting is recommended for vertical illumination on such equipment as control panels, switchgear, and motor control centers. Luminaires with an upward component contribute to improved visual comfort. For high-bay areas (25 ft or higher), it may be appropriate to use either HID or fluorescent luminaires. However, HID high-wattage luminaires are often chosen for high mounting heights.

15.4 CONTROL ROOMS

The control room is the nerve center of the power plant or process plant and must be monitored continuously. Lighting must be designed with special attention to the comfort of the operator; direct and reflected glare and veiling reflections must be minimized, and luminance ratios must be low. Along with ordinary office-type seeing tasks, it is often necessary to read meters 10 to 15 ft away.

Although the practice is not standardized, most control room lighting involves one of two general categories: diffuse lighting or directional lighting. Diffuse lighting may be from low-luminance, luminous indirect lighting equipment, solid luminous plastic ceilings, or louvered ceilings. Directional lighting may be from recessed troffers that follow the general contour of the control board.

A basic conflict exists in trying to light a control room, since some of the tasks are made more visible under reduced illumination, while other tasks require significantly higher levels of illumination. Fortunately, most of the tasks enhanced by light-level reductions are not located at the same place where tasks requiring high illumination are to be found. Therefore, the lighting system can be modified to give a nonuniform distribution of light within the room.

The optimum distribution of illumination for each control room must be determined on a case-by-case basis since the equipment arrangements will vary in each installation. It may be found necessary to use several different types of louvers, lenses, and diffusers, sometimes in combination with one another, to achieve the optimum distribution for each particular room.

15.4.1 Cathode Ray Tube Display

Currently, there is a proliferation of cathode ray tube (CRT) displays in all areas of industry. Lighting for these areas needs special attention. In the case of fluorescent lighting, the first consideration is to eliminate any fluorescent lamp reflection in the face of the CRT. This can be accomplished by using the specular parabolic wedge louver, which has an absolute cutoff of 45 degrees; any lamp images will be reflected below the viewing angle. CRT display is being used more and more in manufacturing operations. Therefore, in addition to the parabolic wedge louver, adequate lighting is needed for these difficult tasks (100 to 200 fc). A diffuser positioned on top of the louver can be useful to avoid lamp reflections on computer input keys and other specular surfaces.

All brightness within the space must be controlled. Walls and other surfaces should be diffuse and have a light reflectance value of less than 50%. Windows must be covered with blackout curtains or other appropriate shielding material. Reflections from shiny plastic keys on CRT keyboards can cause glare problems. Some CRT manufacturers are now producing terminals with matte-finish keys. All desks or work surfaces should be matte, or low-glare, nonreflective surfaces. Another consideration is to position the CRTs so that they are not parallel to each other, to avoid reflections between the two CRTs.

The end result of these precautions is a comfortable workstation with good visibility on the CRT screen, little or no glare from the keyboard or telephones, and easy visibility for occasional references to slick-page printed or written texts. The introduction of CRTs to desktops has created a difficult surface to light, but it has not created new lighting design principles. The techniques regarding the luminous environment are the same today as they were the day before the birth of CRTs. In general, the following points are useful for lighting design involving CRTs and should therefore be followed closely:

1. Test and experience the CRT workstation before design.
2. To avoid reflected images, use the mirror test* for the full range of operator viewing angles and for all types of lighting, up or down.
3. Design to avoid reflected and direct glare from any source, natural or artificial.
4. Illumination levels should not be diminished merely because of the presence of a CRT.
5. Bright clothing and, for critical viewing, reflective jewelry should be avoided.
6. Eyeglasses should be worn with the correct "near" prescription for acuity as well as a range of viewing angles.
7. Design should be critical of luminaire selection.

15.5 MANUFACTURING AREAS

Glass tubing mauufacture is an important part of fluorescent lamp manufacturing operations. Near the end of the glass tubing drawing line, clippers and end-formers are installed to complete the tubing manufacturing process. For this operation, ventilated fluorescent fixtures with high output and center V shield reflectors are chosen to deliver 100 fc on the horizontal work plane. V reflectors provide adequate shielding for visual comfort. An upward component also

---

*The face of the CRT should be traversed with a small mirror. Any bright images reflected in the mirror should then be either moved, removed, or covered, or the finish should be changed to a diffuse finish with a reflectance of less than 30%.

contributes to visual comfort by balance of luminances between luminaires and their backgrounds. Top openings help minimize dirt accumulation. Figure 15.3 shows a fluorescent lighting installation that has proven satisfactory in all aspects essential to a difficult industrial operation.

Fluorescent lighting was widely used for industrial operations prior to the energy crunch in the mid-1970s. During the last decade, HID lighting has been becoming more popular in industrial plants, especially for retrofitting designs, not only because of the higher lumen efficacies of the lamp and consequent lighting energy reduction, but also based on its overall good qualities for industrial applications.

An example of this application was the relighting program at an automotive manufacturing facility. Since the automotive process is continuous and integrated, the fixture layout was designed to satisfy all the various seeing tasks involved in the fabrication and mating of components. The production and maintenance staff has to keep tabs on the logic-programmed machines, run tests, perform adjustments, and assure that the manufactured components and subassembly are in total compliance with the design specifications. The general lighting satisfies these requirements using evenly distributed illuminance.

Figure 15.3 High output fluorescent luminaire with V reflectors for manufacturing area lights.

In high-bay areas, 1000-W HPS luminaires, which provide some uplight component through open and ventilated reflectors are used in an approximate spacing-to-mounting height ratio of 1:1. For low-bay areas, 400-W luminaires with about a 16-ft mounting height provide even candlepower distribution of light without glare using a faceted aluminum reflector and a polycarbonate lens having refractor prism elements. The optical assembly is totally enclosed, gasketed, and filtered to keep contaminants from infiltrating to the lamp and reflector or inside the lens. Since the lens prisms redirect some of the high-angle lumen output toward lower angles, these industrial units provide adequate horizontal beam spread and vertical surface illumination.

Generally, the spacing and mounting-height arrangements of both low- and high-bay units provide approximately 60 fc horizontal maintained on a typical work surface. All the fixtures are wired in a checkerboard pattern on 480 V three-phase circuits. The overall light level can be reduced 50% in a given zone by switching off half the circuits in that zone from a convenient central point. Thus energy savings can be effected during reduced production activity. Each fixture is powered through a fused plug and cord assembly which permits easy replacement of the unit for repair.

15.6 WAREHOUSES

In warehouses with storage racks, the objects to be lighted are on the vertical surfaces of the stacks, not on the floor. Some horizontal illumination is necessary for operators of material handling equipment and for the floor cleaning. Designing a lighting system for a warehouse might appear to be relatively simple. Install enough luminaires to deliver enough light onto the stored material to permit accurate selection, and in the area for operators to see safely where they are going. Over the years, a few rules of thumb have evolved as offshoots of manufacturing area lighting practice. However, in recent years the concept of storage has changed. Warehouse are larger, stacks are higher and deeper, and operations have become more and more automatic and computer controlled. In designing lighting for most new warehouses, illuminating engineers find their job comparable to designing a system for a long, tall, narrow room for which reflectances are unknown and in which luminaires can be placed only in high, inaccessible places.

In the past (before the energy crunch in 1974), fluorescent lighting was commonly used to illuminate vertical surfaces. Figure 15.4 shows high-output fluorescent luminaires operating on 277 V circuits installed at a mounting height of 40 ft in a modern warehouse (built in 1966). It delivers satisfactory vertical illumination, especially near the top of the stacks.

Today, higher mounting heights have necessitated the use of HID lighting equipment. Because the light is emitted from an optically small area, an HID lamp permits finer optical control than do line source fluorescent lamps; luminaires can be designed with better directional characteristics than those of fluorescent luminaires.

Several factors are unique to warehouse lighting. First, there is the necessity of seeing on a vertical surface rather than on a horizontal plane. Second, the warehouse aisle is similar to a narrow, long room with high walls, where interreflectances influence the result to

Figure 15.4 High output fluorescent lighting for warehouse aisles.

a significant degree. Third, the type and amount of material in warehouse aisles are subject to unpredictable fluctuation. A solid wall of material, with 30% reflection today, may become a big black hole tomorrow; or that black hole may allow the beam of the luminaire in the adjacent aisle to penetrate into the subject aisle.

Studies on warehouse aisle lighting have pointed out that luminaires with maximum uplight provide superior vertical as well as horizontal illumination at all locations for several aisle widths. The uplight would be reflected from the ceiling, which would be advantageous in a warehouse aisle setting with narrow aisles and high mounting heights. The study concluded that the amount of uplight appears to be an important factor and that luminaire design appears to be one of the most critical elements in the production of vertical illumination.

Figure 15.5 shows an empty warehouse lighted with 400 W HPS luminaires, resulting in good vertical illumination. The HPS lamps were also chosen because of the cost factor. The trend in warehouses has been toward narrower and higher aisles, in which HID lamps become mandatory.

Figure 15.5 400 W HPS lighting for a warehouse.

15.7 ENGINEERING OFFICES, CONFERENCE ROOMS, AND PLANT HOSPITAL ROOMS

15.7.1 Engineering Offices

Visual requirements for engineering and drafting offices demand high-quality illumination since discrimination of fine detail is frequently required for extended periods. Significant graduation of shadows along T squares and triangles reduces visibility. Harsh directional shadows from drawing instruments may reduce efficiency. Illumination systems that avoid reflections are most important in providing maximum contrast. Studies show that ideal locations for light fixtures were at least 24 in. from either side of the drafting table. For a room of 105 ft by 44 ft that is to be broken into six- or eight-person working groups with 6 ft high partitions; continuous rows of light fixtures mounted 12 ft on centers would be a satisfactory arrangement. These are recessed 2 × 4 ft parabolic fluorescent troffers with 4 in. deep cells and four lamps. Parabolic fixtures are also suited for CRT viewings in the same office. Such a lighting system can

Figure 15.6 Combination of fluorescent and incandescent lighting for a functional conference room.

produce virtually glare-free and adequate illumination at the center of the drafting tables.

15.7.2 Conference Rooms

The diversity of work to be performed in the conference room suggests that the illumination should be flexible and the entire environment should be comfortable and pleasant. The general lighting is provided by the recessed fluorescent in the 2 × 4 grid-type ceiling. Specially chosen reflectors provide high-level illumination without glare.

Incandescent (250 W PAR38) downlights with concentric louvers are provided in the front stage and over the conference table. This, in combination with dimmer-controlled general lighting, provides flexi-

Figure 15.7 Dimmer controlled fluorescent lighting for a engineering manager's office.

bility to create a changing environment as well as eye comfort for all seeing tasks. Figure 15.6 shows a handsomely decorated functional conference room with the lighting system as described above.

15.7.3 Engineering Manager's Office

The fluorescent luminaires for the engineering manager's office are similar to that of the conference room, to provide a maximum of 200 fc. The light fixtures are positioned to minimize both direct and reflected glare. The room is provided with richly paneled walls and dark venetian blinds to reduce glare. The illumination required in this office varies from that necessary for casual sight during visits to higher values for prolonged concentration on difficult seeing tasks. All lights are dimmer controlled to provide flexible, smooth variation in the illumination needed for the diversity of tasks encountered. Figure 15.7 shows a beautifully decorated manager's office with the lighting system as described above.

Figure 15.8 Level controlled fluorescent lighting for a plant hospital room.

15.7.4 Plant Hospital Rooms

A plant hospital room is used for emergency and/or simple treatments only. It is staffed with a nurse and a part-time doctor. In the hospital room a moderately high lighting level is needed for examinations. Ideally, a 50 fc level for local examination is recommended. However, it is advantageous to use three-lamp or four-lamp fluorescent luminaires for the general illumination, switched or dimmed to allow reduction in the light level during certain procedures. This type of room can be better served with general illumination supplemented by a portable or fixed examination light. The general light source should be of the color-improved type. Figure 15.8 illustrates a plant hospital room that is lighted with color-improved lamps and level-controlled switching devices.

BIBLIOGRAPHY

Chen, K., Case Studies: Offices Lighting, *Building Operating Management*, Jan. 1970.

Christensen, Morgan, Lighting Prescription for Areas Containing Cathode Ray Tube (CRT) Displays, *Lighting Design and Application*, May 1981, pp. 23–25.

Feldhaus, Ron, The Look and Feel of Innovative Lighting—One Company's Approach, *Electrical System Design*, Mar. 1987, pp. 34–40.

Goodwin, Patricia, Warehouse Aisle Lighting, *Lighting Design and Application*, June 1985, pp. 30–42.

Lighting Handbook, Sec. 9 of Application Volume, Illuminating Engineering Society, New York, 1987.

McGowan, Terry, and Christensen, Morgan, Recent Findings in Warehouse Lighting Design, *Plant Engineering*, Mar. 18, 1982, pp. 267–271.

Sledz, Roger J., Control Room Lighting: An Application of Human Factors Engineering, IEEE-PES Winter Meeting Record, 1982.

Zekowski, Gerry, Lighting for CRT's: Another View, *Lighting Design and Application*, Nov. 1981, pp. 28–32.

16

Floodlighting Design

16.1 INTRODUCTION

Floodlighting a building, monument, or structure is an effective means of identifying the object at night and thereby calling attention to it and its owner. Thanks to recently developed light sources, luminaires, and techniques, lighting effects can be tailored to the type of building or structure and the significance the owner wants to give it. However, the equipment and techniques must be used intelligently and imaginatively, for it is essential that the building or structure's form, beauty, and architectural identity be neither disturbed nor obscured.

In general, floodlighting should achieve certain objectives. First, the structure surface should have a brightness such that it appears in perspective when viewed from a distance. Shadows cast should look like those cast by the sun; they should not destroy the basic form and depth of the structure's architecture. Walls and other flat surfaces should be illuminated to a level that reveals their texture and the character of the architectural design. Finally, the structure should be identified with the area about it by illuminating sufficient surrounding area; that is, it should not appear suspended but rather oriented with adjacent grounds, slopes, and plazas.

16.2 BASIC FLOODLIGHTING EFFECTS

16.2.1 Flat Lighting

Flat lighting is uniform illumination of a structure. It creates few highlights and shadows and little modeling, but it can be the most economical kind because installation is usually simple and little of the light pattern misses the building. Luminaires can be mounted on the ground, on poles, or on the roof of adjacent building or buildings across the street (see Section 16.7.1).

16.2.2 Grazing Lighting

Grazing lighting dramatically expresses the character of a building by producing strong highlights and shadows. It is achieved by mounting floodlights close to the facade, so it is often used where mounting space is restricted. The best light source for tall buildings is a high intensity discharge (HID) lamp with its arc tube along the axis of a concentrating specular reflector (see Section 16.7.3).

16.2.3 Lighting Patterns

Lighting patterns can be used to emphasize or subdue adjacent architectural elements, strengthen design concpets, or increase the attraction of an otherwise plain surface. The key to success in nonuniform lighting is to create the impression that the effect was planned (see Section 16.7.4).

16.2.4 Color Lighting

Color lighting can supplement the increasing use of bold colors in modern construction, both in general floodlighting and as a means of establishing highlights and focal points. It can be achieved either by use of color filters or by utilizing the inherent color differences among the light sources. Incandescent lighting produces a natural look, clear mercury lighting tends to cast a slight greenish color on neutral colors, fluorescent lighting strengthens white or light blue colors, and sodium lighting is rich in amber color and very effective in adding warmth.

16.2.5 Sparkle or Glitter

Sparkle or glitter, which is achieved with exposed lamps, complements modern architecture with its emphasis on line and plane. The lamp size required for a sparkle pattern depends on the brightness of the area and the effect desired.

16.3 CHOOSING THE FLOODLIGHT SOURCES

The basic categories of light sources are incandescent, fluorescent, and HID lamp. Table 16.1 compares the costs, life and characteristics, color, size, and aesthetic achievement of various light sources.

Table 16.1 Comparison of Light Sources

	Incandescent			*Mercury*		
	Standard	*Quartz-Iodine*	*Fluorescent*	*Standard*	*Metallic Additive*	*High-Pressure Sodium*
Initial Cost	Low	Low	Higher	Higher	Higher	Higher
Annual Operating Cost	Medium	Medium	Low	Low	Low	Low
Service Life	Fair	Fair	Good	Very good	Good	Good
Color Definition	Good	Very good	Fair	Fair	Good	Good
Beam Control	Very good	Good	Poor	Fair	Good	Good
Cold Weather Operation	Very good	Very good	Fair	Good	Good	Good
Long Range Projection (narrow beam)	Very good	Fair	Poor	Fair	Fair	Fair
Medium Range Projection	Good	Good	Fair	Good	Good	Good
Lumen Output	Fair	Fair	Fair	Good	Very good	Best

16.3.1 Incandescent Lamps

Incandescent lamps are perhaps the most useful and versatile flood-light source. Their light can be directed easily by lenses and reflectors in beams of the desired shape, and the color of their light is accepted as "white." Efficacy usually is about 20 lm/W.

Tungsten halogen lamps, the new incandescent sources, have efficacies of about 25 lm/W. They contain halogen that continually removes vaporized tungsten deposits from the quartz envelope and redeposits it on the filament; consequently, their light output remains almost constant overtime instead of diminishing as a result of tungsten depositing on the envelope. The lamps used for building floodlighting are usually about the size and shape of pencils. Most floodlights developed for these linear light sources develop rectangular beam patterns, which are highly efficient for many building floodlighting applications.

Projector lamps are developed for particular needs. The 6 V 120 W PAR64 incandescent lamp produces a thin beam that is very effective for floodlighting tall buildings, columns, steeples, water towers, and the like. Its beam spread of $4\frac{1}{2}$ degrees in one plane by 7 degrees in the other is achieved by masking critical areas of the reflector to prevent refocusing of light.

16.3.2 Fluorescent Lamps

Fluorescent lamps are lower in brightness than the other sources, but more efficient than most. Fluorescent lamps require a large specular reflector for precise control of light, but even with such a reflector, control is limited to the light perpendicular to the length of the lamp. They are sensitive to temperature both in starting and operating, although outdoor-type ballasts ensure reliable starting down to $-20°F$. Regardless of ballast type, however, light output is reduced when the lamp is exposed to low temperature and moving air.

16.3.3 Mercury Lamps

Mercury lamps are almost as efficient as fluorescent lamps and somewhat more compact. Color rendition in general is inferior to that of incandescent lighting, although some mercury lamps have relatively good color rendition.

16.3.4 Metal Halide Lamps

Metal halide lamps furnish approximately 25 to 100 lm/W of white light. In addition to the mercury and argon gas in the arc tube generally are iodine compounds. The advantages of metal halide lamps over mercury lamps are good color without the use of phosphor and a high initial light output. A characteristic of metal halide lamps is that there is some variation in color uniformity from lamp to lamp. This is influenced by fluctuations in line voltage, ballast output characteristics, luminaire design, and ambient temperature. However, a metal halide light source produces a wide range of less subtle colors, including yellow, red, green, and blue.

16.3.5 High Pressure Sodium Lamps

High pressure sodium lamps have a high efficacy of 100 to 130 lm/W of white light with a yellow/orange tone, which provides a rich warm amber color that serves building material well. The appearance of natural surfaces lighted with high pressure sodium lamps is similar to their appearance under warm white fluorescent of low-wattage incandescent lamps, but colors at the "cool" end of the spectrum are substantially grayed down.

16.4 CHOOSING A LUMINAIRE

The first step in determining the type, number, and size of floodlight luminaires required to light a building is to choose a tentative floodlight on the basis of type of light source, shape and size of beam (round or rectangular; wide, medium, or narrow), and wattage or light output (beam lumens) of the source. As a general rule, if a simple requirement must be met, the illuminating engineers simply select the lamp and luminaire best suited for the job. Where there are no clear-cut requirements, the engineers compare the various lamp and luminaire characteristics and weigh the importance of each. If more than one light source is suitable, an economic study must be made to determine which would be the best choice for a number of years of service. The comparison of light source in Table 16.1 can be used effectively as a quick selector.

With the light source chosen, a luminaire can be selected. Floodlight luminaires are usually divided into seven types on the basis of beam spread. In Table 16.2 it is noted that beam efficiencies vary with the type of beam and lamp, as shown. Two popular types of floodlights are shown in Figures 16.1 and 16.2.

16.5 DESIGN PROCEDURES

16.5.1 Determine the Level of Illumination

In Table 16.3 are listed the recommended illumination levels for many floodlighting applications. The illumination should not fall below these levels at any time in the maintenance cycle; therefore, an allowance

Table 16.2 Floodlight Luminaire Types

Beam Spread (degrees)	*NEMA Type*	*Minimum Efficiency (percent)*				
		Incandescent		*Mercury*		*Fluorescent*
		Effective Reflector Area (sq. in.)				
		Under 227	*Over 227*	*Under 227*	*Over 227*	*Any*
10 to18	1	34	35	—	—	20
18 to 29	2	36	36	22	30	25
29 to 46	3	39	45	24	34	35
46 to 70	4	42	50	35	38	42
70 to 100	5	46	50	38	42	50
100 to 130	6	—	—	42	46	55
130 and up	7	—	—	46	50	55

Source: National Electrical Manufacturers' Association. Asymmetrical-beam floodlights may be designated by a combination type designation which indicates horizontal and vertical beam spreads in that order; e.g., a floodlight with a horizontal beam spread of 75 degrees (Type 5) and vertical spread of 35 degrees (Type 3) would be designated as a Type 5×3 floodlight.

Figure 16.1 A typical floodlight for 400 W HID lamps (courtesy of Westinghouse Electric Corp.).

Figure 16.2 A typical floodlight for 1000 W HID lamps.

for reasonable depreciation must be made in the design. In flood-lighting buildings, monuments, and the like, the reflectance of the surface and the brightness of the surroundings must be considered in order to determine the amount of light necessary. If a building is located in an area that is normally crowded, it is advisable to reduce the brightness on the lower portion of the building to prevent possible annoyance to pedestrians and motorists.

16.5.2 Choose Proper Spread

As already discussed in Section 16.4, floodlight equipment is divided into seven types according to beam spread, which is defined as the angle between the two directions in which the candlepower is 10% of the maximum candlepower at or near the center of the beam. Beam efficiency is defined as the percentage of the beam lumens bear to the lamp lumens, the beam lumens being the lumens contained within the beam spread.

Table 16.3 Recommended Levels of Illumination for Floodlighting Design Applications

	Recommended Footcandles (Minimum At Any Time)
Building—	
General Construction	10
Excavation Work	2
Building Exteriors and Monuments, Floodlighted—	
Bright Surroundings—	
Light Surfaces	15
Dark Surfaces	50
Dark Surroundings—	
Light Surfaces	5
Dark Surfaces	20
Bulletins and Poster Boards— (Water Tanks or Stacks With Advertising Messages, Flags)	
Bright Surroundings—	
Light Surfaces	50
Dark Surfaces	100
Dark Surroundings—	
Light Surfaces	20
Dark Surfaces	50
Coal Yards (Protective)	0.2
Dredging	2
Loading Platforms	20
Lumber Yards	1
Parking Lots	5
Self-Parking	1
Attendant Parking	2
Piers, Freight and Passenger	20
Prison Yards	5
Quarries	5
Railroad Yards—Classification	
Switch Points	2
Body of Yard	1
Service Stations (At Grade)—	
Light Surroundings—	
Approach	3
Pump Island Area	30
Service Areas	7
Dark Surroundings—	
Approach	1.5
Pump Island Area	20
Service Area	3
Shipyards—	
General	5
Ways	10
Fabrication Area	30
Storage Yards, Active	20

Although the choice of beam spread for a particular application depends on individual circumstances, the following general principles apply:

1. The greater the distance from the floodlight to the area to be lighted, the narrower the beam spread desired.
2. Since by definition the candlepower at the edge of a floodlight beam is 10% of the candlepower near the center of the beam, the illumination level at the edge of the beam is one-tenth or less of that at the center. To obtain reasonable uniformity of illumination, the beams of individual floodlight must overlap each other as well as the edge of the surface to be lighted.
3. The percentage of beam lumens falling outside the area to be lighted is usually lower with narrow-beam units than with wide-beam units. Thus narrow-beam floodlights are preferable where they will provide the necessary degree of uniformity of illumination and the proper footcandle level.

4. The location of floodlighting equipment is usually dictated by the type of application and the physical surroundings. If the area is large, individual towers or poles spaced at regular intervals may be required to light it evenly; smaller areas may require only one tower with all equipment concentrated on it, or adjacent buildings may be used as floodlight locations.

It is important that the light be properly controlled. Strong light directed parallel to a highway or railroad track can be a dangerous source of glare to oncoming traffic, and light thrown indiscriminately on adjacent property may be a serious nuisance.

16.5.3 Determine the Coefficient of Beam Utilization

To determine the number of floodlights that will be required to produce a specified level of illumination in a given situation, it is necessary to know the number of lumens in the beam of the floodlight and the percentage of the beam lumens striking the area to be lighted. The beam lumens may be obtained from manufacturers' catalogs. The ratio of the lumens striking the floodlighted surface to the beam lumens is called the coefficient of beam utilization (CBU). When an area is uniformly lighted, the average CBU of the floodlights in the installation is always less than 1.0.

The coefficient of beam utilization for any individual floodlight will depend on its location, the point at which it is aimed, and the distribution of light within its beam. In general, the average CBU of all the floodlights in an installation should fall within the range 0.60 to 0.90. If less than 60% of the beam lumens are utilized, it is an indication that a more economical lighting plan should be possible by using different locations or narrower-beam floodlights. On the other hand, if the CBU is over 0.90, it is possible that the beam spread selected is too narrow and the resultant illumination will be spotty. An estimated CBU can be determined by experience or by making calculations for several potential aiming points and using the average figure thus obtained.

To make such calculations the floodlighted area is superimposed on the photometric grid, and the ratio of the lumens inside this area to the total beam lumens is determined. All horizontal lines on a building appear as straight horizontal lines on the grid if the floodlight is so aimed that its beam axis is perpendicular to a horizontal line on the face of the building. All vertical lines except the one through the beam axis appear slightly curved. Figure 16.3 illustrates the superimposed method to determine the coefficient of beam utilization.

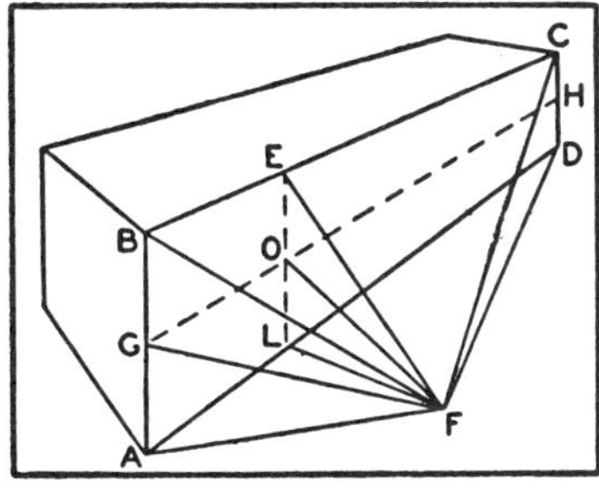

IF: EO = OL = 25
AL = FL = 40
LD = 80

THEN: ANGLE LFO = 32°
EFO = 19°
BFE = 32°
GFO = 41°
AFL = 45°
CFE = 51°
HFO = 60°
DFL = 64°

THE CBU OF THE FLOODLIGHT AT F IS ABOUT .81

NOMINAL LAMP DATA		TEST RESULTS	
WATTS	1500	AVG. MAX. CANDLEPOWER	23,500
VOLTS	115	BEAM LUMENS	18,790
BULB	PS-52 Clear	BEAM EFFICIENCY	57%
SERVICE	General	HOR. BEAM SPREAD	98 degrees
LUMENS	33,000	VERT. BEAM SPREAD	91 degrees

TEST PROCEDURE AND DATA FORM IN ACCORDANCE WITH I.E.S. AND N.E.M.A. STANDARDS

ISOCANDLE CURVES AVERAGE OF RIGHT AND LEFT SIDES

10% MAX. 10, 2350 C.P, 5000 C.P, 10,000 C.P, 15,000 C.P, 20,000 C.P, 23,000 C.P

B, E, C, G, O, H, A, L, D

OUTLINE OF BUILDING

LUMENS AVERAGE OF RIGHT AND LEFT SIDES

	0°–10°	10°–20°	20°–30°	30°–40°	40°–50°		TOTALS - ½ OF HORIZONTAL ZONES
50°–40°	66	58	44	27	14		209
40°–30°	146	125	88	53	26		438
30°–20°	280	228	158	86	40		792
20°–10°	502	388	232	122	53		1297
10°–0°	784	581	310	145	62		1882
0°–10°	884	630	332	151	65		2062
10°–20°	654	481	265	128	56		1584
20°–30°	351	274	167	90	41		923
30°–40°	167	140	98	57	27		489
40°–50°	80	70	50	32	14		246
TOTALS OF VERTICAL ZONES	3914	2975	1744	891	398		9922

Figure 16.3 Superimposed method for determining coefficient of beam utilization.

16.5.4 Estimate the Maintenance Factor

Lighting efficiency is seriously impaired by lamp depreciation and by dirt on the reflecting and transmitting surfaces of the equipment. To compensate for the gradual depreciation of illumination on the floodlit area, a maintenance factor (MF) must be applied in the calculations to make allowance for the following:

1. Loss of light output due to dirt on the lamp, reflector, and cover glass. Under comparable conditions, enclosed floodlights have a higher maintained efficiency than that of open units because the cover glass protects both the reflector and the lamp from the accumulation of dirt.
2. Loss in light output of the lamp with life. Because some of the light must pass through the bulb more than once before leaving the floodlight, bulb blackening also lowers floodlight efficiency, the reduction in beam lumens being about double the reduction in bare lamp output.

Maintenance factors are usually estimated to be from 0.65 to 0.85. However, if the floodlights are cleaned infrequently, or where lamps are replaced only on burnout, it is advisable to use lower maintenance factors. Difference in lumen maintenance among lamp types and sizes should also be taken into consideration. With narrow-beam floodlights, dirt on the reflector and cover glass tends to widen the beam spread, reducing the maximum candlepower more than the total light output. Thus for a small lighted area utilizing only the central part of a beam, a smaller percentage of the beam lumens will strike the target after the floodlight has become dirty. Therefore, the depreciation in foot-candle intensity will be greater than the depreciation in total light output, and it will be necessary to consider this in selecting a maintenance factor.

16.5.5 Determine the Number of Floodlights Required

$$\text{number of floodlights} = \frac{\text{area} \times \text{footcandles}}{\text{beam lumens} \times \text{CBU} \times \text{MF}}$$

where area is the surface area to be lighted in square feet; footcandles are as selected from Table 16.3. For the beam lumens, refer to manufacturers' catalogs for equipment under consideration. Where supply voltage deviates from rated voltage, the lamp lumen output should be adjusted accordingly.

16.5.6 Check for Coverage and Uniformity

After a tentative layout has been made, the uniformity may be checked by calculating the intensity of illumination at a few points on the floodlit surface. This may be done by the point-by-point method described in Chapter 12, using either a candlepower distribution curve or an isocandela diagram. If the uniformity is found to be unsatisfactory, a larger number of units may have to be used.

16.6 APPLICATION GUIDE

16.6.1 Buildings

The floodlighting of a building is primarily a problem in esthetics, and each installation must be studied individually. Under some circumstances, particularly with small, utilitarian buildings or larger buildings that have no special architectural features, uniform illumination is desirable. To create the appearance of uniform brightness over the entire facade of a building, it is necessary to increase the actual brightness appreciably toward the top. Higher brightness at the top of a building increases its apparent height (see Section 16.7.1).

With buildings of classical design or special architectural character, uniform illumination often defeats the purpose of the lighting, which should aim to preserve and bring out the architectural form. Buildings are designed primarily for daytime appearance, when the light comes from above. This effect is almost impossible to duplicate with floodlights, which must be mounted in nearby locations and usually at a height no greater than the elevation of the building. However, it is often possible to achieve a result that is interesting and pleasing, although different from the daytime appearance.

Shadows are essential to relief, and contrasts in brightness levels or sometimes in color can be used to bring out important details and to supress others. Sculpture or architectural details require particularly careful treatment to avoid flatness or grotesque shadows that may distort the appearance as conceived by the architect.

16.6.2 Color

Color can be provided in floodlighting installations in any one of several ways. Amber, blue, and red cover glasses are generally available for standard enclosed floodlights to replace the regular lens, or the floodlight may be recessed in a niche, the opening of which is covered with a filter. Where smaller amounts of color light are needed, 100 W PAR38 colortone (red, pink, yellow, green, blue, or blue-white) lamps are available, or a 300- or 500-W hard glass R-40 lamp may be used with a colored lens. Any color filter absorbs a large amount of light, and this loss must be considered in designing the installation.

16.7 EXAMPLES OF FLOODLIGHTING INSTALLATION

The following examples of floodlighting installations which were designed by the author are chosen to illustrate different themes and methods to achieve somewhat different aims in each case, as discussed in the preceding sections.

16.7.1 Westinghouse Lamp Division Headquarters Building

In the past, incandescent lamps were the most useful and versatile source. However, from an energy conservation and cost-saving standpoint, aesthetic lighting can be achieved with more efficient HID sources wherever feasible. A good example is the floodlighting of the Westinghouse Lamp Division Headquarters Building. The building had been face-lifted with a marble chip finish. The design specified 10 luminaires fitted with 400 W HPS lamps mounted on the roof of a garage some 50 ft away from the building. Each unit was adjusted individually to a proper aiming angle to achieve a uniform illumination of 10 fc over the building face. These floodlights have created a color of warmth and unmistakable identity, viewed by thousands of motorists traveling the Garden State Parkway at night. Figure 16.4 shows the building face floodlighted with HPS luminaires.

16.7.2 Floodlighting a Water Tower

Floodlighting a sign on a circular water tower, without spill or scatter, requires rather more than rule-of-thumb design. The light sources chosen had to be an economical lamp that would produce a narrow

Figure 16.4 Office building floodlighted with HPS light source.

searchlight type of beam and project all of the light exactly where it was aimed.

In this installation, low-voltage 120 W PAR64 very-narrow-spot lamps accomplished this. The bottom part of the tank and the upper tower legs were silhouetted with 500 W PAR64 narrow-spot lamps operating on a 120 V circuit. The 120 W PAR64 lamps operate on a 6 V circuit and can produce a controlled pencil-thin beam which is very effective for uplighting buildings, columns, steeples, towers, and the like. Its beam spread is 4½ degrees by 7 degrees and is achieved by masking critical areas of the reflector to prevent refocusing of the light. A matte black light shield produces an extremely sharp beam cutoff. Mounted precisely in relation to the reflector's focal point, a

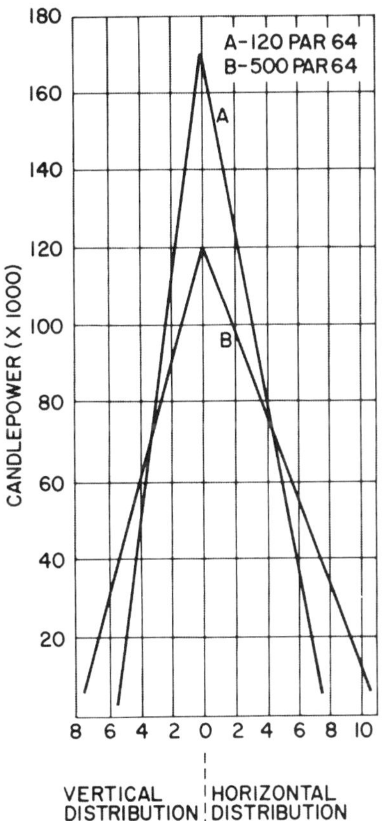

Figure 16.5 Candlepower distribution curves of 120 W and 500 W PAR64 lamps.

specially designed filament yields a maximum beam candlepower of 170,000. Figure 16.5 shows the candlepower distribution of the 120 W PAR64 very-narrow-spot lamp and the 500 W PAR64 narrow-spot lamp.

The water tank stands 100 ft above grade. The overall diameter of the tank is 22 ft. The corporate sign is approximately 12 ft in diameter and 120 ft above grade. The lighting banks pepper the sign from two sides: from the nearby building roof and from an erected pole. Each is located approximately 30 ft above the ground. The bank located on a nearby building roof consists of six 120 W and three 500 W PAR64 lamps. The bank located on the top of a new pole consists of four 120 W and two 500 W PAR64 lamps. Figure 16.6 shows a sketch of the scheme with all essential dimensions of the tower and mounting locations from both field measurements and calculations. Figure 16.7 shows the pole with floodlights mounted on its top.

All lamps are contained in a weatherproof cast-aluminum unit equipped with clear high-heat-treated tempered lenses, and adjustable cast arms permit exact aiming of the units. Raintight aluminum wireways are used to connect the floodlight units at each bank, where the lower row of luminaires is aimed at the bottom of the tank and the upper row at the sign. Figure 16.8 shows the water tower floodlighted

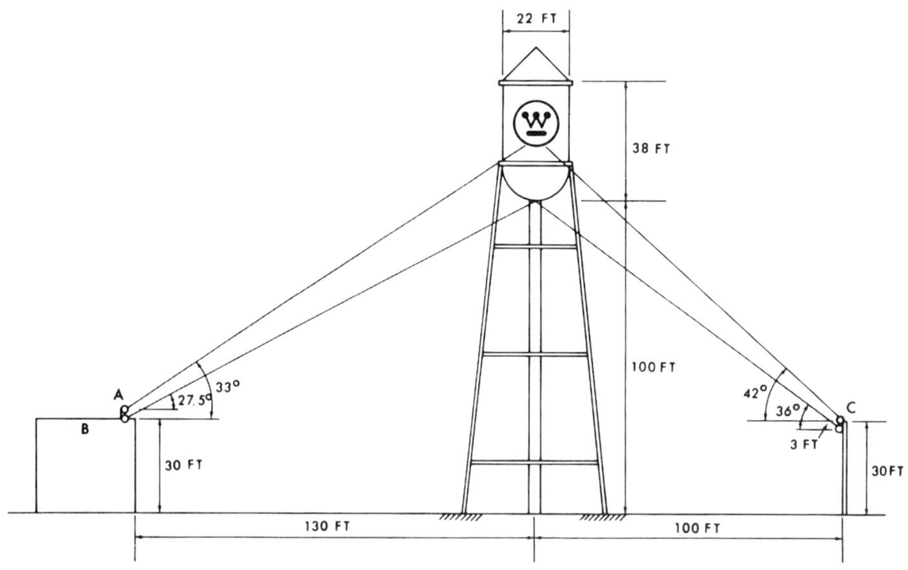

Figure 16.6 Sketch of dimensions and locations of floodlights for a water tower.

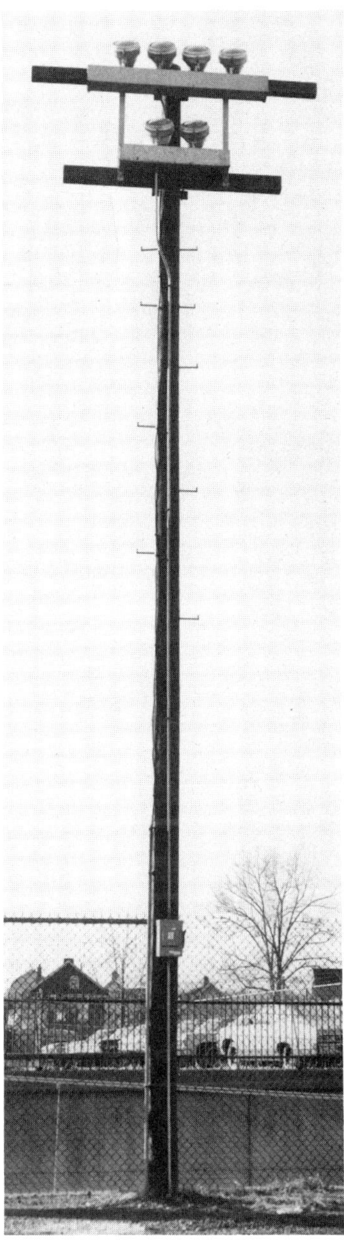

Figure 16.7 Pole-mounted floodlights.

by PAR64 lamps as the light source. The wiring scheme for the entire installation is shown in Figure 16.9. A three-pole starter is used to control power for the lighting units, which can be turned on and off by an astronomical clock. The time settings are adjusted to seasonal variations. The 500 W lamps are wired across a single-phase, three-wire system and controlled by two outdoor disconnect switches. The third contact is used to supply a 120 V source to a 120/12 V transformer rated at 1.5 kVA. The secondary is arranged to form a parallel-series circuit so that two 6 V lamps are series-connected across a 12 V circuit. The main disconnect switch, line starter, timer, and step-down transformer are installed in a nearby shed so that indoor equipment can be used. 120 V feeders are enclosed in steel conduit,

Figure 16.8 The water tower floodlighted with PAR64 light source.

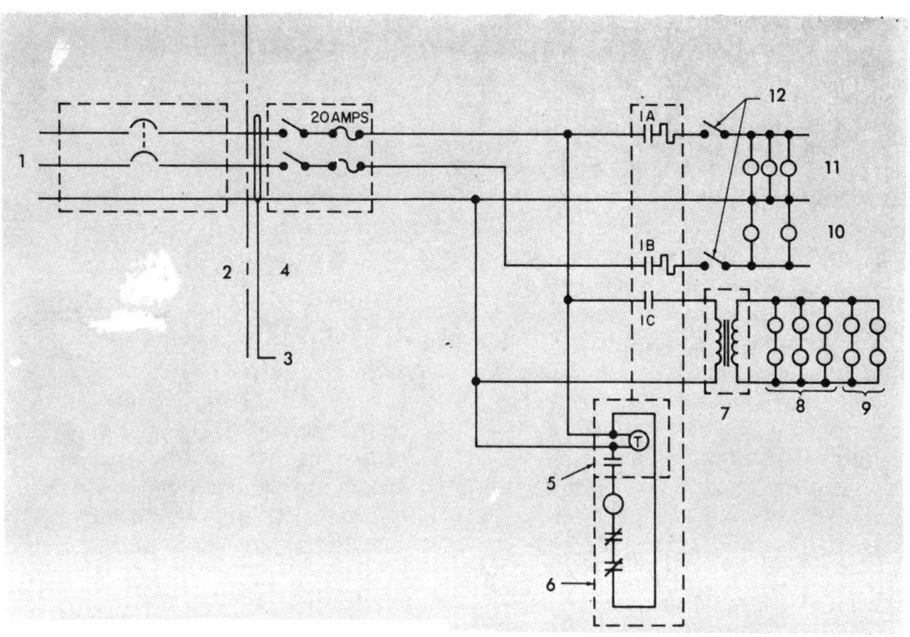

1—230/115-volt, three-wire, 60-cps service.
2—Existing service line in shed.
3—¾-inch–3 No. 10 RHW.
4—Safety switch mounted in shed.
5—Tork timer to control on and off.
6—Line starter.
7—120/12-volt transformer, 1½-kVa, mounted in shed.
8—Six 120-watt, six-volt PAR64 very narrow spots mounted on roof.
9—Four 120-watt, six-volt PAR64 narrow spots on new pole.
10—Two 500-watt, 120-volt PAR64 narrow spots on new pole.
11—Three 500-watt, 120-volt PAR64 narrow spots on roof.
12—S.P. disconnect switch, outdoors at pole and on nearby shed.

Figure 16.9 Wiring scheme for floodlighting a water tower.

while 12 V circuits in the form of underground cables are directly buried in the ground from the control station to the pole.

16.7.3 Pan Am Building

The main faces (north and south) of this New York office building are floodlighted from the tenth floor setback up to the top of the 59th floor, a distance of 550 ft. To achieve such lighting, special searchlight-type luminaires were developed to house a specially designed incandescent lamp.

The lamp had to have the most compact filament possible to enable the luminaire mirrors to project narrow but intense beams to the top of the building. At the same time, enough spread is needed to enable the luminaires to cover the lower areas, and a burning life of more than a year is desirable. The lamp developed has a globular bulb of hard glass 8 in. in diameter and a special 80 V filament operating at 2000 W. A collector grid traps tungsten particles as the filament vaporizes, preventing bulb blackening and thereby assuring good lumen maintenance. The special luminaires have cast aluminum housing with mirrored glass reflectors and clear tempered lenses. Each of the 170 units produces 2,750,000 candelas.

The smaller faces of the building are floodlit with 1000 W quartz-iodine lamps in luminaires that supply the same light distribution provided by the searchlights on the larger sides. As shown in Figure 16.10 the floodlighted building stands out with night city background.

16.7.4 New Fluorescent Lamp Plant

In this case the objectives are to light the front and ends of an office building brightly and flatly so that it would stand out against the background of the manufacturing building behind it, which, although less brightly lighted, is to have enough illumination to define its mass clearly. The office building is illuminated to a minimum intensity of 15 fc by lighting upward from ground level with weatherproof fluorescent floodlights. Special fluorescent lamps that have a low silhouette are used in the fixture.

For the manufacturing building, twenty-seven 1500 W quartz-iodine floodlights are located at ground level, 14 ft from the wall on approximately 40 ft centers, to provide a pattern on the building face and create a "glow" as background for the highlighted office building and the shrubbery. To delineate the pattern of the facade, all floodlights are aimed at 45 degrees from perpendicular to the building face and upward to provide the effect of the sun shining on the building at a 45-degree angle. One floodlight is opposite each column and each corner, with the others located as required to give adequate

Figure 16.10 Pan Am building floodlighted with specially designed luminaires and lamps.

Figure 16.11 A brightly floodlighted office building.

overlapping of patterns. This installation was one of the earliest application of the newly developed 1500 W quartz-iodine floodlight luminaires.

The shrubbery surrounding the office building is illuminated by batteries of floodlights on the building roof along both ends. The best color for most shrubs is provided by mercury lighting, so 100 W PAR38 mercury lamps are used. They are mounted in heavy-duty cast-aluminum outdoor floodlighting luminaires, which are neat and small for good appearance, and so aimed that they do not project light and glare toward viewers of the scene. Figure 16.11 shows a brightly floodlit office building in the foreground of the manufacturing building. Figure 16.12 shows a fluorescent lamp manufacturing plant building that has sunset pattern floodlighting.

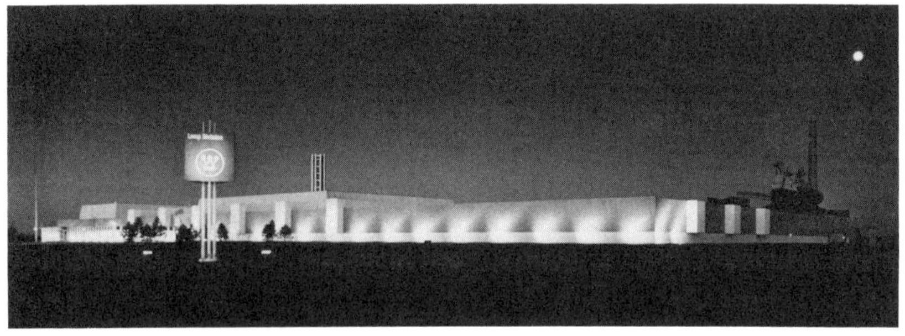

Figure 16.12 A pattern formed floodlighting for the manufacturing plant building.

BIBLIOGRAPHY

Chen, K., Floodlighting Technique for a Water Tower, *Illuminating Engineering*, May 1967, pp. 305–307.

Chen, Kao, Lighting Esthetics with Energy Saving Ideas, *IEEE Transactions on Industry Applications*, Jan./Feb. 1976, pp. 35–38.

Chen, Kao, What's New in Floodlighting, *IEEE Transactions on Industry Applications*, July/Aug. 1977, pp. 343–347.

Chen, K., and Karns, E. B., New Techniques Enhance Effectiveness of Building Floodlighting, *Westinghouse Engineer*, July 1968, pp. 118–121.

Floodlighting with High-Pressure-Sodium, Data Sheet, Lighting Design and Application, June 1972.

Lighting Handbook, Westinghouse Electric Corporation, Bloomfield, N.J., 1976.

17

Energy Conservation in Illuminating Systems

17.1 INTRODUCTION

Continuing improvements in the efficiency of generation and distribution equipment during the early years of incandescent lighting helped drive energy costs down steadily. Significant reductions in energy costs have been achieved which would have more than halved the cost of electric lighting since 1913 even if lamp performance had not improved at all. However, this trend has been reversed since the energy shortage in 1974.

The average cost of obtaining a given amount of light takes into account the reduction in energy cost, the reduction in lamp price, the increase in lamp efficacy, and the increase in lamp life. The record of the Edison Electric Institute shows that lighting costs in 1945 were 1.3% of what they were in 1882. About 60% of the savings since 1923 is attributable to reductions in the cost of electric energy, about 30% to increase in lamp efficacy, and about 10% to reductions in lamp prices. Cheap energy in the United States has been a most influential factor in the flourishing incandescent lighting industry for several decades. However, since the energy crisis the lighting industry has been facing one of its toughest challenges to date. On the one hand, lighting plays an important role in every walk of life and should be allowed to perform its intended function. On the other hand, the lighting industry has been called upon to make contributions to energy conservation. To put the role of lighting as it relates to energy conservation in perspective, we must look at the impact of lighting on total energy resource consumption in the United States. Currently, 80% of the resources used in this country are fossil fuels (coal, oil, and natural gas), the most critical fuels in terms of estimated resources consumed; approximately 25% are used to generate electricity. Of that 25%, about 20% ends up as lighting. In other words, approximately 5% of the total energy resources consumed in this country ends up in the form of lighting.

With these facts, why is lighting a target for energy conservation? The reason is that lighting is "visible." In addition, in terms of the end user, lighting represents 30 to 50% of the operating cost of a building. Lighting energy conservation is important in terms of the total resources and operating cost for the building owner. As utility rates continue to increase, the impact of lighting on operating cost will become more apparent.

17.2 ENERGY-EFFICIENT LIGHT SOURCES

Although it constitutes only one component in an illuminating system, the light source is often the major factor determining the overall efficiency of the entire illuminating system. Typical light sources cover a wide range of efficacies, that is, how well the light source converts electric energy to visible light. The expression used to measure the efficacy of light sources is lumens per watt (lm/W).

17.2.1 Incandescent Light Source

Recent technical advances have made possible a line of energy-saving incandescent lamps that use rare krypton gas as a fill gas. The krypton allows hotter filament operation without the undesirable short life. The result is a slight increase in the efficacy. For example, a 100 W, 2500 h life extended service lamp can be replaced with a 93 W energy-saving lamp with krypton fill which produces the same light output and same life.

Energy-saving potential also exists for incandescent lamps where reflector lamps can be used. Incandescent lamps using built-in reflectors offer better utilization of the light produced by the lamp compared to a nonreflector type. In this family of lamps there are R lamps, indoor reflector lamps; and PAR lamps or parabolic aluminized reflector lamps. A new line of indoor reflector lamps called ER, or elliptical reflector lamps, allow reduction of 50% or more in energy consumption in many installations.

In addition, incandescent lamps with special radiation covering envelopes are being developed. Compact high-emissivity filaments are required to absorb infrared radiation reflected from specially coated envelopes. The energy-saving potential for the new lamps could be as much as 60% compared with equivalent conventional incandescent lamps.

17.2.2 Fluorescent Light Source

Since the early 1970s there has been available a line of energy-saving replacements for standard fluorescent lamps. Krypton gas is again employed to improve the efficacy, up to a level of 100 lm/W. Energy-saving lamps are now available in all popular sizes and colors for most applications. Limitations of energy-saving reduced-wattage lamps are:

1. Should be used only where ambient temperature does not drop below 60°F
2. Should be used only on high power factor fluorescent ballasts
3. Not to be used where drafts or cold-air ducts would be directed

Typical energy savings are about 17 W per lamp for the popular 4-ft 40 W replacement and 17.5 W per lamp for the popular 8-ft slimline 75 W replacement. Savings in energy cost normally pay back the new lamp cost in a year at typical power rates and lamp costs.

A recently developed arc-discharge lamp, called a compact fluorescent, can replace an incandescent light source in particular applications. These compact, twin-tube lamps are gaining popularity because they are energy efficient and fit into a small enclosed housing. A number of the compact fluorescent lamps use trichromatic phosphor coating, which provides peak response to the human eye in the three primary colors, red, green, and blue.

The so-called PL lamps are available in 7-, 9-, and 12 W ratings, which provide equal-light-output alternatives to 40-, 60-, and 75 W incandescent lamps. They are suitable for exit signs, information signs, and entrance and hallway security fixtures. Another recent compact lamp developed is the SL lamp (see Figure 14.2), which is a double-folded, bent-tube assembly that can be retrofitted to a standard medium-base socket.

17.2.3 HID Light Source

Mercury light sources are available in many popular wattages, ranging from 40 to 1000 W, in several phosphor colors. However, the newer HID sources offer higher efficacy and superior color-rendering properties compared to mercury lamps. This has practically eliminated standard mercury from new installations where energy conservation is a deciding factor.

Metal halide lamps in 325-, 400-, and 1000 W ratings have been designed for direct replacement of mercury lamps in existing mercury fixtures to gain up to 42% more light output. However, these retrofit lamps should be applied carefully, since they are compatible with only a few mercury ballasts. If used with existing reflector, the metal halide lamp may cause undesirable light distribution.

One recently developed high pressure sodium lamp eliminates the high-voltage pulse starting by using a tungsten wire heater wrapped around the arc tube. When power is applied to the lamp, the tungsten coil heats the thin arc tube to the required 300°C and, when full light output is reached, the heater circuit is opened. The heated-tube lamp is available in 150-, 215-, 310-, and 360 W sizes for use on existing mercury reactor or lag-type autotransformer ballasts rated 175, 250, 375, and 400 W, respectively. The changeover to a more efficient light source in an industrial fixture can be done by a simple lamp replacement. This changeover can be practical if the existing fixture is still in good condition. The average rated life of the HPS in many popular sizes is 24,000 h. This long life plus excellent lumen maintenance (90% mean lumens over the lamp life) gives HPS lighting systems an advantage in applications in industrial facilities.

Low pressure sodium lamps have a decidedly yellow color, which limits their application, but they are widely used as "night lights" in retail stores and wherever long burning hours are a factor and color discrimination is not important. They are the most energy-efficient general light source available today (see Figure 14.7).

17.3 ENERGY-EFFICIENT BALLASTS

To assure optimum performance, ballasts are manufactured within a rigid tolerance to supply a lamp with specified voltages and currents. ANSI specifies the standards as well as the methods used to test the ballast (C82) and lamp (C78) performance. The Certified Ballast Manufacturers Association (CBM) sets the performance criteria in accordance with the ANSI standards. These ballasts are designated as CBM ballasts.

17.3.1 Intrinsic Ballast Parameters Affecting Efficiency

Ballasts are designed to operate fluorescent lamps over a range of ±10% about the rated center voltage. Over this operating range, the power input and light output tend to decrease with decreasing input voltage. The system efficiency increases slightly as the input voltage decreases over this range by 2 to 4%.

In general, the relative ballast losses are less for a 40 W ballast than a 20 W ballast. A two-lamp ballast is more efficient than a one-lamp ballast. With the recent increase in energy costs, ballast manufacturers have introduced an energy-efficient ballast that minimizes ballast losses. Ballast efficiencies have improved, which has resulted in an 8 to 10% increase in system efficiency.

It is well known that fluorescent lamps driven at high frequency are more efficient. Electronic ballasts are now available for the F40,

T12, the slimline, the new F8 lamps, and other energy-saving fluorescent lamps on both 120- and 277 V circuits. Operation of an electronic ballast involves the use of transistor circuitry to rectify the 60-Hz ac branch-circuit supply to a dc component and then invert it back to an ac sine-wave component having the frequency range 10 to 30 kHZ. When the frequency of the on-off operation of the mercury arc within the lamp is increased from 60 Hz to many times that value, the lamp efficecy can be raised by nearly 12%. At the same time, with the absence of the magnetizing losses within a core-coil ballast, the relative efficiency of the ballast is increased. Although the electronic ballast costs more than the standard core-coil ballast, the operating factors should reflect an appreciable reduction in life-cycle costs for a lighting system.

17.3.2 Extrinsic Ballast Parameters Affecting Efficiency

High- and low-core ballasts are available with means to operate fluorescent lamps at either 100% or 50% light output. These ballasts can be used in simple low-cost dimming systems where the need to change illumination levels is infrequent.

There are two types of dimming ballasts: core and electronic. Core ballasts can dim fluorescent lamps over a wide range of light levels. Auxiliary switching equipment is required to reduce the duty cycle or limit the current. These systems are energy effective where a substantial portion of the time lamps can be operated well below 100%. High-frequency ballasts can readily be used to dim fluorescent lamps over a wide range of light levels. No major auxiliary equipment is required, as the light levels are controlled through the ballasts' internal reactance. All external control wiring is low-voltage wiring. The great advantage of electronic systems is the ease of controlling the dimming circuitry with external sensors.

New fluorescent lamps rated at slightly less wattage have been introduced that present a new load to the ballasts. For example, 34/35 W lamps are available that can be used in place of 40 W lamp with 40 W ballasts. In this application, both the power and light levels are proportionally reduced with no net change in efficiency. Some manufacturers have introduced ballasts that are optimized for 35 W lamps; the systems are slightly more efficient than with the 40 W ballasts. However, in applications where ballasts need not be replaced, the slight increase in efficiency does not justify the cost of refitting these ballasts. Table 17.1 summarizes the input watts for typical fluorescent lamp ballasts. The reduction in input watts for the new energy-saving ballasts versus standard ballasts are shown clearly in this table.

Table 17.1 Typical Fluorescent Lamp Ballast Input Watts

Lamp Type	Nominal Lamp Current	Nominal Lamp (W)	System Input (W)				Circuit Type
			Standard Ballasts		Energy-Saving Ballasts		
			One-Lamp	Two-Lamp	One-Lamp	Two-Lamp	
F20T12	0.380	20	32	53	—	—	Rapid start, preheat lamp
F30T12	0.430	30	46	81	—	—	Rapid start
F30T12, ES	0.460	25	42	73	—	—	Rapid start
F32T8	0.265	32	—	—	37	71	Rapid start
F40T12	0.430	40	57	96	50	86	Rapid start
F40T12, ES	0.460	34/35	50	82	43	72	Rapid start
F48T12	0.425	40	61	102	—	—	Instant start
F96T12	0.425	75	100	173	—	158	Instant start
F96T12, ES	0.455	60	83	138	—	123	Instant start
F48T12, -800 ma	0.800	60	85	145	—	—	Rapid start
F96T12, -800 ma	0.800	110	140	257	—	237	Rapid start
F96T12, — ES, 800 ma	0.840	95	125	227	—	207	Rapid start
F48 - 1500 ma	1.500	115	134	242	—	—	Rapid start
F96 - 1500 ma	1.500	215	230	450	—	—	Rapid start

4/22/83 RWW

17.3.3 Electronic Ballasts for HID Lamps

Electronic ballasts can be designed to have a steady, constant wattage output with changes in the source impedance as well as excellent regulation. The circuit of one such ballast, called an "energy-efficient" ballast, is shown in Figure 14.11. The electronic ballast is designed to maintain a light level at the minimum power level. During the life of a high pressure sodium lamp, electronic ballasts can save 20% more energy by maintaining a constant wattage output in addition to the 15% intrinsic energy savings compared to the equivalent core-coil ballasts.

A number of manufacturers offer HID dimming systems for such applications as meeting rooms, assembly halls, auditoriums, and so on. This equipment differs in construction and function, since not all system can control all HID lamp types.

With automatic level control, a photocell is used to read average illumination in an area similar to the method used in a fluorescent system. As lamps age and luminaires become dirty, the control unit gradually increases power to the system's dimming ballasts. Also, if daylight enters the space, the control system reduces the lamp power in proportion to the amount of daylight received. Important factors in HID dimming system are energy costs, burning hours, quantity of supplemental daylight, and the lamp wattage selected. Table 17.2 summarizes the input watts for typical HID lamp ballasts. No data for the electronic ballasts are included in the table.

17.4 NEW LUMINAIRES FOR ENERGY-EFFICIENT LIGHT SOURCES

Proper luminaire design is the key to lighting efficiency. For high-pressure sodium, it is extremely important that luminaires with precise optical design and control be matched to the smaller arc sources. Using reflectors originally designed for phosphor-coated mercury lamps can result in poor light distribution and lack of footcandle uniformity. Newly developed luminaires use prismatic glass reflectors especially made for high pressure sodium lamps. In addition to achieving maximum light utilization, they redirect the intense HPS light source with excellent light cutoff and high-angle brightness control. Luminaire manufacturers recommend aluminum reflectors for all general-purpose industrial applications, and glass-coated reflectors where environmental conditions are severe and where maintenance practice is compatible with servicing glass.

Table 17.2 Typical HID Lamp Ballast Input Watts

				Ballast Type			
Lamp Type	ANSI Designation	Watts	Reactor	High Reactance Autotransformer (LAG)	Constant Wattage Autotransformer (CWA)	Constant Wattage Regulated (CW)	High Reactance Regulated (Regulated Lag)
Mercury	H46	50	68	74	74	—	—
	H43	75	94	91–94	93–99	—	—
	H38/44	100	115–125	117–127	118–125	127	—
	H39	175	192–200	200–208	200–210	210	—
	H37	250	272–285	277–286	285–300	292–295	—
	H33	400	430–439	430–484	450–454	460–465	—
	H36	1000	1050–1070	—	1050–1082	1085–1102	—
Metal-halide	M57	175	—	—	210	—	—
	M58	250	—	—	292–300	—	—
	M59	400	—	—	455–465	—	—
	M47	1000	1050	—	1070–1100	—	—
	M48	1500	—	—	1610–1630	—	—
High-pressure sodium	S76	35	43	—	—	—	—
	S68	50	60–64	68	—	—	—
	S62	70	82	88–95	95	—	105
	S54	100	115–117	127–135	138	—	144
	S55	150 (55 V)	170	188–200	190	—	190–204
	S56	150 (100 V)	170	188	188	—	—
	S66	200	220–230	—	245–248	—	254
	S50	250	275–283	296–305	300–307	—	310–315
	S67	310	335–345	—	365	—	378–380
	S51	400	463–440	464–470	465–480	—	480–485
	S52	1000	1060–1065	—	1090–1106	—	—

17.4.1 Construction

The amount of dirt that will accumulate on the lamp and luminaire reflector is affected by luminaire construction. IES recognizes five degrees of ambient dirt conditions: very clean, clean, medium, dirty, and very dirty. Different manufacturers have different philosophies regarding the best way to keep dirt from accumulating on the critical surfaces of luminaires. Construction of HID luminaires for lighting of manufacturing areas can be placed into four general classes: open bottom, open bottom with open top, enclosed sealed and gasketed, and sealed and filtered.

17.4.2 Beam Spread

IES classifies luminaires by light distribution as highly concentrating, concentrating, medium spread, spread, and wide spread. Luminaire manufacturers will provide candle power distribution curves and/or isolux curves for the luminaires in line. Most HID luminaires designed for indoor applications have symmetrical light distribution patterns. But asymmetrical versions are available and may be applied to the solution of special lighting problems. For lighting the perimeter of manufacturing areas, luminaires are available that throw almost all light downward and inward, so that it is not wasted on the wall surfaces.

17.4.3 Types of Reflector Material

HID luminaires are offered with prismatic glass, anodized aluminum, and bonded glass-to-aluminum reflectors.

1. *Prismatic glass reflector.* This reflector has excellent durability, and strong cleaning agents can be used repeatedly. Optical control is provided primarily by the prism constructions on the reflector surface rather than by the specific shape of the reflector.
2. *Anodized aluminum reflector.* Most HID luminaires utilize anodized aluminum reflectors produced by the Alzak process.
3. *Bonded glass to aluminum reflector.* Glass is bonded to an anodized or chemically brightened aluminum reflector to provide a more durable and cleanable finish.

17.5 COST ANALYSIS

Energy management and energy-efficient design have a tremendous impact on cost. The final decision as to which lighting system to install depends heavily on the cost. These costs should include not only the initial cost of the installation, but also the operating and maintenance costs. With the rise in energy cost and inflation in all sectors of economy, an inexpensive system (low initial cost) could cost the owner many times more to operate and maintain.

Initial Cost

Energy-efficient design will result in an increase of initial cost over the traditional approach of "make it cheap." Many factors are involved, so a decision as to which system to use should not be made on the basis of initial cost alone.

Maintenance Cost

Maintenance cost is tied directly to the selection of lighting equipment, which affects initial cost. In general, the initial cost will increase for equipment with better maintenance characteristics. A good maintenance program will minimize light loss from dirt accumulation and surface deterioration, which would avoid increasing light to compensate for such losses.

Operating Cost

Operating cost is tied into the amount of power consumed. System design in terms of light source efficacy and overall system efficiency will determine the operating cost.

Cost Summary

There are a number of methods of economic analysis by which the choices available may be reviewed. Types of analysis precently in use are the payback period, the internal rate of return, precent value, and the savings investment ratio. With the rapid increase in the cost of energy in recent times, an inflationary factor is critical to an analysis of operating costs.

17.5.1 Payback Period Method

A major lamp manufacturer designed an energy cost management analyzer that contains essential lamp information and fixture data, and provides parallel working spaces for the present lighting system and the new energy-saving lighting system, side by side. Each contains items such as annual energy charges per fixture, annual lamp replacement cost per fixture, annual cleaning cost per fixture, and so on. The total of these items is equal to the annual operating cost for the present system or the proposed system. From the difference between the two systems, one will be able to calculate return on investment (ROI). Figure 17.1 shows such an analyzer.

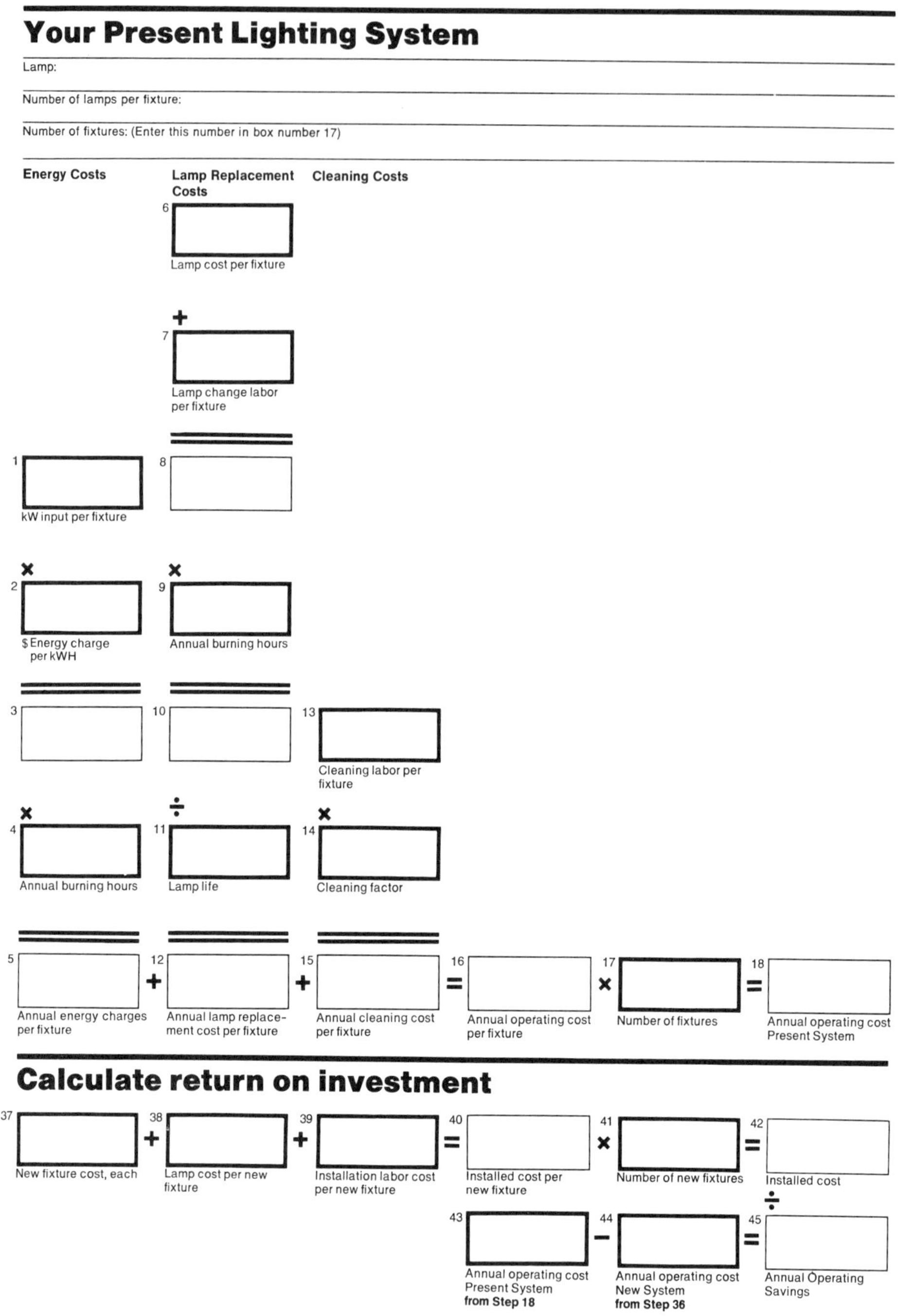

Your Present Lighting System

Lamp:

Number of lamps per fixture:

Number of fixtures: (Enter this number in box number 17)

Energy Costs

1 kW input per fixture
× 2 $ Energy charge per kWH
= 3
× 4 Annual burning hours
= 5 Annual energy charges per fixture

Lamp Replacement Costs

6 Lamp cost per fixture
+ 7 Lamp change labor per fixture
= 8
× 9 Annual burning hours
= 10
÷ 11 Lamp life
= 12 Annual lamp replacement cost per fixture

Cleaning Costs

13 Cleaning labor per fixture
× 14 Cleaning factor
= 15 Annual cleaning cost per fixture

5 + 12 + 15 = 16 Annual operating cost per fixture × 17 Number of fixtures = 18 Annual operating cost Present System

Calculate return on investment

37 New fixture cost, each + 38 Lamp cost per new fixture + 39 Installation labor cost per new fixture = 40 Installed cost per new fixture × 41 Number of new fixtures = 42 Installed cost

÷

43 Annual operating cost Present System **from Step 18** − 44 Annual operating cost New System **from Step 36** = 45 Annual Operating Savings

a

Figure 17.1 Lighting energy cost management analyzer.

Your New Energy Saving Lighting System

Lamp:

Number of lamps per fixture:

Number of fixtures: (Enter this number in box number 35)

Energy Costs

19 kW input per fixture

× 20 $ Energy charge per kWH

= 21

× 22 Annual burning hours

= 23 Annual energy charges per fixture

Lamp Replacement Costs

24 Lamp cost per fixture

+ 25 Lamp change labor per fixture

= 26

× 27 Annual burning hours

= 28

÷ 29 Lamp life

= 30 Annual lamp replacement cost per fixture

Cleaning Costs

31 Cleaning labor per fixture

× 32 Cleaning factor

= 33 Annual cleaning cost per fixture

23 + 30 + 33 = 34 Annual operating cost per fixture × 35 Number of fixtures = 36 Annual operating cost New System

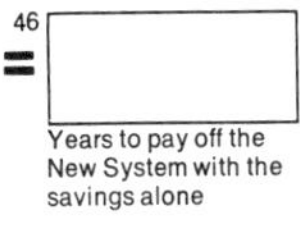

46 Years to pay off the New System with the savings alone

b

Figure 17.1 (Continued)

There are many other forms that one can use to make a relighting study. However, none is more complete and thorough than the cost analysis form proposed by IES and/or major lighting manufacturers. Figure 17.2 shows a typical lighting cost analysis form. A number of lamp and luminaire manufacturers offer computer services for evaluating lighting system alternatives.

17.5.2 Life-Cycle Costing

Life-cycle costing (LCC) is simply the evaluation of a proposal over a reasonable time period considering all pertinent costs and the time value of the money. The evaluation can take the form of a present value analysis or uniform annual cost analysis. More sophisticated analysis would include sinking fund and rate of return on extra investment. This method of analysis is not intended to be a detailed study of various systems. It is to be used by practicing engineers as a guide for comparing the advantages of alternative design cases. A more detailed analysis taking into account other items, such as future costs, could be performed. The method given here utilizes differential costs to provide a direct comparison of systems. Figure 17.3 shows an outline of one method of determining costs, called "life-cycle cost analysis."

17.5.3 Examples of Cost Analysis

A comprehensive cost analysis comparing an existing mercury lighting system with four alternative systems was made for an industrial plant. It served as a basis for selecting a best system which is not only more efficient, but delivers better-quality light at the same time. Table 17.3 exhibits the comparative cost analysis. Figure 17.4 exhibits an economic analysis for two lighting systems in a 30 ft by 30 ft classroom, to illustrate the use of the life-cycle cost method; it is not intended to show the benefits of one lighting system over another.

17.6 ENERGY-SAVING LIGHTING TECHNIQUES

17.6.1 Using Incandescent Systems

As discussed in Section 17.2.1, recent technical advances have made possible a line of energy-saving incandescent lamps that use rare krypton gas as a fill gas. In addition, energy saving can be accomplished with incandescent lamps, and a variety of choices are available: (1) use of lower-wattage lamps where less light is acceptable, (2) use of shorter-life high-efficacy lamps, (3) use of reflectorized lamps in place of standard lamps, and (4) use of transformer fixtures

		Lighting System Parameter	Base	II
Basic Data	1.	Rated initial lamp lumens per luminaire		
	2.	Rated lamp life (hours) at ________ hours per start		
	3.	Group replacement interval (hours)		
	4.	Average watts per lamp		
	5.	Input watts per lumaire (including ballast losses)		
	6.	Coefficient of utilization		
	7.	Ballast factory (fluorescent)		
	8.	Lamp depreciation factor		
	9.	Dirt depreciation factor		
	10.	Effective maintained lumens per luminaire (1 • 6 • 7 • 8 • 9)		
	10A.	Average footcandles on work surface (10 ÷ ft^2/luminaire)		
	11.	Relative number of luminaires needed for equal maintained footcandles (10 of base system ÷ 10 of system compared)		
Initial Costs	12.	Net cost of one luminaire		
	13.	Wiring and distribution system cost per luminaire		
	14.	Installation labor cost per luminaire		
	15.	Net initial lamp cost per luminaire		
	16.	Total initial cost per luminaire (12 + 13 + 14 + 15)		
	17.	Annual owning cost per luminaire (15% of 12 + 13 + 14)		
	18.	Relative initial cost for equal maintained footcandles (16 • 11 of system compared ÷ 16 of base system)		
Operating Costs	19.	Burning hours per year		
	20.	Number of lamps group replaced per year (19 • = lamps/unit ÷ 3)		
	21.	Number of interim spot replacements (20 • = burn outs in GR interval)		
	21A.	Number of lamps spot replaced per year — No group relamping (19 • lamps/unit ÷ 2)		
	22.	Replacement lamp cost per year (20 or 21A • net lamp cost)		
	23.	Labor cost for group replacements (20 • group labor rate/lamp) at $ ________/ lamp		
	24.	Labor cost for spot replacements (21 • spot labor rate/lamp at $ __________ / lamp		
	25.	Cost of cleaning per luminaire per year		
	26	Annual energy cost per year (5 • 19 • ¢/kWH ÷ 100 000 at __________ ¢ kWH		
	27.	Total annual operating cost per luminaire (22 + 23 + 24 + 25 + 26)		
	28.	Relative annual operating cost for equal maintained footcandles (27 • 11 of system compared ÷ 27 of base system)		
Total	29.	Total annual cost — owning and operating — per luminaire (17 + 27)		
	30.	Relative total annual cost for equal maintained footcandles (29 • 11 of system compared ÷ 29 of base system)		

Figure 17.2 A typical lighting cost analysis form.

Life cycle cost analysis for _______ ft^2 _______________.

	Luminaire _______ Layout _______	Luminaire _______ Layout _______
A. Lighting and air conditioning installed costs (initial)		
1. Luminaire installed costs: luminaire, lamps, material, labor	$_______	$_______
2. Total kW lighting:	_______kW	_______kW
3. Tons of air conditioning required for lighting: (3.41 × kW/12)	_______tons	_______tons
4. First cost of air-conditioning machinery: @ $_____/ton	$_______	$_______
5. Reduction of first cost of heating equipment:	$_______	$_______
6. Other differential costs:	$_______	$_______
	$_______	$_______
	$_______	$_______
	$_______	$_______
7. Subtotal mechanical and electrical installed cost:	$_______	$_______
8. Initial taxes:	$_______	$_______
9. Total costs:	$_______ (A1)	$_______ (B1)
10. Installed cost per square foot:	$_______	$_______
11. Watts per square foot of lighting:	_______watts	_______watts
12. Salvage (at *y* years):	$_______ (As)	$_______ (Bs)
B. Annual power and maintenance costs		
1. Lamps: burning hours × kW × $/kWh	$_______	$_______
2. Air conditioning: operation-hours × tons × kW/ton × $/kWh	$_______	$_______
3. Air conditioning maintenance: tons × $/ton	$_______	$_______
4. Reduction in heating cost fuel used: _____	$_______	$_______
5. Reduced heating maintenance: MBtu × $/MBtu	$_______	$_______
6. Other differential costs:	$_______	$_______
	$_______	$_______
	$_______	$_______
	$_______	$_______
7. Cost of lamps: (No. of lamps _____ @ $_____/lamp per *N*) (Group relamping every *N* years, typically every one, two or three years, depending on burning schedule.)	$_______	$_______
8. Cost of ballast replacement: (No. of ballasts _____ @ $_____/ballast per n) (n = number of years of ballast life.)	$_______	$_______
9. Luminaire washing cost: No. of luminaires _____ @ $_____ each. (Cost to wash one luminaire includes cost to replace *or* wash lamps.)	$_______	$_______
10. Annual insurance cost:	$_______	$_______
11. Annual property tax cost:	$_______	$_______
12. Total annual power and maintenance cost:	$_______ (Ap)	$_______ (Bp)
13. Cost per square foot:	$_______	$_______

Notes on analysis

A. 1. An estimate is prepared for material and labor of the installation.

2. In the example that follows a 40-W rapid-start lamp with ballasts loss is considered one 48-W load, and the 150-W HPS with ballasts is considered 175-W.

4. First cost of machinery will vary from $1000 to $2000/ton. Use the same value for both systems.

Figure 17.3 Life cycle cost analysis form.

Table 17.3 Comprehensive Cost Analysis for Five Illuminating Systems

	Existing 400 W Mercury System	250 W Metal-Halide System	400 W Metal-Halide System	250 W HPS System	400 W HPS System
Number of luminaires required	108	97	68	70	42
Luminaire spacing (square grid), ft	9.62	10.15	12.13	11.95	15.43
Initial lamp lumens per lamp	22,500	20,500	34,000	30,000	50,000
Lamp lumen depreciation factor	0.78	0.83	0.75	0.90	0.90
Estimated lamp life, hr	24,000	10,000	15,000	24,000	24,000
Average lamp replacements per year	18	38.8	18.13	11.67	7
Lamp net cost, dollars per lamp	10.23	23.55	22.35	38.40	36.00
Luminaire input watts	450	285	460	300	475
Average watts per sq ft	4.9	2.8	3.1	2.1	2.0
Total connected load, kw	48.6	27.65	31.28	21	19.95
Luminaire per unit cost, dollars	-0-	68	85	145	150
Installation labor per unit, dollars	-0-	36	36	36	36
Installation cost summary					
Luminaire cost, dollars	-0-	6,596.00	5,780.00	10,150.00	6,300.00
Initial lamp cost, dollars	-0-	2,284.35	1,519.80	2,688.00	1,512.00
Installation labor cost, dollars	-0-	3,492.00	2,448.00	2,520.00	1,512.00
Total installation costs, dollars	-0-	12,372.35	9,747.80	15,358.00	9,324.00
Annual operating cost summary					
Lamp cost, dollars	184.18	913.74	405.28	448.00	252.00
Maintenance labor, dollars	180.00	388.00	181.33	116.67	70.00
Energy cost, dollars	7,776.00	4,423.20	5,004.80	3,360.00	3,192.00
Total annual operating cost, dollars	8,140.14	5,724.94	5,591.41	3,924.67	3,514.00
Relative operating cost, percent	100.00	70.33	68.69	48.21	43.17
Total annual cost summary					
Annual owning cost, dollars	-0-	1,513.20	1,234.20	1,900.50	1,171.80
Owning and operating cost, dollars	8,140.14	7,238.14	6,825.61	5,825.17	4,685.80
Relative owning and operating cost, percent	100.00	88.92	83.85	71.56	57.56
Annual cost per fc per sq ft, dollars	0.8107	0.7216	0.6832	0.5819	0.4681
Lighting investment payback summary					
Annual operating cost, dollars	8,140.14	5,724.94	5,591.41	3,924.67	3,514.00
Operating cost savings, dollars	-0-	2,415.20	2,548.73	4,215.47	4,626.14
Total new investment, dollars	-0-	12,372.35	9,747.80	15,358.00	9,324.00
Simple investment payback interval, years	-0-	5.12	3.82	3.64	2.02
Simple return on investment, percent	-0-	19.52	26.15	27.45	49.62
Adjusted discounted investment payback interval, months	-0-	90.4	60.7	57.0	28.3
Summary of costs over next 20 years					
Net lamp costs, dollars	3,682.80	17,132.62	7,345.70	7,616.00	4,284.00
Lamp replacement labor costs (at $10 per lamp), dollars	3,600.00	7,760.00	3,626.67	2,333.33	1,400.00
Energy consumption, kwh	3,888,000	2,211,600	2,502,400	1,680,000	1,596,000
Total energy costs, dollars	155,520.00	88,464.00	100,096.00	67,200.00	63,840.00
Total initial costs, dollars	-0-	12,372.35	9,747.80	15,358.00	9,324.00
Total 20 year life-cycle costs, dollars	162,802.80	125,728.97	120,816.17	92,507.33	78,848.00

*Basis: 10,000 sq ft manufacturing area illuminated to 100 fc, 4000 burning hours per year, effective electrical energy rate (including demand and other charges) 4¢/kwh, average dirt conditions, 20 year amortization at interest rate of 10 percent.

that accomodate low-voltage lamps. More detailed discussions of each method are given below.

Use of Lower-Wattage Lamps

Substitution of lower-wattage lamps will result in loss of light output. The effects on task performance may not be acceptable. However, the variety of wattages available with direct interchangeability makes this approach a consideration that should not be passed over.

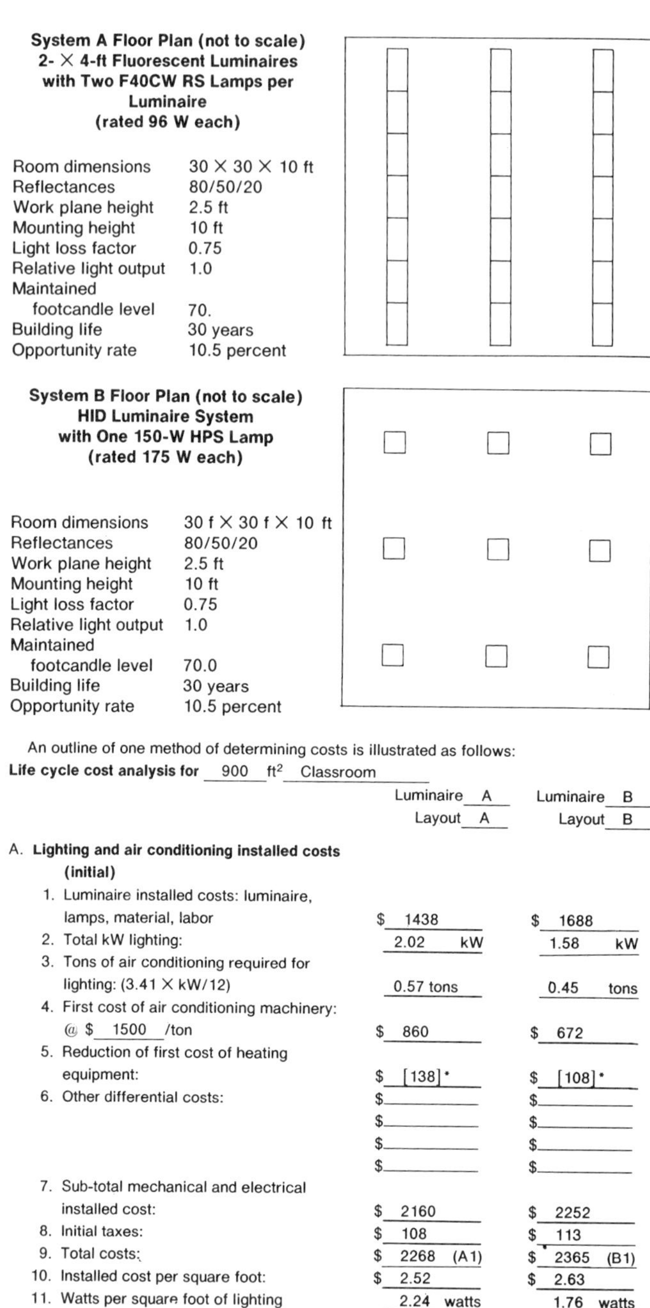

System A Floor Plan (not to scale)
2- × 4-ft Fluorescent Luminaires
with Two F40CW RS Lamps per Luminaire
(rated 96 W each)

Room dimensions	30 × 30 × 10 ft
Reflectances	80/50/20
Work plane height	2.5 ft
Mounting height	10 ft
Light loss factor	0.75
Relative light output	1.0
Maintained footcandle level	70.
Building life	30 years
Opportunity rate	10.5 percent

System B Floor Plan (not to scale)
HID Luminaire System
with One 150-W HPS Lamp
(rated 175 W each)

Room dimensions	30 f × 30 f × 10 ft
Reflectances	80/50/20
Work plane height	2.5 ft
Mounting height	10 ft
Light loss factor	0.75
Relative light output	1.0
Maintained footcandle level	70.0
Building life	30 years
Opportunity rate	10.5 percent

An outline of one method of determining costs is illustrated as follows:

Life cycle cost analysis for 900 ft² Classroom

	Luminaire A Layout A	Luminaire B Layout B
A. **Lighting and air conditioning installed costs (initial)**		
1. Luminaire installed costs: luminaire, lamps, material, labor	$ 1438	$ 1688
2. Total kW lighting:	2.02 kW	1.58 kW
3. Tons of air conditioning required for lighting: (3.41 × kW/12)	0.57 tons	0.45 tons
4. First cost of air conditioning machinery: @ $ 1500 /ton	$ 860	$ 672
5. Reduction of first cost of heating equipment:	$ [138]*	$ [108]*
6. Other differential costs:	$	$
	$	$
	$	$
	$	$
7. Sub-total mechanical and electrical installed cost:	$ 2160	$ 2252
8. Initial taxes:	$ 108	$ 113
9. Total costs:	$ 2268 (A1)	$ 2365 (B1)
10. Installed cost per square foot:	$ 2.52	$ 2.63
11. Watts per square foot of lighting	2.24 watts	1.76 watts
12. Salvage (at *y* years):	$ 227 (As)	$ 237 (Bs)

Figure 17.4 An economic analysis for two lighting systems in a 30 × 30 classroom.

B. Annual power and maintenance costs

Item	System A	System B
1. Lamps: burning hours × kW × $/kWh	$ 314	$ 246
2. Air conditioning operation hours × tons × kW/Ton × $/kWh	$ 72	$ 56
3. Air conditioning maintenance: tons × $/ton	$ 86	$ 67
4. Reduction in heating cost fuel used: coal	$ [23]*	$ [18]*
5. Reduced heating maintenance: MBtu × $/MBtu	$ [14]*	$ [11]*
6. Other differential costs:	$	$
	$	$
	$	$
	$	$
	$	$
7. Cost of lamps: (No. of lamps 42 / 9 @ $ 1.08 / 34.00 /lamp per *N*) 3 yrs. / 6 yrs (Group relamping every *N* years, typically every one, two or three years, depending on burning schedule.)	$ 15	$ 51
8. Cost of ballast replacement: (No. of ballasts 21 / 9 @ 12 yrs / 12 yrs $ 12 / 63 /ballast per n) (n = number of years of ballast life.) *Labor A = 0.8 hrs, B = 1.0 hrs; Rate = 14.50/hr*	$ 41	$ 58
9. Luminaire washing cost: No. of luminaires 21 / 9 @ $ 7.25 / 7.25 each. 1yr / 1yr (Cost to wash one luminaire includes cost to replace *or* wash lamps.) *Labor A & B = 0.5 hr; rate = 14.50/hr*	$ 152	$ 65
10. Annual insurance cost:	$ 27	$ 28
11. Annual property tax cost:	$ 136	$ 142
12. Total annual power and maintenance cost:	$ 806 (Ap)	$ 684 (Bp)
13. Cost per square foot:	$ 0.90	$ 0.76

* Bracket indicates negative values.

PAYOUT (Example)

(Sys. A) = ($2268 − $227) = $2041

(Sys. B) = ($2365 − $237) = $2128

$X_1 = (\$2128 - \$2041) = \$87$

$X_2 = (\$806 - \$684) = \$122$

$$X_3 = \frac{122}{(.105 \times 87)} = 13.36$$

$a = 1.105$

$$b = \frac{13.36}{12.36} = 1.081$$

$$y = \frac{\ln 1.081}{\ln 1.105}$$ = 0.78 years before Sys. B begins to be the more economical system. □

Figure 17.4 (Continued).

Use of Shorter-Life High-Efficacy Lamps

A basic characteristic of incandescent lamps is the interrelationship between life and lumens. This seesaw effect is demonstrated in Table 17.4. When a 100 W lamp of 3500 h life design is replaced, an equal amount of light can be obtained from a 75 W lamp of 850 h life design. An energy saving of 25 W and a cost saving of $4 per year will result. A more important aspect of the energy-saving ideas beyond energy and dollar savings may be the necessity to comply with government regulations.

Use of Reflectorized Lamps

Reflector (and PAR) lamps are designed to put light on the work surface efficiently. The filament is positioned optically within the reflector during lamp making to further maximize the efficacy of this source. In many instances, a reflector lamp of only half the wattage of a standard lamp can be substituted quickly with very nearly the same lighting results on the work surface. Table 17.5 provides a direct substitution guide for this energy-saving scheme.

Use of Low-Voltage Lamps

Low-voltage lamps are inherently more efficient than lamps of standard voltage design with equal life values. These gains vary by wattage and range in the vicinity of 10 to 30%. A relatively simple method is available to switch to low voltage. A change to track lighting could be the answer. Many track lighting fixtures accommodate reflector low-voltage lamps, thus providing efficiency gains in two ways.

Table 17.4 Lumen Comparison Chart

	Average Life in Hours		
Wattage	Standard	2500 Hours	3500 Hours
40	480	415	400
60	890	740	670
75	1210	1000	—
100	1710	1480	1280
150	2850	2350	2150
200	3900	3250	2890

Table 17.5 Direct Substitute Guide for Incandescent Lighting

Present Lamp	60A (1000 hr.)	60/99IF (2500 hr.)	75A (850 hr.)	75R30/FL (2000 hr.)	100A (750 hr.)	100/99IF (2500 hr.)	150PAR FL (2000 hr.)	150R/FL (2000 hr.)	200A (750 hr.)
Substitute	30R20 (2000 hr.) 50R20 (2000 hr.)	50R20 (2000 hr.) 55PAR/FL (3000 hr.)	50R20 (2000 hr.) 60A (1000 hr.)	55PAR/FL (2000 hr.)	75R30/FL (2000 hr.) 75PAR/FL (2000 hr.) 75ER30 (2000 hr.)	75PAR/FL (2000 hr.) 75A (850 hr.)	100PAR/FL (2000 hr.)	100PAR/FL (2000 hr.)	150PAR/FL (2000 hr.)
Light Level*	100% 200%	200+% 200+%	175% 75%	150%	125% 200+% 200%	200+% 80%	70%	150%	200+%
Watts Saved	30 10	10 5	25 15	20	25 25 25	25 25	50	50	50
Annual Energy Cost Savings Per Socket**	$4.80 $1.60	$1.60 $ .80	$4.00 $2.40	$3.20	$4.00 $4.00 $4.00	$4.00 $4.00	$8.00	$8.00	$8.00

*Expressed as % of present level. Specific cases may vary depending on conditions. **Based on 4,000 hours/year and $0.04/KWH

New Developments

Recent developments in incandescent sources prompted primarily by the energy crunch have occurred in the following areas:

1. *New wattage sizes:* introduction of 55-, 100-, and 200 W PAR lamps
2. *New bulb shapes:* introduction of ER lamps
3. *New coatings:* the use of special coatings in a number of new incandescent lamps
4. *Basic design modifications:* the use of krypton fill gas instead of argon

17.6.2 Using Fluorescent Systems

There are some 900 million fluorescent sockets in place in the United States, so short-range reduction in the use of lighting energy must depend on modifications that can be made to existing systems without major retrofit and conversion costs. With this in mind, lighting manufacturers have developed a line of energy-saving lamps and various types of energy-efficient ballasts and luminaires in the last decade. Today numerous possible combinations can be selected to suit any particular application.

Energy-Efficient Lamps

As discussed in Section 17.2.2, reduced wattage fluorescent lamps use krypton fill gas as a first response to the need for energy savings. However, because fluorescent ballasts tend to operate lamps at constant current, there is a limit to the lamp manufacturers' ability to reduce lamp wattage without compromising system reliability. In practice, light output and wattage reductions have amounted to 18 to 20% for instant-start circuits, but only 10 to 14% for rapid-start circuits, which make up almost three-fourths of sockets installed.

Energy-Efficient Ballasts

The intrinsically higher efficiency of reduced-wattage lamps suggested a low-loss high-performance ballast specifically designed for them. For various energy-efficient ballasts, reference should be made to Section 17.3.

Fluorescent System Considerations

Fluorescent lamps are sensitive to ambient temperatures. By using reduced-wattage lamps or low-loss ballasts, less heat will be generated and the operating temperature point of the lamps will probably change. The critical area is the coldest spot on the bulb surface.

Most fluorescent lamps will peak in light output at around 100°F cold-spot temperature. For enclosed luminaire types which ordinarily operate the lamps at higher temperature, replacing standard lamps with high-efficacy reduced-wattage lamps may result in a net increase in luminaire output even though the reduced-wattage lamps are rated for less output than that of standard lamps.

Economics of Energy-Saving Fluorescent Lamps

For the purpose of cost comparison between 40 W and energy-saving lamps, although 34 W lamps can be operated on low-loss ballasts to further reduce energy consumption, conventional ballasts are assumed, which makes the analysis applicable to retrofit as well as new installations. Within limits of conventional design practices, low-wattage lamps can provide illuminating engineers with a new flexibility to more closely tailor illumination levels to specific standards or requirements while simultaneously reducing owner's life-cycle costs. Figure 17.5 illustrates a fluorescent substitution guide for energy-saving schemes.

17.6.3 Using High Intensity Discharge Systems

The HID light source can provide many energy-saving opportunities, due to their high efficacy. Retrofitting or relighting with a light source of higher efficacy is an essential technique to achieve energy savings in a HID illuminating system.

Retrofit Options for a Mercury System

The operating cost of a mercury lighting system is second only to that of an incandescent system. For maximum savings, a mercury system can be replaced by a high pressure sodium system, or if color discrimination is important, a metal halide system. Two basic retrofit options can be applied to a mercury system. The first option is to replace the luminaire ballasts with metal halide or high pressure sodium lamps. The second option is to replace with special retrofit MH or HPS lamps designed to operate on mercury ballasts. Ballast change-out permits conventional lamps to be used in the luminaire. Replacing ballasts require a larger cash expenditure, but the long-term benefits are greater. Although retrofit lamps produce substantial savings when applied to an existing mercury system, they generally cost more and have a shorter life, or lower efficacy or lumen maintenance than those of standard MH or HPS lamps operated on companion ballasts. Most ballasts manufacturers and some luminaire manufacturers offer remote-mounting retrofit ballasts that can be mounted outside the luminaire for conversion. However, retrofit lamps cannot be used on all mercury systems. Each prospective

To Save Energy Use This Lamp	Instead of	Light Level Will Be	Watts Saved
Type 1 Low Energy Lamp 35 Watts	Standard 40 Watt 4 Foot Rapid Start	Same or Slightly Less	5 Per Lamp
Type 1 Low Energy Lamp 60 Watts	Standard 75 Watt 8 Foot Slimline Lamp	Same or Slightly Less	15 Per Lamp
Type 1 Low Energy Lamp 95 Watts	Standard 115 Watt 8 Foot High Output Lamp—800 MA	Same or Slightly Less	20 Per Lamp
Type 2 "33" Low Energy Lamp	1 Standard 40 Watt Lamp Operating on a 2 Lamp Ballast	33% Less Light Than 2 Standard Lamps	30 Per Pair
Type 2 "50" Low Energy Lamp	1 Standard 40 Watt Lamp Operating on a 2 Lamp Ballast	50% Less Light Than 2 Standard Lamps	46 Per Pair

Low Energy Lamp Designation

Type 1—Low Energy Lamp: Available from 34 to 36 watts. Does not affect light output of other lamp on a 2 lamp ballast.

Type 2—Low Energy Lamp: Replaces one *standard* 40 watt lamp of a two lamp pair operating on a 2 lamp series lead ballast. Light output of the pair is reduced by either 33% (Type 2, "33") or 50% (Type 2 "50").

Figure 17.5 A energy-saving substitute guide for fluorescent lighting.

retrofit project should be checked with the lamp manufacturer for advise.

At the present time, retrofit HPS lamps in wattages of 150, 215, and 360 can be used on mercury systems using 175-, 250-, and 400 W lamps respectively. One manufacturer offers an 880 W HPS lamps for use in existing 1000 W mercury luminaire.

Retrofit metal Halide lamps are available in 400- and 1000 W versions which are intended for use with CW/CWA type mercury ballasts. The 1000 W metal halide can reduce wattage per luminaire by either

35 or 85 W, depending on the type of ballast it is used with. A new addition is the 325W MH lamp which saves about 70 W per luminaire while delivering 40% more light than 400 W mercury lamp it replaces. Figure 17.6 illustrates some sample retrofit schemes as discussed in this section.

Relight with High Pressure Sodium System

The most effective way of achieving energy savings with a HID system is to replace existing incandescent or mercury system, even fluorescent lighting, with a properly chosen HPS light source. This replacement operation is called a "relighting program" in most industrial plants. In making a relighting study, the most difficult and

Use This Lamp	Instead of	Light Level Will Be**	Watts Saved Per Socket*
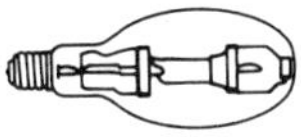 325 Watt High Efficiency Metal Halide Lamp	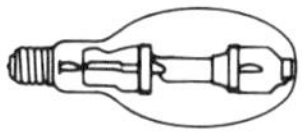 400 Watt Mercury (Single-lamp CW/CWA Regulator Ballasts)	Higher	70
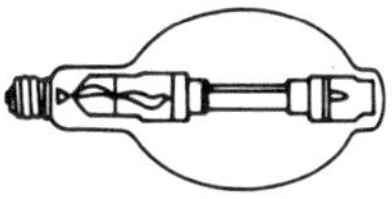 1000 Watt High Efficiency Metal Halide Lamp	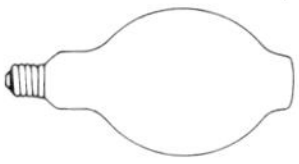 1000 Watt Mercury (CW/CWA Regulator Ballasts)	Much Higher	85
	 1000 Watt Mercury (lag or rector ballasts)	Much Higher	35
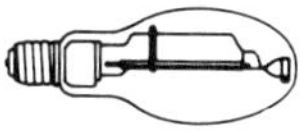 150 Watt High Pressure Sodium Lamp	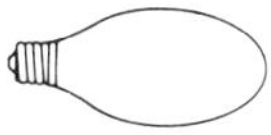 175 Watt Mercury	Higher	40
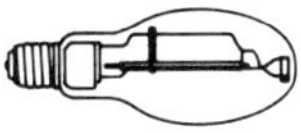 215 Watt High Pressure Sodium Lamp	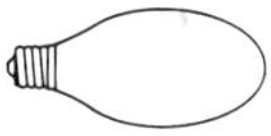 250 Watt Mercury	Higher	65

Figure 17.6 A energy-saving substitute guide for mercury lighting.

perhaps the most controversial determination is the approach that one would take to tackle the problems. A necessary first step is to obtain a plot plan of the plant property, showing all buildings with structural dimensions, and lighting layout drawings and construction bills of material to help identify and confirm the number, type, and wattage of all light sources. The next step is to identify and categorize by wattage and lamp type each light source in all areas, working within one defined area at a time. Information on the efficacies of various light sources should be assembled both for the types of lights used in the plant and for other types that will be proposed. Data on lamp life of various light sources should be available from lamp manufacturers. Ballast losses should also be determined. If they are not available, 10% of lamp watts can be used as an approximation. Tasks to be performed in each area should be established so that appropriate lighting level can be determined accordingly.

After all necessary data are gathered, the information can be plugged into various study forms (see Figures 17.1 and 17.2). In going through the forms, one will soon be confronted with some puzzling questions as to the best criteria for conducting the study. In making a relighting study, the following basic guiding principles must be observed:

Fixture for Fixture Replacement. This pattern of relighting represents one of the simplest conditions. A typical case for such a system is to replace 1000 W mercury lights with 400 W HPS on a one-for-one basis. Usually, the spacing-to-mounting height ratio is suitable for the new light source. The existing wire sizes are usually adequate and connections can easily be made. The overall installation costs of the relighting program will be low, and the ROI will be about 50%. The lighting level will stay practically the same as before. However, the energy savings realized in this program amount to 60% of the original consumption. Figure 17.7 shows a typical industrial relighting program of this nature. On the left, one hundred seventy-eight 1000 W mercury luminaires were used to light the plant; on the right, the same number of 400 W HPS new luminaires are in place. The ROI in this case is 58%.

Energy Savings Plus Improvement of Lighting Quality. In some cases the relighting program can result in not only energy savings, but also improved lighting quality, such as increased illumination levels and greater visibility. Warehouses lighted with high-wattage incandescents usually fall under this category. Due to higher efficacy of the HPS light source and usually ample mounting height, the number of new fixtures that will deliver a desirable higher illumination level can be reduced. Consequently, this type of relighting program can result in energy savings as well as operating cost sav-

Figure 17.7 A typical industrial relighting program—400 W HPS replacing 1000 W mercury.

ings. In addition, improvements in lighting quality are achieved as a bonus.

Lumen for Lumen Replacement. The lumen output of one 400 W HPS luminaire is approximately equal to that of two 400 W mercury, two super-high-output fluorescent, or six 500 W incandescent units. These simple relationships can provide a convenient replacement scheme as long as the spacing-to-mounting height ratio of the new HPS luminaires remains satisfactory. The replacement scheme can result in the reduction of luminaires required to deliver a lighting level equivalent to that of an existing system. Energy savings can readily be achieved.

Unfavorable Conditions. Favorable situations may not always exist. If the relighting program cannot achieve both energy savings and improvements in lighting quality, a fair standard must be set as a guideline for the program, to determine whether it can be justified. The fair standard would be that the relighting program be judged on its capacity to give an equal illumination level and visibility as the existing system, yet it can result in considerable energy and cost savings.

Identifying Problems and Possible Solutions in a HPS Lighting System

In recent years, most of the old problems that existed in applying HID light sources to industrial plants have been overcome. However, despite the many significant improvements, premature component failures, glare, and color rendition, and so on, remain as the major concern of the HPS lighting system. The more important problems are discussed below.

Low Lighting Level. Lamp lumen output is nearly proportional to lamp wattage, and lamp wattage is dependent on the ballast design and the lamp voltage. Therefore, it follows that reducing the lamp voltage will, in turn, reduce the light output. Low lighting levels for HPS lighting system usually result from (1) low lamp arc voltage, (2) a steep slope of the ballast characteristic curve, (3) low capacitance in the lead ballast circuit, (4) low line voltage, and (5) dirt accumulation on the luminaire. ANSI specifications allow a production range of 85 to 115 V. If the lamps are made with their arc voltages near the lower limit, they will deliver a low lighting level when installed.

Excessive Voltate Rise. HPS lamp voltage increases during lamp life. When a lamp starts, the amalgam is fully condensed and the lamp voltage is low. As the lamp warms up, the amalgam vaporizes, causing the voltage to increase. If the voltage stabilizes at a value below the drop-out voltage, the lamp will operate properly. However, if the lamp voltage exceeds the drop-out voltage, the lamp will extinguish. One of the more common causes of voltage rise results from operating the lamp above its wattage value, which produces arc tube blackening and sodium loss. Another problem arises from using ballasts that are not matched to the input voltage.

Starting Difficulties. When a HPS lamp does not start, one generally looks at the starter. There are devices available to detect whether there is a voltage pulse of sufficient magnitude to start the lamp. Starters fail from overheating due to component failure from the heat within the luminaire, or from repeated attempts to light an inoperative lamp previously used in that socket. The pulse circuit utilizes the ballast windings to develop the high starting voltage and it must be connected directly to the eyelet of the socket, not to the shell or ground. If both the lamp and starter are functioning, it may be that the supply voltage is too low or the output voltage of the ballast is insufficient.

Color Rendition. HPS lamps have a spectral distribution richer in golden white than other types of lamps. Because tests show that there is little difference in the ability of the human eye to identify safety colors under incandescent, metal halide, or high pressure sodium lighting, the overall color-rendering properties of HPS lighting are satisfactory. Thus any concern about or negative reactions to the color of these lights are more psychological than factual. In a very rare case, the yellow color may actually cause difficulty in color distinction. If this happens, supplementary lighting such as fluorescent or even incandescent should be used to assist the task. HPS lamps are now available with substantially better color rendition than the standard line.

Glare. Due to the inherent nature of HID light sources, regardless of the types of fixtures, there is a hot spot that can cause a glare sensation when one views from a specific distance. However, at a normal 30-in. working plane, one will not be bothered by the hot spot. Diffused HPS lamps can reduce glare; however, their light output is somewhat lower. Nevertheless, they do offer a choice for the industrial users should glare become a serious problem.

Future Prospects of HPS Lighting System

Electric energy saved from lighting with HPS light source can have a significant impact on the national economy. Some of the problems discussed in previous sections can be minimized or eliminated in the planning stages by careful examination of the components selected for the lighting system. While short-term goals tend to concentrate on improvements in lamp performance, a longer-range perspective suggests some interesting possibilities with the system. Shorter arc source lengths, more sophisticated optical design, more efficient ballasts, and better color offer an attractive alternatives to present-day spot and floodlighting system. As the use of HPS lighting expands to include commercial and eventually residential establishments, the luminaire manufacturers will be encouraged to develop new lines of decorative and compact luminaires.

17.6.4 Using Daylighting

Daylighting should be dealt with by first analyzing it and then establishing a design technique to integrate it with the electric lighting system. In Section 13.2.2 we discussed some basic aspects of daylighting.

Analysis of Daylight

The primary difficulty in daylighting is the variability of daylight with respect to the time of day and year with respect to environmental conditions. These variations in the quality and quantity of light result in variations in the interior environment.

Interior Environment. The daylight entering a space may be analyzed in terms of the quantity and quality of the light. Daylight may be adequate in quantity to reduce the electric lighting level and result in false energy conservation if the quality of the light is not analyzed. Poor quality of daylight may lead to discomfort and a loss in visibility that may result in a decrease in human performance and productivity.

Quality of Daylight. The quality of daylighting in a space affects the physical and psychological impact on the human occupants. The physical effects of daylighting can be investigated in terms of the visual, thermal, and acoustical environment.

Daylighting Design from Windows

Assuming suitable sun control, a southern exposure is preferred to optimize the daylight contribution into a space. Sufficient daylighting must be provided to replace electric lighting if energy conservation is to be realized. Longhand calculational procedures are based on the data and procedures presented in the IES Lighting Handbook. Although the computer program may be more complex, it offers a more complete and accurate profile of the daylight contribution. The longhand design procedure involves two steps:

1. Determine the quantity of illumination coming to the window surface.
2. Use that quantity to determine the daylight contribution to the interior part of the space.

Once the contribution of illumination to the window surface has been calculated, two longhand methods are available for determining the illumination contribution to the space. Two methods of calculation can be applied. The first method is to follow the point-by-point procedure, which makes two assumptions: (1) interreflected component is ignored, and (2) the window is a uniform diffuse emitter. The second method is a lumen method that calculates illumination values at three points defined as the maximum, midway, and minimum. The second procedure includes both the direct and interreflected components of illumination. It also assumes that the illumination coming to the window is uniformly distributed across the surface of the window. References should be made to Sections 12.3 and 12.4 for the detailed procedures of both methods.

17.6.5 Lighting Controls

Electrical and electronic controls can be used to conserve energy by reducing light levels by turning off lamps or by dimming. The inclusion of switching provisions for reduced lighting is a cost item in an initial installation, but can result in relatively large savings. Dimming is a more sophisticated manner of control that can most readily be justified when variable light levels are desired in addition to energy conservation. Chapter 18 will be devoted entirely to the latest lighting control schemes and techniques.

17.7 LIGHTING AND ENERGY STANDARDS

Early in 1974, the document "Energy Conservation Guidelines for Existing Office Buildings" was published by the General Services Administration and the Public Building Services (GSA/PBS). At approximately the same time, the National Bureau of Standards (NBS) prepared the document "Design and Evaluation Criteria for Energy Conservation in New Buildings" for the National Conference of States on Building Codes and Standards (NCSBCS) covering lighting for new construction. NCSBCS asked the American Society of Heating, Refrigeration, and Air Conditioning Engineers (ASHRAE) to prepare a standard on energy conservation for new buildings based on this NBS document. ASHRAE turned to the Illuminating Engineering Society (IES), and the IES Task Committee on Energy Budgeting procedures was formed. The work of this committee culminated in recommendations that were published by ASHRAE as Chapter 9 of the ASHRAE 90-75 and by IES as "IES Recommended Lighting Power Budget Determination Procedures EMS-1." NCSBCS also adopted it as part of its Model Building Code. Both documents established a procedure for determining a "lighting power budget."

In 1975, Congress passed Public Law 94-163, entitled "Energy Policy and Conservation Act of 1975," which was amended by Public Law 94-385, "Energy Conservation Standards for New Building Act of 1976." Public Law 94-164 makes mandatory certain lighting efficiency standards set forth in Public Laws 94-163 and 94-385.

In 1976, the Energy Research and Development Association (ERDA) contracted with NCSBCS to codify ASHRAE 90-75. The resulting document was called "The Model Code for Energy Conservation in New Building," or more simply, the "Model Code." The Model Code has been adopted by a number of states to satisfy the requirements of Public Laws 94-163 and 94-385.

In June 1976, the IES Board of Directors adopted important revisions to Chapter 9 of ASHRAE 90-75 that tightened the lighting power budget procedure to assure energy conservation. These revisions were included in 90-75R but did not become part of the NCSBCS Model Code.

ASHRAE 90-75R was cosponsored by IES and ASHRAE and submitted to the American National Standards Institute (ANSI) in late 1977 for adoption as an ANSI standard. The document that resulted will be known as ANSI/ASHRAE/IES Standard 90, "Energy Conservation in New Buildings."

In 1975, work was begun by IES on a series of six documents that deal with energy standards for existing buildings. The documents cover low-rise residential, high-rise residential, commercial, industrial, institutional, and public assembly occupancies. Several of these documents have since become ANSI standards.

There have been several revisions on the ANSI/ASHRAE/IES 90-75R since then. All were included in the lighting portion of ANSI/ASHRAE/IES 90A-1980, "Energy Conservation in New Building Design," and in EMS-1-1981, "IES Recommended Lighting Power Budget Determination Procedure."

As a means for simplifying and shortening the EMS-1 procedure, the IES developed a Unit Power Density (UPD) Procedure and published it as EMS-6 in 1980. The present LEM-1-1982, "Lighting Power Limit Determination," is a further refinement which combines EMS-6 and EMS-1 and, as such, supersedes both. LEM-1 is concerned only with the determination of lighting power limit; lighting control guidelines are contained in LEM-3, "Design Considerations for Effective Building Lighting Energy Utilization."

Since 1982, IES has published the LEM series. In addition to LEM-1 and LEM-3 are LEM-2, "Lighting Energy Limit Determination," LEM-4, "Energy Analysis of Building Lighting Design and Installation," and LEM-6, "IES Guidelines for Unit Power Density (UPD) for new Roadway Lighting Installation."

ANSI/ASHRAE/IES Standard 90-75R was by far the most popular energy conservation design standard. It has been adopted as energy codes by most states within the United States and serves as a model standard in many other countries. Although this standard underwent a revision in 1980, major revisions were not made to incorporate new practices and technology. The proposed standard 90.1P takes a totally new approach compared to earlier versions of standard 90.

Three parallel and alternative paths for compliance are provided: prescriptive, system performance, and building energy methods. Each of these methods may be used selectively and interchangeably when determining building subsystems compliance. Standard 90.1P will be the standard of the future for energy-conserving building design and operations. The interaction between the three compliance methods fulfills needs that arise during the various phases of the building process. Although the document is in its infant stage, ongoing analysis and reevaluation will assure that the standard will mature into a useful, practical standard.

In the meantime, the U.S. Department of Energy published in the *Federal Register* of May 6, 1987, note of a proposed interim rule entitled "Energy Conservation Voluntary Performance Standards for New Commercial and Multi-family High-Rise Residential Buildings." This rule, when issued, will be mandatory for all federal buildings and a voluntary recommendation for nonfederal buildings. There is industry concern that these standards may be adopted by states and thus would become the standard for all buildings. Section 3 on lighting is similar to the proposed ANSI/ASHRAE/IES Standard 90.1P second review draft, dated August 22, 1986. The major exception is the inclusion of the 1992 Unit Power Density Values for lighting which become mandatory on January 1, 1992. Therefore, all lighting professionals should seriously review and take a close look at these values before the interim rule becomes mandatory.

BIBLIOGRAPHY

ANSI/IEEE 739-1984, Energy Conservation and Cost-Effective Planning in Industrial Facilities.

Benton, C. C., Daylighting Can Improve the Quality of Lighting and Save Energy, *Architectural Lighting*, Nov. 1986, pp. 46–48.

Chen, Kao, Lighting Esthetics with Energy Saving Ideas, *IEEE Transactions on Industry Applications*, Jan./Feb. 1976, pp. 35–38.

Chen, Kao, The Energy Oriented Economics of Lighting Systems, *IEEE Transactions on Industry Applications*, Jan./Feb. 1977, pp. 62–68.

Chen, Kao, and Guerdan, E. R., Resource Benefits of Industrial Relighting Program, *IEEE Transactions on Industry Applications*, May/June 1979.

Chen, Kao, and Kane, R. M., Achieving Optimum Performance in a High-Pressure-Sodium Lighting System, *IEEE Transactions on Industry Applications*, July/Aug. 1982, pp. 416–423.

Chen, Kao, and Lally, William, Update: Fluorescent Lighting Economics, *IEEE Transactions on Industry Applications*, May/June 1983, pp. 328–333.

Chen, Kao, and Main, George, Jr., Industrial Relighting Program: Its Purpose, Progress, and Prospects, *IEEE Transactions on Industry Applications*, Mar./Apr. 1981, pp. 217–221.
Chen, Kao, and Murray, W. A., Energy, Incandescent Lighting, and 100 Years, *IEEE Transactions on Industry Applications*, May/June 1980.
Chen, K., Unglert, M. C., and Melafa, R. L., Energy-Saving Lighting for Industrial Applications, *IEEE Transactions on Industry Applications*, July/Aug. 1978.
Hart, A. L., Cutting Lighting Costs by Applying Energy Efficient Lighting Sources, *Plant Engineering*, Mar. 8, 1979.
IES Committee Report on Life Cycle Cost Analysis, *Lighting Design and Application*, May 1980, pp. 43–48.
1986 IES Progress Report, *Lighting Design and Application*, Nov. 1986, pp. 25–42.
Peery, R., Daylighting and Energy Conservation, *Lighting Design and Application*, Oct. 1974, pp. 27–29.
Recommended Procedure for Lighting Power Limit Determination, Illuminating Engineering Society Publication LEM-1-1982.
Tao, William, Lighting Compliance Procedures of ANSI/ASHRAE/IES Standard 90.1P, *Lighting Design and Application*, Sept. 1985, pp. 36–40.

18

Lighting Controls

18.1 INTRODUCTION

Lighting controls have become increasingly sophisticated in the recent years, mainly due to dire need for energy savings and advancements in the solid-state control devices. As the cost of energy has continued to rise, increasing effort has gone into minimizing the energy consumption of lighting installations. This effort has evolved along three major directions: the development of new energy-efficient lighting equipment, the utilization of improved lighting design practice, and finally, improvements in lighting control systems. The lighting industry has been introducing more efficient lighting components and systems, which cost more to install but result in a lower total cost (operating plus initial). End users begin to base their decisions on the ROI or the life-cycle cost. These decisions should lead to much greater application of cost-effective lighting control systems.

In the last decade, many new lighting control hardwares have appeared on the market. In this chapter we review all important types of lighting controls for industrial illuminating systems and offer some guidelines for selection to suit individual needs and achieve optimum energy savings.

18.2 TYPES OF CONTROLS

All lighting controls, whether a simple switch or a sophisticated programmable controller, can normally be classified in two basic categories: (1) on-off controls and (2) level controls. In its simplest form, lighting control can be accomplished manually by means of a switch located on a wall, in a luminaire or a panel box. Even though manual on-off switches for lighting control are used in commercial and industrial facilities, the current trend is toward greater use of lighting contactors. However, this type of control cannot be used conveniently to change lighting levels. Many circumstances require a varied level of illumination. Dimming devices are the most used means of providing the level controls.

Lighting controls can also be grouped into two general categories: centralized controls and local controls. The main difference between them falls into the realm of function of the area considered. Centralized controls are used in buildings where it is desirable to control large areas of the building on the same schedule. An example of centralized controls is a microprocessor that turns all lights on and off on a preprogrammed schedule. Localized controls are designed to affect only specific areas. Examples of localized controls are a personnel detector or a photocell controlling each office within a suite of offices.

Since a centralized system can be utilized to activate local controls, a time clock (centralized control) can be used to energize the building's entire lighting system on a time-of-day schedule, while a personnel detector (local control) located in a specific office overrides the centralized control to turn lighting in the specific offices on and off as demanded by occupancy. Figure 18.1 shows a matrix of the functions required for lighting control systems, with the type of control designed to perform that function or functions. No attempt is made to differentiate the actual technology utilized by manufacturers.

18.3 ON-OFF CONTROLS

As discussed previously, lighting control can be accomplished by means of a switch or a breaker, which will not be repeated here. However, various other types of on-off controls are discussed below.

18.3.1 Power Lighting Contactors

Power contactors can be divided into three general categories: power lighting contactors rated up to 1200 A, multipole contactors with up to 12 poles, and single-pole contactors with low voltage control. More popular sizes range from 60 to 225 A. Since small contactors can be used for control stations and an unlimited number of stations can be used for each power lighting contactor, the illuminating engineers can be liberal with the number of control stations used. A typical installation is the Twin Towers of World Trade Center in New York City. Four 225 A power lighting contactors are used for each floor. They are in turn controlled by two master controllers. Being

Type of Control	Total Building Control	Localized Control	Programmable	On/Off	Manual Override	Telephone Control	Time of Day	Holiday/Weekend Schedule	Dimming	Co-ordination with Daylight	Remote Control	Co-ordination with HVAC	Over-the-wire Control	Low Voltage
	Functions*													
Centralized Controls:														
Computer operated	X	X	X	X	X	X	X	X	X	X	X	X	X	X
Over-the-wire controller clock	X	X	X	X	X		X	X			X		X	X
Programmable control	X	X	X	X	X		X	X			X	X		X
Localized Controls:														
Daylight control (dimmable)[1]		X			X				X	X				X
Daylight control (non-dimmable)		X			X					X	X			X
In-fixture daylighting control		X			X				X	X	X			
Solid state dimming (no dimming ballast)		X	X		X				X	X			X	
People sensors		X		X	X									X
Ambient light co-ordinators[2]	X	X	X						X	X			X	
Timer control	X	X		X	X		X						X	

[1]Requires dimming ballast in each luminaire
[2]Compensates for light loss factors
*Many functions can be added; check with manufacturer

Figure 18.1 Lighting control matrix.

mechanically held, they can be controlled by:

1. A manually operated three-position switch with a center-off position
2. Auxiliary relays as a function of photoelectric cells
3. A time switch with a single-pole, double-throw contact
4. Control relays in an energy management system (e.g., programmable controllers)

18.3.2 Multipole Lighting Contactors

Multiple contactors are available up to 12 poles with contacts usually limited to a 20 A rating. The latest multipole contactors can be provided with optical solid-state control modules that provide two-wire control, three-wire control, or stop/start control. A typical six-pole lighting contactor with a solid-state two-wire control module is shown in Figure 18.2.

Today's multipole lighting contactors are of shallow construction, permitting them to be mounted in panelboards and in some cases mounted in a 4-in. stud-construction wall (Figure 18.3). Magnetically held 20 A multipole lighting contactors are also available. They are often furnished, assembled and prewired, in enclosures with electronic transceivers for connection to programmable controllers that utilize card readers, telephone interfaces, card printers, and other external components as needed.

18.3.3 Low-Voltage Relays

Low-voltage relays are usually single-pole and used for individual branch-circuit or luminaire control. These relay contacts are normally rated for a 20 A tungsten filament load at 125 V ac and are mechanically latching, requiring only a momentary 24 V rectified ac switch circuit pulse to either open or close the local contacts. A step-down transformer is required to provide low-voltage power for relay actuation. With a 40 VA rating, one transformer can supply power for up to 15 relays with No. 14 AWG wiring.

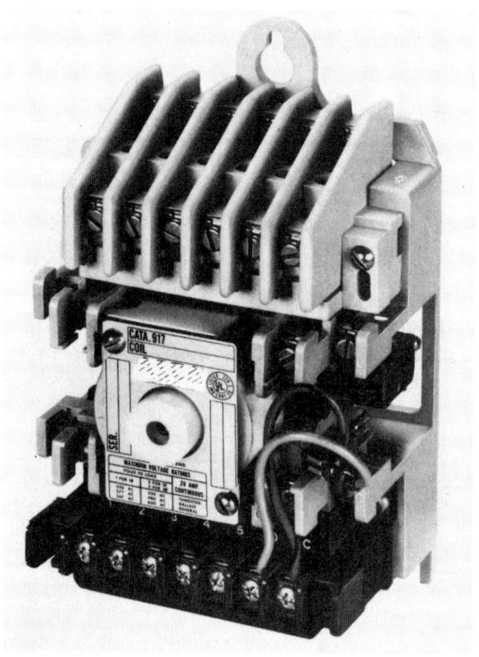

Figure 18.2 Six pole 20 A lighting contactor (Courtesy of Automatic Switch Company).

Figure 18.3 Multi-pole lighting contactor for flush mounting (Courtesy of Automatic Switch Co.).

18.3.4 Timing Controls

A broad variety of time switches are available for controlling lighting loads. They may be used for direct on-off control of lights or for control of lighting contactors. Twenty-four-hour dial timers are available with one or more sets of on/off trippers. A day-omitting device is optional for applications in which lighting is not used on weekends. Astronomic dial timers turn lights on at sunset and off at a prescribed time. Seven-day-calender dial timers are used when the on/off program is set on one dial. Operation can be omitted on selected days. Tripper will require periodic resetting to conform to seasonal changes.

Program dial timers are used for multiple daily operations or for short-duration needs, such as turning lights on and off for cleaning crews and security guards. A total of 48 on/off operations can be programmed daily. They are available in multidial versions and with day-omitting capability. Reset timers incorporate a manual control and a timer. They are typically used by security guards on patrol; the guard can turn on selected lights when entering the area, and the timers turn off the lights after a prescribed time has elapsed.

Digital timers are used where multiple circuits, accuracy, and more sophisticated programming are required. They can be scheduled a year in advance for holidays and so on. However, they have a limited load-carrying capacity, typically 3 A at 24 V ac per circuit. This requires interfacing with a lighting contactor or relay and a step-down transformer. Figure 18.4 shows the front-panel illustration of digital timers for controlling eight lighting circuits.

Phototimers combine a remotely mounted photo-control and a seven-day dial timer. It controls three circuits, each with its own program. One circuit provides on and off operation by photocontrol; in the second, photocontrol turns the circuit on and a time switch turns it off; and in the third, on and off control are both provided by a time switch. Built-in manual bypasses maintain any circuit in on and off operation for prolonged periods without affecting the other circuits or the master program.

18.3.5 Sensors

Sensors for lighting controls can be divided into two categories: photoelectric sensing and presence detectors.

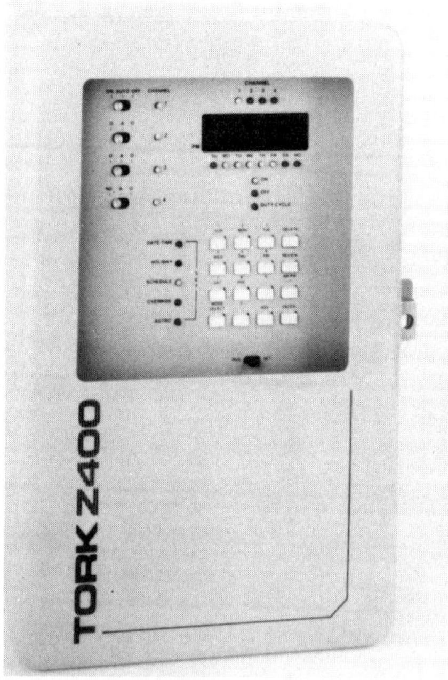

Figure 18.4 Front panel of a digital timer with self-prompting LED features (Courtesy of Tork Co.).

Photoelectric Sensors

When used for control of outdoor lighting, photoelectric sensors are normally set to switch lighting on at dusk and off at dawn. Adjustments can be made to change response to higher or lower light levels. Built-in time delay helps to eliminate nuisance switching in response to sources other than natural ambient light.

Personnel Sensors

A variety of personnel sensors have recently been developed. Personnel sensors are complete control systems that sense when a space or room is occupied and automatically turn the lights on for a preset period. If no further occupancy is sensed during this time interval, the lights are then turned off automatically. There are many types of personnel sensors; they differ only in their method of sensing occupancy. Some of these methods include passive and active infrared, ultrasonic, and acoustic sensing. Two of the most commonly used methods are passive infrared and ultrasonic.

Passive Infrared Sensors. These detect and respond to changes in radiated heat within a room caused by the presence and movement of a human body. They consist of two components, the sensor and the control unit. The sensor, which contains the passive sensing-element optics system and the electronic logic circuitry, is designed to be mounted in the ceiling panel. The control unit, which contains the low-voltage power supply for the system and the load relay used to switch the lighting load, is typically mounted above the ceiling. The sensor includes a timer circuit that keeps the lights on as long as changes in infrared energy are detected. If no changes are detected for a certain period, the lights are turned off automatically.

The most common passive infrared sensors have a coverage area of approximately 200 ft^2. If a larger area needs to be covered, multiple sensors can be connected to a single control unit, as shown in Figure 18.5, which indicates placement of three 200-ft^2 sensors for coverage of a 20 ft by 20 ft L-shaped office.

There are many design considerations in the use of personnel sensors. The sensors must be placed so as to cover areas of the room where occupants are expected to be. Care must be taken to ensure that the sensors cover all potential occupant locations. The sensor should not be placed too close to the entry; otherwise, people walking in an outside corridor past an open door will activate the lights. Passive infrared sensors should also be placed where they will not sense any nonhuman heat sources, such as an HVAC register or baseboard heater.

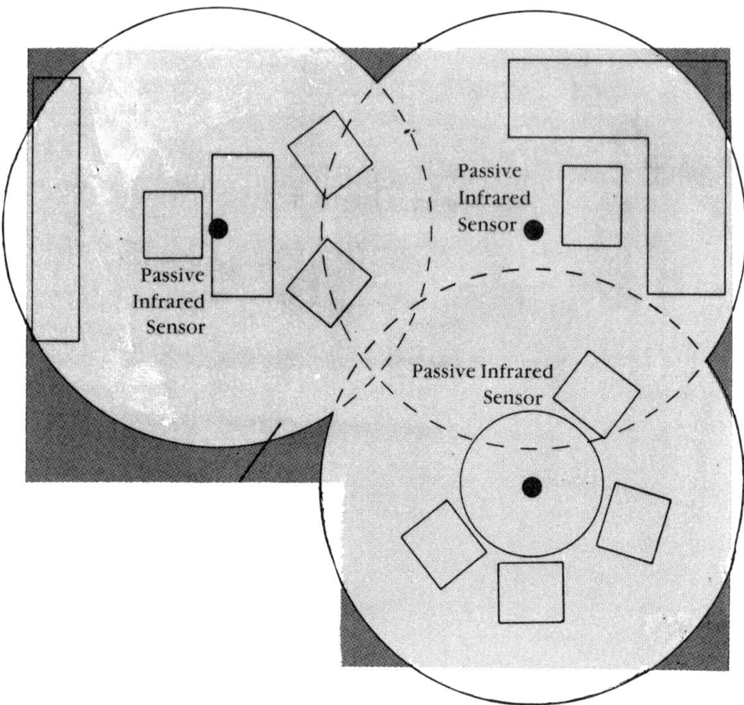

Figure 18.5 Placement of sensors for coverage of 400 ft^2 office.

Ultrasonic Sensors. An ultrasonic sensor creates an ultrasonic field in the room being monitored. When a person enters the room, the field is disturbed; the sensor detects the disturbance and activates the room lights. A typical ultrasonic sensor unit contains all the sensing and local control equipment, is mounted on the ceiling, and can provide up to 900 ft^2 of coverage. A single unit often contains multiple sensors.

Each type of ultrasonic sensor has a coverage pattern that depends on the number of sensors mounted in the unit. The pattern may include two different coverage areas for the same unit: an area within which large body motions, such as walking, are detected, and a smaller coverage area, within which small body motions, such as the movement of an arm, are detected. Careful placement of the ultrasonic sensors ensures proper operation of the system. Sensors must be placed such that they do not detect motion outside the room being controlled. Another consideration in the application of ultrasonic sensors is the acoustics of the room. Room acoustics can affect the coverage pattern and therefore the appropriate number and placement of sensors.

Ultrasonic sensors are available in many different styles, each providing specific coverage patterns. There are also ultrasonic sensors designed for wall mounting as a direct replacement for a doorway light switch. Personnel sensors are ideal for controlling lights in any space with random and intermittent occupancy patterns. Some spaces that meet this criterion include enclosed offices, rest rooms, storage rooms, library stacks, conference rooms, and classrooms.

18.3.6 Programmable Control System

Control systems are now available that employ microprocessor logic to replace hard wiring with soft wiring. Coded commands can be multiplexed to control points over a pair of low-voltage wires. Control points have receiver/switches, the latter component generally a low-voltage relay or lighting contactor. Logic functions can be programmed into a control device to turn lighting on and off over a 24-h or 1-week period. Overrides are available, and some systems can be accessed with Touchtone telephones. Such systems, which control lighting in both time and space, save considerable energy compared with past control practices. The heart of such systems is the programmable controller, which holds in memory a series of on-off instructions to the lighting circuits throughout a building. They can provide minute-by-minute control of an entire lighting scheme according to a user-determined schedule, with pulse initiation of the control signal generated from its internal clock.

Multiplexed signals from the programmable controller can be sent throughout a building via twisted low-capacitance wire. Signals can also be sent via existing power wiring to the receiver control modules (see the section below on power-line carrier systems). The codes from the controller are transmitted to the transceivers, where they may be combined with other signals (e.g., photocontrol relay output). In turn, signals are provided to low-voltage relays or lighting contactors, which require only a momentary electrical pulse to operate its latching or mechanical mechanism to the on or off position.

Control of the lighting pattern is not limited to the programmable controller keyboard. In addition, a manually operated, momentary contact switch connected to the transceiver can provide signals to operate not only those relays, or contactors connected to it, but also relays or contactors located anywhewre throughout the building via the data line back to the programmable controller.

Power-Line Carrier System

The power-line carrier system contains electronic transmitters and receivers that use the building power wiring system as a communications pathway. The transmitter accepts a control signal input, converts the control signal into digital form, and injects it onto the power wiring system. The low-voltage digital signal is transmitted at a frequency anywhere from 25 to 250 kHz, depending on the specific transmitter design. The control signals can be inserted onto 120 to 480 V lines. It is important to note that host signals cannot pass through transformers; bypass devices are usually required at each transformer.

One interesting use of power-line carrier technology is a high intensity discharge (HID) dimming ballast that has a receiver built into it. Control signals carry dimming information to all ballasts equipped with integral receivers. This system can be connected to a photosensor to provide for either daylighting compensation or lumen maintenance control. A power-line carrier system can be economically used in large facilities with many control points, such as factories and warehouses where control cabling would be lengthy and expensive. The technology is also suitable for existing buildings, where the cost of installing new lighting control wiring can be prohibitive.

18.4 LEVEL CONTROLS

18.4.1 Dimmers

Many circumstances require a varied level of illumination. Dimmers are the most used means of providing lighting level control. The original dimmers were of the resistance type. In the last decade, solid-state dimmers have taken over 90% of the market. By means of an electronic switch, the electronic dimmers turn off the current to the load for a portion of the cycle, thus delivering less power to the load. Electronic dimmers are now available for incandescent, fluorescent, and HID lighting.

Types of Dimmers

Conventional Dimmers. The modern SCR (silicon-controlled rectifier) dimmer operates on the principle of switching on the current a proportional distance through each half-cycle. An SCR is nothing more than a very fast switch. Dimmers operate simply by delaying the turning on of these switches by an amount of time inversely proportional to the incoming control voltage. In the simplest of examples, a dimmer set at 50% will delay halfway through each half-cycle and then turn on.

Digital Dimmers. The digital dimmer uses the same very fast switch (SCR), but instead of using a voltage to change a capacitor and turn on the switch, the digital dimmer counts a number of steps through each half-cycle, then turns on the switch.

Another term that is often used when referring to new technology dimmers is multiplexing. Multiplexing refers to the use of a single cable to carry data to a group of dimmers. Multiplexing is a concept that can be used on both analog and digital dimmers.

Comparison of Digital Dimmers and Analog Dimmers. The digital dimmer would have some significant advantages over a conventional analog dimmer:

1. The digital dimmer would communicate with the controller over a digital communication link, sending numbers rather than analog voltages. This communication technique has the distinct advantage of having much higher immunity to noise and distortion.
2. Analog dimmer components are very sensitive to changes in temperature. The digital dimmer is counting in reference to a crystal and will maintain its accuracy.
3. Analog dimmers using components that have manufacturer variations of as much as 2% will vary greatly from dimmer to dimmer. Thus sophisticated trimming is required to balance dimmer to dimmer.
4. Analog dimmers use a series of components to curve the output of the dimmer to approximate accepted standards of controller setting versus apparent lumens. A digital dimmer looks up these curve data in a table and can perform within 1/2 of 1% of the accepted standards. The digital dimmer has the added advantage of being able to reproduce any standard or nonstandard curve.

Intelligent Dimmers. Intelligent dimmers communicate with the controller in the same way as does the digital dimmer except that it has the intelligence to recognize data other than the setting of the controller. The scope of what these data could include is as broad as the scope of what the intelligent controller can produce.

Technology of Patching. As dimmer per light system becomes more and more clearly the wave of the future, moving the job of patching out of the controller and into the dimmer makes more and more sense. If in this data stream the intelligent dimmer is being informed that it is patched to channel 5 and that whenever it sees data for channel 5 it should use them, the number of dimmers that can be patched to any control channel will be limitless.

One problem when lights are patched together in a single control channel is that the light produced when that channel is active may not be uniform across an area. If a proportional level is part of the data that the intelligent dimmer stores and uses, all the dimmers patched to one channel can be balanced to make the lighting uniform. It is possible for an intelligent dimmer to communicate information gathered back to the controller so that the controller can make adjustments and/or communicate this information to the operator. For years, conventional analog dimmers have used feedback information to adjust the operation of dimmers—line regulation, load regulation, and current limiting being most common.

The intelligent dimmer will be able to do as many of these functions as engineers deem desirable. The intelligent dimmer will run a program like any other computer, and this program can be changed. The scope and capabilities offered by it are going to be important parts of control technology in the future.

18.4.2 Computer/Microprocessor

More sophisticated lighting control now consists of a microcomputer with an oscillator as its controller and a receiver switch as decoder. The microcomputer is programmed for various lighting patterns and addresses of the luminaires. The address and condition codes generated are modulated by the oscillator and superimposed on the building electrical system line frequency. At the receiver/switch, a decoder takes the message off the line, and if the address code corresponds to the one given that switch, the condition codes are executed and turn the luminaire fully on, halfway on, or off. One such system contains all programming, logic circuitry, and hardware. Discretionary control and override functions can also be incorporated. Printers can be connected to the microcomputer to provide hard-copy records of all control activities. This feature gives information to the system operator about the frequency and time of any local override activities. Video display terminals and recent software have made the system much easier to use, which promotes prompt revision of control schedules. Transceivers are available in 16- and 32-output versions. The control relays are capable of switching a 20 A inductive load. Figure 18.6 shows a typical scheme for such a control system as described above.

18.4.3 Daylighting

Contemporary lighting design practices require that illumination meeting a specified design criterion be provided whenever building is occupied. The sole use of daylighting does not meet such a criterion, because daylighting illumination levels change continually with time, season, and sky conditions. To exploit daylighting as a source of illumination, it is necessary to establish an interactive link between ambient lighting conditions and the electric lighting system. This can be achieved with a photoelectrically controlled lighting system that adjusts the output of the electric lighting system based on the amount of prevailing daylight. Lighting control hardware that links

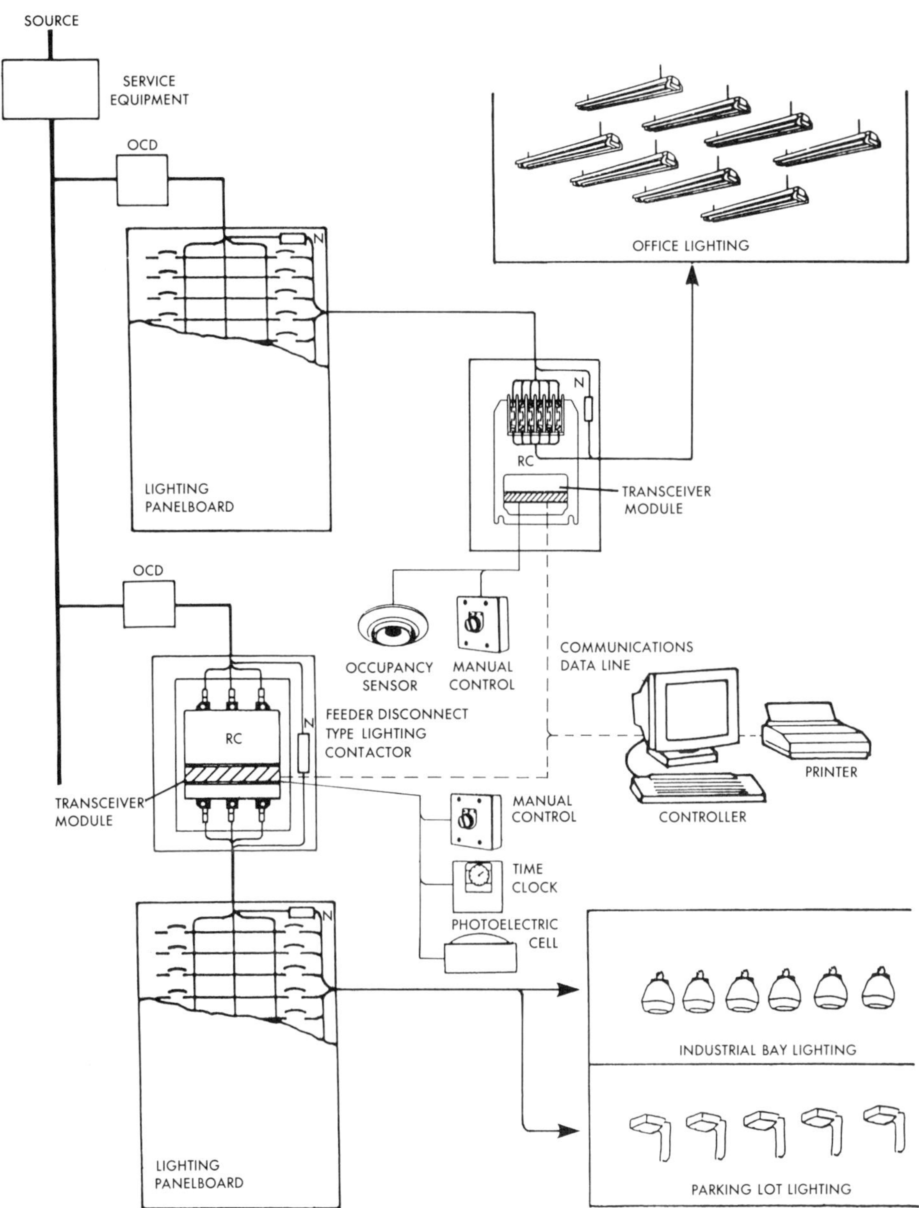

Figure 18.6 Programmable lighting control scheme (Courtesy of Automatic Switch Co.).

electric lighting to available daylighting generally falls into two categories:

1. Continuously dimmable electric lighting systems controlled by photosensors, which continually adjust the electric lighting level in response to the amount of daylight striking the control photosensors. At the heart of the system is the controllable output ballast or dimming ballast.
2. Photo-relay-based systems, which automatically switch off perimeter lighting when daylighting is sufficient to meet lighting needs.

The lighting control system shown in Figure 18.6 can be used to achieve the above described lighting level control scheme with the aid of photosensors located within the task zones. Photosensor information is used by the microcomputer to control zone lighting levels, which can be controlled from 100% to 0%. In zones where natural light is available, the system automatically takes advantage of this resource and incorporates it into the space. The electric lighting in daylight zones is used as fill-in lighting to provide even illumination across the space. The amount of usable daylight is dependent on the windows and skylight types and sizes, window and skylight treatments (drapes, blinds, glazing, etc.), building design, interior design, and furnishings.

Another scheme of utilizing daylighting to conserve energy is shown in Figure 18.7. Indicated herein is a method of wiring two luminaires with a two-lamp ballast to achieve three-step control. As more daylighting is available, more rows near the window are switched off. This may appear to be a crude method, but it is an economical approach that requires no elaborate control hardware. The wiring method responds to daylighting in the following manner:

Sequence	Illumination (%)
Power on to ballasts X and Y	100
Switch off power to either X or Y	50
Switch off power to both X and Y	0

18.5 ENERGY-SAVING STATISTICS FROM DIFFERENT TYPES OF LIGHTING CONTROLS

Data collected from various projects that have been successfully completed and put into operation in the recent years are presented in Table 18.1. Although each project is different in its space, lighting control, system installed cost, and resultant energy savings, the

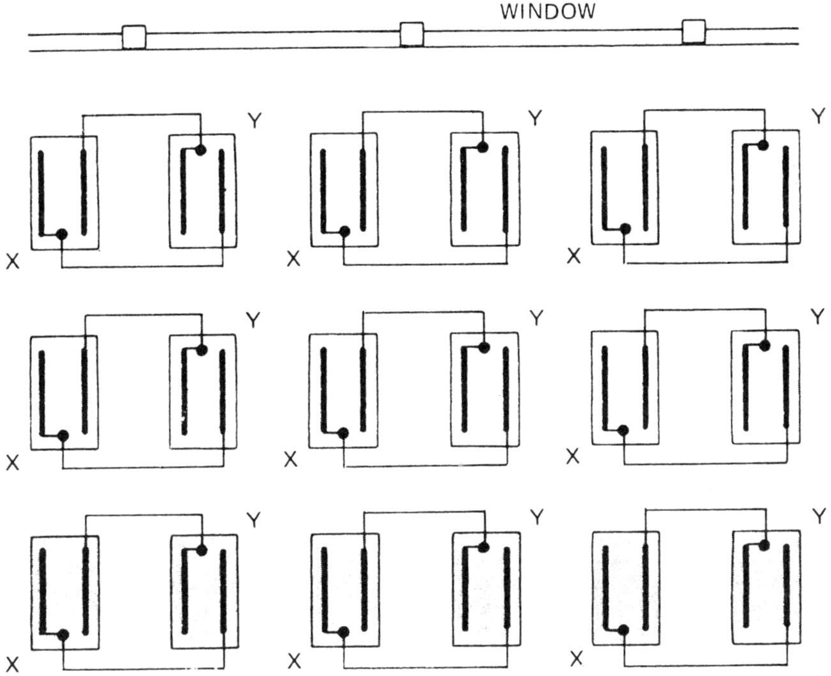

One two-lamp ballast controls one lamp in the luminaire and one lamp in the adjacent luminaire. Illumination of each row can be adjusted in 3-steps in response to daylighting in the following manner:

SEQUENCE	ILLUMINATION
Power on to ballasts X and Y	100%
Switch-off power to either ballast X or Y	50%
Switch-off power to both ballasts X and Y	0%

As more daylighting is available, more rows near the window are switched off.

Figure 18.7 Method of wiring for three-step control.

Table 18.1 Energy-Saving Statistics from Lighting Controls[a]

Project	Type of building	Space (ft^2)	Lighting control	System installed cost	Resultant energy savings	Payback (years)	Cost ($\$/ft^2$)
A	Headquarter offices	120,000	CCS (central control system)	$ 19,000	$40,000 at $0.06/kWh	0.5	0.158
B	Offices and manufacturing	384,000	HP computer and telephone interface	$ 59,140	$23,000 at $0.015/kWh	2.6	0.154
C	Offices (commercial building)	250,000	GE-PLC and telephone interface	$ 37,000	$32,300 at $0.04/kWh	1.1	0.148
D	Offices	800,000	Microprocessor transceivers and relays	$154,400	$56,000 at $0.10/kWh	3	0.193

[a]No elaborate dimming controls are involved in the projects listed.

unit costs in dollars per square foot are fairly consistent. None of the projects listed include elaborate dimming controls, which would certainly have increased the cost of the project.

The equation below shows a simple relationship between the electricity costs and control system unit cost in dollars per square foot. The figures are derived from a simple payback formula expressed in terms of most relevant factors involved in evaluating an effective cost an end user could be expected to pay for a lighting control system.

$$\text{Payback years} = \frac{CA}{R \times UPD \times HR/1000 \times PA}$$

where

A = area, ft^2
UPD = unit power density for interior lighting design, W/ft^2
P = electricity cost, dollars/kWh
C = unit cost of lighting control system, $dollars/ft^2$
R = percent energy reduction
HR = annual operating hours

18.6 BASES FOR SELECTING LIGHTING CONTROLS TO ATTAIN OPTIMUM SAVINGS

18.6.1 Guidelines for Minimum Number of Lighting Controls

Table 18.2 represents an attempt to establish some guidelines for a minimum number of lighting controls to be installed with reference to the area size and unit power density of a lighting system. The lighting controls used are either on-off devices or dimming controls.

18.6.2 Medium-Sized Offices

An assessment of the cost-effectiveness of photoelectric control equipment was made for three medium-sized offices:

1. A fully automated mixed control system would be unlikely to be cost-effective for single offices.
2. A partially automated on-off system would be cost-effective in new buildings at present energy cost.
3. A partially automated mixed system for new buildings would probably be cost-effective only if designed to control luminaires in several offices.

Table 18.2 Recommended Minimum Number of Lighting Controls[a]

	UPD (W/ft^2)		
A (ft^2)	<1.5	1.5–3	>3
<125	1	2	2
125–250	1	2	2+
251–500	1	2+	4
501–1000	2	3+	5+
1001–2000	3	4+	6
>2000	3+	5+	6+

[a]When multiple controls are used, it is generally installed to permit reducing the general lighting in the space by at least one-half in either a uniform pattern or by zones, as most appropriate.

The results of this study are summarized in Table 18.3. From a close examination of Table 18.3 it would seem sensible to include a dimming line to each luminaire during its installation in new buildings. This would involve only a small extra capital cost, but would keep open the option of installing a dimming system later without the necessity for complete rewiring. In another study, when due considerations were given to the daylight factor, it appeared that it is generally best to control only those luminaires nearest the windows.

18.6.3 Factors Affecting Selection of Lighting Controls

In general, there do not appear to be any general rules or guidelines that conveniently lead one to select specific controls. The following factors will have a bearing on the selection of lighting controls.

Size of the Facility

A large facility may justify a building management system or programmable controllers that provide centralized lighting control. On the other hand, a small facility may obtain optimum savings by selecting a simple time switch control. From experience, large computers often cost around $500 per hour to operate, whereas a minicomputer costs no more than $10 per hour to operate.

Table 18.3 Cost-Effectiveness of Photoelectric Control Systems[a]

Types of control	New buildings		Existing buildings	
	Single office	Multioffice	Single office	Multioffice
Fully automated 1 on-off + 1 dimming	2	1	2	2
Partially automated				
1 on-off	1	1	3	3
1 on-off + 1 dimming	2	1	2	3

[a]1, Cost-effective within 15 years; 2, not cost-effective within 15 years; 3, depending on energy costs.

Size of Individual Lighting Area

If the current drawn by an individual lighting area exceeds 20 A, a power contactor may be the choice. Small areas such as individual offices would be candidates for low-voltage relays. In either case, these devices would be controlled by timers, photoelectric sensors, and the like. In recent years, the trend has been to control smaller individual lighting areas.

Availability of Daylighting

As discussed previously, energy savings from daylighting depends on many factors: climate conditions, building form and design, and the activities within the building. Only a portion of a building can be daylit; however, in most cases, 30% of the floor space is sufficiently close to the perimeter to be daylit. There is little documented research in this field. As a general guide, the average energy savings from daylighting for an entire building will be in the neighborhood of 15%.

Type of Usage in the Facility

If the facility provides commercial rental of office space, consideration is often given to flexible controls, such as low-voltage relays. In the institutional facilities where lighting requirements are more fixed, other types of controls should be considered.

New Installation or Modification of Existing Facility

From Table 18.3 it becomes evident that a more elaborate photoelectric control system or even a sophisticated lighting management system can be easily justified for a new building. However, for an existing building, an extensive lighting control system may not be cost-effective. In this case a control system that utilizes existing power wiring for signal transmission would be preferred (see Table 18.1, project A).

18.6.4 Comparison of Lighting Control Systems

Table 18.4 shows a simple comparison of different systems according to how well each meets the needs of both the occupant and the building management. The relative performance rating of each system is often a judgement call. Individual manufacturers for a single system may vary sharply in system performance and cost. The objective here is to show a process, not pass final judgement or provide hard pricing guidelines. Some of the reasoning that went into the relative rating for each function shown in Table 18.4 is given below.

1. *Occupancy sensitivity.* From an occupant's perspective, individual wall switches work just fine. Contactors on most building automation systems can be a real disadvantage, since they do not normally allow the occupant to override for after-hour usage. Programmable lighting control caters to the occupant. When he or she enters the building after-hours, the person's particular working area can be lit in anticipation of his or her arrival with a single phone call. Similarly, when a person is staying late, his or her office and related work space can both be kept on with a phone call or switch override.

2. *Occupant-level selection.* This is a function affected by both control system capacity and floor layout. Individual office layouts with manual switching of split-wired fixtures or several lighting sources within the space give most occupants the degree of control they need. The dimmable solid-state ballast approach provides an even better method for allowing the occupant to adjust the overhead lighting.

3. *Energy-saving potential.* The only surprise here is in the "good" rating for switches in individual offices and the "poor" rating for the same devices used to control a zone. In practice, what happens is that when more than one person is in a zone, the first in turns it on and the last out never looks back.

4. *Management data.* This term reflects system monitoring, analysis, and reporting capabilities. These require a communications capability not normally inherent in switches or occupancy sensors.

Table 18.4 Comparison of Various Lighting Controls

	Occupant Sensitivity	Light Level Selection	Energy Savings Potential	Management Data	Integration Capability	Space Adaptability	Cost
Standard Wall Switches (Individual offices)	Good	Fair	Fair	No	No	Poor	Medium
Contactor Control Via Building Automation System	Poor	Poor	Poor	Yes	Yes	Good	Low
Programmable Lighting Control Relay Based	Excellent	Fair	Good	Yes	Yes	Good	Medium
Programmable Lighting Control Dimming Based	Excellent	Excellent	Excellent	Yes	Yes	Good	High
Occupancy Sensor	Fair	Poor	Good	No	No	Poor	High

5. *Space adaptability.* The important point here is that devices physically linked to the occupant's walls or ceiling pose problems when it is time to rearrange a space.

6. *Costs.* These are for illustration purpose only. The solid-state ballast cost, in particular, represents a combination of functions not presently available in the market. Perhaps the biggest surprise is that individual office switches are not cheap. They typically cost $0.50 per square foot. Occupancy sensors may reduce the installation labor, but the added hardware content still means a relatively high total cost. Both switches and sensors incur added cost for office rearrangements.

In conclusion, regardless of the type and/or size of the facility illuminating engineers may deal with, there will always be a suitable type of control for them to choose from. The engineers must become knowledgeable in the field of lighting controls and exercise their sound judgment in the selection of control schemes to achieve a high-quality lighting system with optimum energy savings.

BIBLIOGRAPHY

Alling, W. R., The Integration of Microcomputers and Controllable Output Ballast—A New Dimension in Lighting Control, *IEEE Transactions on Industry Applications*, vol. IA-20, no. 5, Sept./Oct. 1984.

Chen, Kao, New Concepts in Interior Lighting Design, *IEEE Transactions on Industry Applications*, vol. IA-20, no. 5, Sept./Oct. 1984.

Chen, Kao, and Castenschiold, Rene, Selecting Lighting Controls for Optimum Energy Savings, IEEE-IAS Annual Meeting Proceedings, Oct. 1985.

Crisp, V. H. C., Preliminary Study of Automatic Daylighting Control of Artificial Lighting, *Lighting Research & Technology*, vol. 9, no. 1, 1977.

Dunlop, R., Management of Lighting Loads with Controls, Electric Power Research Institute Publication EPRI EM-3866, Project 2285-7 Proceedings, Jan. 1985.

Hunt, D. R. G., and Crisp, V. H. C., Lighting Controls: Their Current Use and Possible Improvement, CIE Seminar, Electricity Council Research Center, Capenhurst, England, July 1978.

Pearlman, Gordon W., The Emergence and Future of Intelligent Dimmers, *Lighting Design and Application*, June 1982, pp. 22–23.

Peterson, D., Integrated Lighting Management, *Lighting Design and Application*, Sept. 1988, pp. 14–19.

Peterson, D., and Rubinstein, F., Effective Lighting Control, *Lighting Design and Application*, Feb. 1983.

Index